AF301206

Linking nutrition security and agrobiodiversity:
the importance of traditional vegetables for nutritional
health of women in rural Tanzania

Dissertation
submitted for the degree of
Doctor of Agricultural Sciences
of the Faculty of Agriculture, Nutritional
Sciences and Environmental Management,
Justus-Liebig-Universität Gießen, Germany

By
Gudrun B. Keding
Born in Berlin, Germany

Gießen, December 2010

Bibliografische Information der Deutschen Nationalbibliothek

Die Deutsche Nationalbibliothek verzeichnet diese Publikation in der Deutschen Nationalbibliografie; detaillierte bibliografische Daten sind im Internet über http://dnb.d-nb.de abrufbar.

1. Aufl. - Göttingen : Cuvillier, 2010
  Zugl.: Gießen, Univ., Diss., 2010

  978-3-86955-598-0

Dissertation im
Institut für Ernährungswissenschaft
Fachbereich Agrarwissenschaften, Oecotrophologie und Umweltmanagement
Justus-Liebig-Universität Gießen

Referee: Prof. Dr. Michael Krawinkel, Justus-Liebig-Universität Gießen
Co-referee: PD Dr. Brigitte L. Maass, Georg-August-Universität Göttingen

Date of Examination: 3 December 2010

Printed with the financial support of the *fiat panis* Foundation, Ulm, Germany

Table of Contents

List of Tables

List of Figures

List of Abbreviations

AGP	Acid Glucoprotein
BIA	Bioelectric Impedance Analysis
BMI	Body Mass Index
CIC	Conjunctival Impression Cytology
CRP	C-Reactive Protein
DBS	Dried Blood Spots
DDS	Dietary Diversity Score
DGE	Deutsche Gesellschaft für Ernährung
DS	Dry Season
ELISA	Enzyme-Linked ImmunoSorbent Assay
EV	Exotic Vegetable
EVDS	Exotic Vegetable Diversity Score
FANTA	Food and Nutrition Technical Assistance
FAO	Food and Agriculture Organization of the United Nations
FVS	Food Variety Score
Hb	Haemoglobin
HH	Household
HORTI Tengeru	Horticultural Research and Training Institute in Tengeru, Tanzania
IDA	Iron Deficiency Anaemia
IDE	Iron Deficiency Erythropoiesis
IOM	Institute of Medicine (USA)
K-S test	Kolmogorov-Smirnov test
K-W test	Kruskal-Wallis test
KCCO	Kilimanjaro Centre for Community Opfthalmology
LR	Long Rainy season
PA(L)	Physical Activity (Level)
PCA	Principal Component Analysis
ProNIVA	Promotion of Neglected Indigenous Leafy and Legume Vegetable Crops for Nutritional Health in Eastern and Southern Africa
PUFA	Polyunsaturated Fatty Acids
RBP	Retinol-Binding Protein
RDA	Recommended Daily Allowance
RDR	Relative Dose Response
S-W test	Shapiro-Wilk test
SPSS	Statistical Package for the Social Sciences
SR	Short Rainy season
SUA	Sokoine University of Agriculture
TfR	Transferrin Receptor
TV	Traditional Vegetable
TVDS	Traditional Vegetable Diversity Score
UNICEF	United Nations International Children's Emergency Fund
USDA	United States Department of Agriculture
VAD	Vitamin A Deficiency
VARF	Vitamin A Rich Foods
VAS	Visiual Analogue Scale
VDS	Vegetable Diversity Score
VIF	Variance Inflation Factor
WHO	World Health Organization of the United Nations

1 Introduction

Currently, about one in three people worldwide, mostly women and children, suffer from diseases associated with malnutrition. Simultaneously, diseases associated with affluence, such as obesity, type 2 diabetes mellitus and heart disease, are on the rise not only in developed but also in developing countries, creating a double burden of nutrition. While the causes of malnutrition are complex, a leading cause is suggested to be a general simplification of diets; in short, declining diversity may lead to a decline in nutrition quantity (Bioversity International 2009).

In fact, it was observed that the so-called "food baskets" of Zambia (Luapula Province) and Tanzania (Rukwa Region), where there is a high degree of market-oriented maize mono-cropping, are, at the same time, the regions with the highest prevalence of malnutrition (Bellin-Sesay 1995). It is generally acknowledged that malnutrition – either micronutrient deficiencies or an imbalance in nutrient intake resulting in obesity – is, amongst others, caused by declining diversity in traditional food systems when these are being replaced by mono-cropping agriculture (Frison 2007). Yet, in many African farming systems, polyculture is the norm. The latter can be defined as a traditional strategy to achieve yield advantages as well as yield stability, an optimal exploitation of resources, reduction of pests and disease occurrence and, thereby, minimisation of risk, efficient use of labour and dietary diversity (Liebman 1995).

A high level of agrobiodiversity, besides providing nutritional diversity, is considered as a valuable resource in supplying the necessary nutrients for a health-oriented human diet. In fact, nutrition security addresses not only the access to but also the utilisation of food. One important underlying factor of nutrition security is a certain degree of diversity in the available food with its macro- and micronutrients (Bouis and Hunt 1999).

The project 'Promotion of Neglected Indigenous Leafy and Legume Vegetable Crops for Nutritional Health in Eastern and Southern Africa' (ProNIVA), in which the present study was integrated, focused on the improvement of household food security of resource-poor groups in Tanzania, Rwanda, Uganda and Malawi. Thereby, safeguarding biodiversity of indigenous vegetables, the promotion of production and consumption of the latter, and the establishment of improved cultivation practices together with the provision of higher quality seeds of indigenous vegetables, in order to stabilise farmers' incomes and nutritional health, were the main goals. Next to the evaluation of nutraceutical values of indigenous vegetables, nutritional health of people in the researched areas was assessed; associations between the availability of indigenous vegetables, their diversity, production, consumption and the nutritional status of people was investigated; the latter was the central topic of the present study in Tanzania and was expanded beyond indigenous vegetables, including all other vegetables as well as all foods in general.

Several studies have investigated the link between farming or home-gardening and nutritional health, with positive results. A study in Kampala, Uganda, for example, suggests that child nutritional status (height-for-age) was significantly higher among households that farmed. Furthermore, among lower-socioeconomic status households, there was a significantly higher prevalence of moderate to severe malnutrition among children from non-farming households compared to those from farming households (Maxwell 1998). In Bangladesh, a large-scale home-gardening project found that families growing both more fruits and vegetables and a larger variety of fruits and vegetables, were likely to have a higher intake of vitamin A (Helen Keller International/Asia Pacific 2001). In several Asian countries, it was calculated that vegetables are an efficient source of micronutrients, both with respect to their unit cost of production and per unit of land required (Ali and Tsou 1997). Even where there is no evidence of a direct nutritional impact, farming and especially home-gardening can indirectly impact on nutritional status by increasing the income of the female farmers (Reynaud 1989).

Increased fruit and vegetable consumption has been widely promoted because of the provision of micronutrients as well as of many bioactive phytochemicals associated with health maintenance and prevention of chronic diseases (Steinmetz and Potter 1996). Greater fruit and vegetable consumption can, in fact, help address the double burden of micronutrient deficiencies and chronic diseases. Furthermore, diets rich in micronutrients and antioxidants are strongly recommended to supplement medicinal therapy in fighting HIV/AIDS. (Friis 2002)

Within the present study, firstly, the hypothesis was tested, that a high vegetable diversity available ("production") results in high dietary diversity ("consumption") in different districts of rural Tanzania. It was also of interest whether especially a high vegetable diversity was of advantage for or had a direct influence on nutrition, and which linkages existed between production and consumption. As other studies have already established these links for different regions, it was also considered important to investigate possible reasons for these links or reasons why there were no associations. Furthermore, as the integration of nutrition and agriculture is necessary but often highly neglected, to understand the linkages between the diversity of foods in the field and on the plate was one important focus.

Secondly, in order to test the hypothesis that a high cropping diversity and a high dietary diversity resulted in a good nutritional health status, production as well as consumption were linked to nutritional health of participating women, namely to their body mass index (BMI), vitamin A and iron status.

While this study focused, on the one hand, on single food groups, health problems and nutrients, on the other hand, a more holistic view on dietary diversity/food variety and dietary patterns and their relation with nutritional status and vegetable production was taken. For some time already, nutritionists have been advocating for a change in the focus of the dietary recommendations from nutrients to patterns of dietary intake (Sichieri 2002). Especially dietary intake in relation to diseases is highly complex, as foods contain many nutrients and foods are eaten in combinations and, thus, it is difficult to credit effects to single dietary components (Schulze and Hoffmann 2006). Therefore, investigating the links between nutritional health and food intake is a challenge, yet, it is an important and necessary step in order to gain more knowledge about nutrition security and its underlying causes.

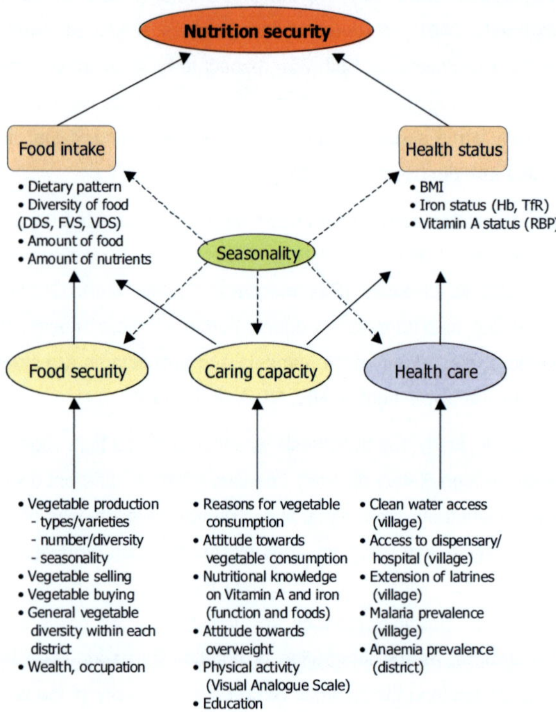

Figure 1.1   Model of nutrition security and underlying factors showing available data from the present study in Tanzania (modified after UNICEF (1998) and Krawinkel (2006))

When organising the different areas of this study, a model of nutrition security (modified after UNICEF(1998) and Krawinkel (2006)) was applied (Figure 1.1). This model gives a holistic view of all factors influencing nutrition security. However, as not all of these factors were surveyed within the present study (e.g. "health care" variables only for whole villages, not for single participants), the model was simplified for this study showing the three areas explored in Tanzania (Figure 1.2). Thereby, "production" includes all activities related to vegetables "in the field", while "consumption" stands for vegetable and general food intake. The socio-economic parameters as well as other influencing factors, grouped under "caring capacity" in the nutrition security model, are expected to influence all three areas directly and are clustered as socio-economic factors, knowledge and attitudes. The arrow from "production" to "nutritional health" is only dotted, as a direct link was expected to be less likely, yet, the link was tested with the present data.

Though being highly complex, the aim of this study was that each discussion chapter could be understood independently and, therefore, other chapters are strongly cross-referenced to. The present study can not give representative results for the whole of Tanzania, yet, it identified relevant aspects of the investigated themes and emphasises the importance of true interdisciplinary research.

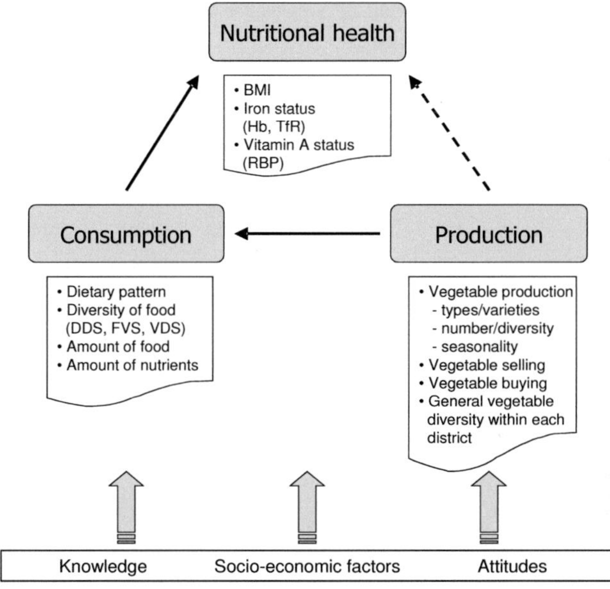

Figure 1.2   The three main areas investigated within the present study and influencing factors

2 Materials and Methods

A cross-sectional sequential study was performed with different survey methods. The main part of the survey was an individual interview on different topics; further, the health status of study participants was measured by various means. While some measurements were performed and questions asked only once, others were repeated three times within one year. Thus, the design can also be called cohort or follow-up study (Lang and Secic 2006) as always the same participants were included. Ethical clearance for this study was granted by the ethical clearance committee of the faculty of medicine at Justus-Liebig-University of Giessen, Germany. Furthermore, the study was approved and permission for the research was given by Sokoine University of Agriculture, Morogoro, Tanzania.

2.1 Research location and participants

The present study was part of the ProNIVA project and built i.a. on its baseline survey in 2003 (Keller 2004), e.g. regarding the selection of districts and villages. The time for conducting the surveys was chosen as to cover three different seasons within a whole year, namely first the dry season (DS) during June/July; second the short rainy season (SR) during November/December; and third the end of dry and beginning of the long rainy season (LR) during March/April. Yet, rain fall patterns are different for different areas of Tanzania (see below). While the year – here 2006 and 2007 – obviously will be crucial for the results, especially those regarding vegetable cropping, in the following only the abbreviations of the months are applied together with the abbreviation for the seasons, thus, Jun/Jul (DS), Nov/Dec (SR) and Mar/Apr (LR).

2.1.1 Selection of villages

The districts and villages for this study were those already visited within the ProNIVA baseline survey (Keller 2004; Weinberger and Msuya 2004), however, a smaller sample was chosen out of all participating villages. Selection criteria were the location to cover best the whole district, how cooperation of village members was during the baseline survey and, in terms of Kongwa, if irrigation was available (Table 2.1).

Rainfall patterns were said to be bimodal in all chosen districts, however, Singida is located in a transition zone between a bimodal and a unimodal rainfall regime. Still, it was assumed that all chosen areas receive short *vuli* rains from October to December with a minor cropping season, and long *masika* rains from March to May providing the main cropping season (USDA 2003).

Table 2.1   Characteristics of chosen villages according to baseline survey in 2003

Village	Ward	GPS data			Irriga-tion
Kongwa		Alt. (m asl)	Lat.	Long.	
Mlali-Iyegu	Mlali	1326	S 06°17	E 36°45	yes
Sagara B	Sagara	1251	S 06°14	E 36°32	no
Ibwaga	Ugogoni	1165	S 06°13	E 36°20	no
Chamae	Hogoro	1137	S 05°59	E 36°26	no
Tubugwe A	Chamkoroma	1050	S 06°21	E 36°37	yes
Manungu	Sejeli	893	S 06°09	E 36°20	no
Muheza					
Sakale	Misalai	1037	S 05°04	E 38°35	n.a.
Potwe-Ndondondo	Bwembwera	428	S 05°12	E 38°37	n.a.
Nkumba Kisiwani	Nkumba	260	S 05°09	E 38°43	n.a.
Kwemsala	Kilulu	240	S 05°15	E 38°47	n.a.
Mashewa	Kisiwani	220	S 05°09	E 38°40	n.a.
Mlingano / Mzambarauni	Pandedarajani	201	S 05°08	E 38°51	n.a.
Singida					
Mwakiti	Ilongero	1615	S 04°34	E 34°53	n.a.
Itamka I	Ilongero	1606	S 04°42	E 34°52	n.a.
Ilongero	Ilongero	1585	S 04°44	E 34°48	n.a.
Iseke	Ihanja	1516	S 05°04	E 34°40	n.a.
Ihanja	Ihanja	1484	S 05°02	E 34°40	n.a.
Nkuninkana	Puma	n.a.	n.a.	n.a.	n.a.

n.a.=not available

2.1.2   Selection of women

It was not the aim of the study to estimate representative prevalences, but rather to gather information from as many women as possible within a given time and budget frame. About 300 participants should be included, whereby, a major challenge was that participants should be available three times within one year.

During the first phase in each of the six villages per district, exactly 20 women took part in the study, thus, 120 per district and 360 overall. These women were randomly selected by the responsible village extension officer together with the *bwana shamba* or *bibi shamba,* the respective male or female agricultural extension worker for this village. The selection criteria included age (between 15 and 45 years, i.e. the reproductive age) and the performance of vegetable cropping. Oral informed consent from each woman enrolled was obtained during recruitment.

During the second phase, it was not possible to meet all women again who had been selected during the first phase. This was mainly due to rain, which made women being busy in their gardens and fields, but which also made it impossible for some women to reach the meeting point. A few women had moved to other places and few were travelling or in hospital. In the district of Kongwa the large number of women missing was also due to the horticultural extension officer not being present and his representative not being able to remind all women on time. Thus, misinformation was also a reason for some women not being present and overall 291 women attended the study during this phase.

During the third phase, similar reasons such as travelling, illness or having moved prevented some women from attending the study and, thus, only 299 were available. Finally, women who attended all three meetings and who consequently formed the study cohort were 252 in number, 72 in Kongwa, 76 in Muheza and 104 in Singida. If not stated otherwise all participants (n=252) were used for the analysis. The words "women" and "participants" are used as synonyms within the present study.

2.2   Data collection

2.2.1   Individual interviews

For the first phase of data collection, the questionnaire was divided into two parts: the "health part", which contained yes/no questions on the general health status of women and the "vegetable part", which contained questions on both production and consumption. For the second and third phase, the vegetable part of the questionnaire was further divided into two parts, a "production part" and a "nutrition part". During the first phase the interview on vegetable production and consumption was performed by two nutritionists and a horticultural researcher from the horticultural research and training institute in Tengeru (HORTI Tengeru). When the questionnaire was divided into two parts for the 2nd and 3rd phase, the nutrition part was carried out by the nutritionists only, while the production part was conducted by the horticulturalist. In this way, the quality of interviews was improved as the horticultural researcher was much more familiar with vegetable production and the identification of different vegetable types, while the nutritionists were more experienced with the methods of 24-h recall and 7-d recall. The questionnaires of all three phases are shown in the appendix.

Health

The health part contained questions on supplementation and factors influencing iron and/or vitamin A status. Further, four standard WHO questions on night blindness were included as well as ten standard questions for three major and seven minor signs for HIV/AIDS. The aim of this questionnaire was to ensure that no other major factors than food would influence vitamin A and/or iron status of women – or if influencing factors occurred, e.g. a respondent had taken iron supplements recently, she was excluded from the analysis of iron status. This part of the questionnaire remained overall the same for all three phases of data collection; only one additional question on smoking was asked during the last phase. While during the first phase, the questions on health were asked by both a nurse who also performed the health check-up and the two nutrition students, they were mainly asked by the nurse during the second and third phases.

Production

During the first phase within this part of the questionnaire ("vegetable part") socio-economic data of participants were obtained. These questions were not repeated during the second and third phase. The questionnaire further contained questions on cultivating, collecting, selling and purchasing vegetables (purchase only second and third phase) as well as on the attitude of women towards consuming different vegetables. Additionally, during the second and third phase it was asked why other vegetables which were cropped during the preceding phase were not cropped at the moment. The vegetables mentioned by women were identified with a picture book containing pictures and descriptions of most traditional vegetables of Tanzania (Vegetable Picture Folder by Patrick Maundu, Bioversity International, unpublished).

Consumption / nutrition

Both a 24-h recall concerning all food stuffs and a 7-d recall concerning vegetables only were performed during all three seasons. To identify the amount of food eaten during the last 24 hours, three containers of different sizes were shown for estimation of portion sizes; other measures were local cups and spoons. During the second phase two questions on physical activity were included as well as questions on nutritional knowledge and the attitude of participants towards obesity.

2.2.2 Health status

Anthropometrical data

During the first survey, heights of women were measured to the nearest 0.1 cm with a person-check (Kirchner & Wilhelm, Asperg, Germany; max. 2 m) fixed to a portable wooden measuring board with a foot rest on which women had to stand barefoot and without headgear. During all three surveys, weights of women were measured with a calibrated person standing scale (seca 862, Seca Co., Hamburg, Germany; min. 0.1 kg – max. 200.0 kg) on which women were examined barefoot and with minimal clothing according to the FANTA protocol (Cogill 2003). From body height and weight, the body mass index (BMI) was calculated for each participant (BMI = weight in kilograms / height in meter^2).

If the health risk, which can be caused by obesity, is of concern, the location of fat storage in the body is in fact more important than the amount of fat itself. To check if somebody is overweight because of high muscle mass or if he/she is, in fact, obese, it is necessary to measure besides body weight and height also body fat and its distribution in the body. This is possible with bioelectric impedance analysis (BIA), which measures the proportion of body fat and further substances with an extremely weak electric current send through the body. Until now, this method is only approved for measurements in healthy subjects and patients with a steady water and electrolyte balance because of a missing standard method and quality controls (Kyle et al. 2004). BIA is neither invasive nor costly and the instrument for the measurement is portable, still its use is restricted due to power requirement, especially in rural areas of developing countries.

Another non-invasive and cheap method to determine the distribution of fat in the body is to measure the waist-hip-ratio. However, this method is not suitable for all ethnic groups (Rush et al. 2007). The best alternative to measure the complete body fat percentage is the dual energy x-ray absorptiometry, which is, however, highly expensive. Thus, as an easy, cheap, non-invasive method, which is more or less applicable everywhere, the measurement of BMI remains to determine the degree of under-, overweight or obesity. Due to certain limitations of this study, such as restricted budget and rural location of target villages, the BMI was consequently used solely in the present context.

Iron status: Haemoglobin (Hb)

Iron nutritional status can be assessed by different biochemical and hematological tests, whereby either serum iron concentration, total iron binding capacity, transferrin saturation, protoporphyrin, serum ferritin or transferrin receptors are measured (Vijayaraghavan 2004). Mainly because of cost implications, in most population studies on iron deficiency anaemia (IDA), haemoglobin (Hb) is measured to indicate the iron status.

As this is a rather easy and cheap tool for the field, also in the present study haemoglobin (Hb) of all participating women was measured as a proxy indicator of iron deficiency during all three seasons. Yet, its sensitivity and specificity is low if used alone as there are many causes of anaemia besides iron deficiency (Zimmermann 2008). Additionally, serum transferrin receptor (TfR) was measured in dried blood spots (DBS), however, only during the March/April survey, mainly because of cost implications.

Measurement of Hb was carried out with a HemoCue Hemoglobin system (© HemoCue AB Sweden). A retired nurse, familiar with taking blood samples from the finger tip, was trained how to use the HemoCue Safety Lancets (2.25 mm deep), the Microcuvettes and the HemoCue 201 Analyser itself. After either the middle or ring finger was pricked, the first few drops of blood were wiped away and the following blood absorbed by a microcuvette. The results were given after 20 to 60 seconds by the HemoCue 201 Analyser, which functions as a dual wavelength photometer. The HemoCue 201 Analyser was calibrated with the HemoCue calibration fluids and cleaned after every 20 to 30 samples with the HemoCue Cleaner.

Vitamin A status: Conjunctival Impression Cytology (CIC)

The vitamin A status of a person can be assessed by clinical, biochemical or dietary parameters. Clinical parameters include all signs of xerophthalmia. Biochemical parameters include the measurement of the relative dose response (RDR), modified RDR (MRDR) or serum 30-day dose response (+S30DR), which are indirect assessments of liver stores; further, serum retinol can be measured as well as serum retinol-binding protein (RBP) and both values correlate very closely with each other (Sommer 1995). Dietary parameters are not an actual measure of vitamin A status, yet they can provide complementary and useful information, e.g. to identify and monitor individuals or groups at risk for suboptimal vitamin A intake. Intake information can be obtained by different dietary assessments such as the 24-hour dietary recall or food frequency questionnaires (Sommer 1995). Impression cytology is a histological technique for identifying keratinizing metaplasia of the conjunctiva, which is also responsible for Bitot's spot (see also Focus C).

As a method of choice for the present study, impression cytology with transfer was selected mainly for its feasibility in the field, non-invasiveness and low costs. Both the technique of obtaining the cell sample from the eye as well as of fixing the cells were discussed and pre-tested before data collection in the field.

A filter paper (Whatman OE66 Membrane Filters, 0.2 µm, Ø47 mm) was applied to the lower bulbar conjunctiva. The round filter paper was cut into 16 similar pieces in shape of a cone. The paper was pressed gently to the conjunctiva with the blunt and soft end of a pencil with rubber. With this and an even pressure on the paper, it was aimed to cover always a more or less similar-sized surface of the conjunctiva and, thus, obtain a more or less similar amount of cells.

After applying the paper to the bulbar conjunctiva for at least one second, the paper was laid directly on a glass slide and pressed several times gently to it with the rubber. After removing the paper the slide was sprayed directly with the fixative spray (Microscopy Merckofix ® Fixationsspray) and left to dry for at least 15 minutes.

The nurse, a retired ophthalmic nurse experienced in trachoma and general eye research, was trained at Kilimanjaro Centre for Community Ophthalmology (KCCO) by an eye doctor how to apply the paper to the eye and also how to transfer the cells to the glass slide. In the field, the procedure of the examination was explained to each participant. First, they were checked for the existence of Bitot spots, then the smear was taken from both eyes. Usually, tears were wiped away before sample taking, yet, sometimes it was only possible to obtain few cells with many tears.

<u>Iron, vitamin A and infection status: Dried Blood Spots (DBS)</u>

The dried blood spots (DBS) analysis was applied during the third phase (Mar/Apr (LR)), to measure retinol binding protein (RBP) as an indicator of the Vitamin A status of participants. As nearly all vitamin A in the blood is associated with RBP, the latter can be used to measure retinol content and, consequently, vitamin A status (Erhardt et al. 2004). Besides RBP also transferrin receptor (TfR) to check for the iron status, as well as two parameters for the infection status, namely c-reactive protein (CRP) and acid glucoprotein (AGP), were measured. While CRP is used to check for transient effects of morbidity in serum retinol, AGP is applied to check for more chronic infections. Because plasma retinol is reduced by clinical and subclinical infections, overestimations of vitamin A deficiency are possible (Thurnham et al. 2003). Therefore, all participants with either CRP>10 mg/L and/or AGP>1 mg/L were excluded from the study when vitamin A was the dependent or target value.

While TfR concentrations seem to be a specific indicator of iron deficiency erythropoiesis (IDE), its value in diagnosing iron deficiency anaemia (IDA) in children is uncertain in regions where Malaria, inflammatory conditions and infections are endemic (Zimmermann 2008). However, it was also found that TfR is not affected by infections or chronic diseases (Cook et al. 1993; Baynes 1996; Massawe 2002). Also, Zimmermann (2008) suggests that in adults elevations in TfR can be a definitive indicator of iron deficiency.

For taking the DBS samples, filterpaper (Whatman 903) was used. After the finger was pricked and a capillary blood sample was taken for determining the Hb, blood was dripped on two circles printed on the filter paper in order to obtain two samples for the DBS analysis. Yet, sometimes not enough blood could be secured so that no analyses could be performed.

The filter paper attached to a card was placed on a table so that the blood was able to dry from both sides for about three hours before it was packed into a ziploc plastic bag. A pack of silicagel was used in each bag as a humidity indicator. After travelling from the village back to the place of accommodation, papers were unpacked and let to dry over night. Unlike retinol, RBP is considerably more stable with respect to temperature and light; additionally, the technique used to quantify RBP, namely a sandwich ELISA, is much easier and less expensive than the one used for retinol (Baeten et al. 2004; Erhardt et al. 2004).

The DBS were analysed by DBS-Tech/Dr. J. Erhardt (Willstaett, Germany) with a sandwich ELISA. For this, a small part with a diameter of 1.5 mm was cut out from the filter paper equalling an amount of 0.75 µl blood. These papers were put into a flat-bottomed microtitre plate, and filled by extraction fluid composed of water, salt and detergent "Tween20". Overnight the protein fraction was triggered, and the solution was pipetted on an ELISA plate and then analysed.

2.3   Data analysis

Data generated within this study was analysed at different levels: the data was usually summarised for all three districts and – if a variable was collected during each season – also for all three seasons. Additionally, data was viewed individually for each district to compare the different study locations and also for each season to check for seasonal differences.

The summarised data from all three seasons, e.g. for food consumption, is simply the sum of the three seasons divided by three. If categories were formed with this data such as food intake below, within or above a recommended daily allowance (RDA), the results sometimes differ from the single season analysis. In fact, it is possible due to rounding that a value for the seasonal mean is lower or higher than one for the three seasons seen alone and does not reflect the mean or median value of the single seasons. Groups were formed with the aim to get equal shares of women in each of them; as reference usually the mean of all districts was used if not stated otherwise. The specific classification of categories for each parameter is given in chapter 3 "Results".

All data generated through individual interviews were first tested for normal distribution with both Kolmogorov-Smirnov (K-S) and Shapiro-Wilk (S-W) tests. Thereby, the requirement for normal distribution applied was for the test statistic of the K-S test to be below 0.1 and for S-W test above 0.95 (Hollenhorst 2006). Then, descriptive statistics were accomplished. To compare means of measured variables between the different districts (independent samples), the t-test for equality of means was applied. The critical t-value for df=100 and 0.975 is 1.984 and when the test statistics were greater, a significant difference between the means was assumed (Weiß 2002). However, in most cases data was not normally distributed, and the Kruskal-Wallis (K-W) test, as a non-parametric alternative that does not rely on an assumption of normality, was applied. For comparison between seasons (paired samples), the Friedman test or Friedman one-way ANOVA was used, which is also a non-parametric test (Lang and Secic 2006). If it was of interest whether an ascending or descending trend existed e.g. between quintiles of a dietary pattern, the non-parametric Jonckheere-Terpstra test for trend was used.

2.3.1   Individual interviews

A main part of the individual interviews was the 24-h recall, which was repeated three times. The amounts of foods that women stated to have consumed during the last 24 hours were entered into NutriSurvey for Windows (Erhardt 2004). With this tool, the amount of nutrients as well as amount of single foods and food groups consumed by each participant were calculated. The amount of nutrients per food item were taken from different databases, which are available at www.nutrisurvey.de, namely the database from Kenya, Senegal and USDA Agricultural Research Service, as there was no database available for Tanzania when analysing the data. Additionally, nutrient amounts for some indigenous vegetables were added from Protabase (http://database.prota.org/search.htm).

The databases usually apply a conversion factor of six for the bioconversion of carotene into vitamin A for all foods; yet, this factor has been questioned (Castenmiller and West 1998) and, thus, it was changed for carotene from vegetables and fruits into the factors 24 and twelve, respectively, according to suggestions of the International Vitamin A Consultative Group (IVACG 2002) and the US Food and Nutrition Board (IOM 2001). It must be alluded to the fact that the databases provided nutrient amounts of some foods in processed form e.g. cooked, yet, for several foods, especially the indigenous vegetables, data only for raw foods were available. Consequently, the amount of nutrients consumed by each participant within 24 hours is only a rough estimate.

Besides vitamin A, the analysis was carried out for another seven major vitamins and seven major minerals. Yet, for this group of nutrients it is especially important to remember that the analysis was done on the basis of mean amounts of nutrients in foods. Therefore, the data can give only a rough estimate on the nutrient intake of the group of participants and is not sufficient to look at single cases.

Besides single nutrients and foods, the consumption of food groups were focused on. There is no international agreement on certain food groups to use for standardised nutritional analysis (FAO, 2007). Therefore, according to different sources and various considerations, it was agreed to use 14 food groups for dietary diversity score and food variety score (Table 2.3.1) and twelve food groups for dietary pattern analysis (Table 2.3.2).

In developed countries, dietary variability must be taken more into account. The reported energy intake, nutrient intake and recorded numbers of food items were found to decrease when dietary intake was recorded only on four or less days. Consequently, for industrialised countries there are limitations on the usefulness of short-term dietary recording methods when the objective is to characterize usual intake in individuals (Patterson and Pietinen 2004). Conversely, in less developed countries, where individuals rely heavily on locally grown foods, diets do not change that much on a day-to-day basis, although there may be wide seasonal shifts in dietary patterns depending on local harvests (Patterson and Pietinen 2004). Therefore, this study considers mean values of three non-consecutive 24-h recalls during different seasons in order to gain a minimum level of representativeness, especially when considering dietary patterns. At the same time, each season is viewed individually to compare them, assuming that food intake during one season is rather similar on every day in the study districts of Tanzania.

<u>Elimination of outliers and data cleansing</u>

Before calculating the mean and median nutrient intake of participants, standard deviation (SD), minimum and maximum levels as well as the number of participants falling within the range of the RDA of each nutrient, it was important to first exclude any outlier. There is a common rule saying that one out of ten measured values can be discarded if it falls without the range of $x \pm 4s$, whereby x is the mean and s the SD. However, this rule is highly controversial (Lozán and Kausch 1998) and, therefore, it was decided to determine lower and upper levels for each nutrient, which were then used as a cut-off point to exclude any participants with values below or above. These lower and upper values were not only defined from a nutritional point of view but also taking into account that respondents often under- or overestimate their food intake.

In fact, it is not unusual that energy intakes less than 500 kcal/day or greater than 5000 kcal/day are reported (Patterson and Pietinen 2004). Combining these data with extreme values for energy intake that can be possible on single days in Tanzania (assuming a "celebratory meal" with unusual high intake or a "sick day" with unusual low intake), the cut-off points were set at 600 kcal/day as the lower level and 3,500 kcal/day as the upper level. From the overall 252 participants, 16 had to be excluded according to these levels, six for falling below 600 kcal/day (Kongwa two, Muheza one, Singida three), and ten for having intakes above 3,500 kcal/day (Kongwa and Muheza three each, Singida four).

Outliers were also excluded for vitamin A and iron. According to Bender (2002) an amount of 6.7 µg retinol equivalents/kg body weight is needed to maintain a concentration of 70 µmol/kg of liver. Thus, depending on her weight, an adult women would need about 600 – 800 µg retinol equivalents per day. An upper limit for adult women is 7500 µg/day, which was used as the upper cut-off, and all participants with values above were excluded. These were four cases for all districts and all seasons.

While there is a recommendation on upper iron levels, which is 45 mg/day for both men and women from 14 years of age onwards (IOM 2001), it is possible to consume more than this amount on single days. Therefore, it was assumed that consumption of iron could be as high as 90 mg/day, while all participants with values above this cut-off point were excluded from the study. These were 23 cases, while further reasons for being excluded from analysis were taking of supplements, pregnancy or the fact that a participant most likely had HIV/Aids or tuberculosis (TB).

<u>Analysing vegetable production and consumption data</u>
"Production" is used in this study as an umbrella term for both cultivation and collection of vegetables. The 20 most common vegetables in production and consumption were chosen and used for cluster analysis. A hierarchical cluster analysis was performed with SPSS Version 16.0 using the Squared Euclidian Distance as the measure of distance and Linkage Between Groups as the method of fusion. The clusters can be derived from directly clustering the 20 vegetables or from clustering the results from a PCA which is performed in advance.

While the clusters from the factor scores derived by PCA looked rather similar, those from the original production and consumption data varied from each other. Therefore, it was decided that only the latter was further used for comparison between vegetable production and consumption and that within the cluster – through PCA before clustering – data were probably too much concentrated so that they could not be interpreted any more.

<u>Analysing dietary diversity data</u>
Dietary diversity scores were calculated by summing up the number of food groups consumed by an individual over the 24-h recall period (FAO 2007). However, currently no international consensus exists on which food groups to include in the scores. The 14 food groups assumed and counted for this study are summarised in Table 2.3.1. Besides, the Food Variety Score (FVS) was calculated, whereby all single foods consumed on one day were summed up (Table 2.3.1). Thereby, some food items such as wheat products or leafy vegetables were grouped as one food

type. In terms of vegetables a detailed analysis, also on vegetable diversity consumed, was carried out in addition and results are summarised in chapter 3.3.4.

A Vegetable Diversity Score (VDS) was calculated with data from a 7-d recall. All different vegetables consumed on each day were counted and then all seven days summed. Besides a number for a whole week, the scores were divided by seven to get a comparable figure per day. In addition, the VDS was split into a score for traditional vegetables and one for exotic vegetables as it was of special interest to detect any difference between the consumption of these two vegetable groups. In general, the recall period of seven days instead of 24 hours was chosen because, when considering a single food group (vegetables), it is uncertain that this group is consumed every day; therefore, a short recall period would give an improper picture.

Table 2.3.1   Seventy-six different dietary items within 14 different food groups according to West et al. (1988) and FAO (1970) and adapted to food items found during the survey 2006/2007 in Tanzania

No	Food group	Foods within this group
	Plant origin	
1	Cereals and grain products	Maize, millet, rice (incl. *vitumbua*), sorghum, wheat (incl. *chapati*, *andazi*, scones etc.)
2	Starchy roots, tubers and fruits	Cassava, plantain, potato, sweet potato, taro/cocoyam, yam
3	Grain legumes and legume products	Bambara groundnut, beans, chickpea, cowpea, mungbean, green pea (dried), pigeon pea, soya bean
4	Nuts and seeds	Cashew nut, coconut (and products e.g. milk), groundnut, melon seeds, pumpkin seeds, sunflower seeds
5	Vegetables and vegetable products	Bulbs (onion, leek, garlic), cabbages, cucumber (wild or exotic), eggplant (African, brinjal), leafy vegetables (dark green), legumes/pulses unripe (e.g. fresh peas), mushrooms, root vegetables (carrot), okra, pumpkin fruit, tomato
6	Fruits	Avocado, baobab, banana, carambola, jackfruit, lemon, mango, melon, orange, papaya, pear, pineapple, tangerine, watermelon, wild fruits (*lade*, *matogo* etc.)
7	Sugars and syrups	Sugar/sugar cane, honey
	Animal origin	
8	Meat	Beef, goat, mutton, pork, offal
9	Poultry	Chicken, duck, offal of poultry
10	Eggs	Eggs (from chicken)
11	Fish and fish products	Cod-type fish (saltwater/sea), salmon-type fish (freshwater/lake), anchovy-type (dried) fish (*dagaa*)
12	Milk and milk products	Milk (cow, goat), sour milk
	Others	
13	Oils and fats	Ghee/clarified butter, lard/animal fat, (red) palm oil (*korie*), vegetable oil (sunflower)
14	Beverages	Carbonated soft drink, non-alcoholic hot drink (tea, coffee)

To compare districts and seasons with regard to their availability of vegetables consumed as well as produced, the Sørensen coefficient was calculated according to the formula

$$S = 2c\,/\,a + b + 2c * 100$$

Thereby *a* equals the number of vegetables that occurred only in district/during season A; *b* equals the number of vegetables that occurred only in district/during season B; and *c* equals the number of common vegetables that occurred in both districts/during both seasons (Dierßen 1990).

Generating and analysing dietary patterns

Dietary patterns cannot be measured directly, yet, different statistical methods are available (Figure 2.3) to characterise dietary patterns by using collected dietary information (Hu et al. 2002). *A posteriori* approaches include both factor or principal component analysis (PCA) and cluster analysis. For these exploratory approaches, study-specific data is used (Schulze and Hoffmann 2006). Dietary patterns are derived through statistical modelling of the dietary data at hand. Consequently, they do not necessarily represent optimal patterns, and one has to carefully check if the generated patterns fit into common eating habits of the population (Hu et al. 2002).

In contrast, the dietary index approach is *a priori* and generates eating patterns based on previous knowledge, which is summarised, e.g. in dietary guidelines or recommended diets, indexes and scores (Hu et al. 2002; Hoffmann et al. 2004). This hypothesis-oriented approach can be based on available scientific evidence for specific diseases (Schulze and Hoffmann 2006). The latter method is limited by current knowledge and, for Tanzania for instance, no dietary guidelines exist up to date (only for some districts within the country e.g. Morogoro and Iringa districts (TARP II – SUA Project 2004)) against which surveyed data could be tested.

A useful method postulated for both approaches is the reduced rank regression (RRR), which includes, next to a person's food intake in g/d, further information like nutrient intake considering a certain research question. It can then be tested directly which dietary pattern is associated, for instance, with diabetes (Hoffmann et al. 2004). This method could have been suitable for the present study to check directly the association between dietary patterns and the measured health status of participants. However, it could not be applied for logistical reasons.

Data for this study on food consumption was generated by three 24-h recalls, which were accomplished during three different seasons within one year. Consequently, here *a posteriori* approaches were applied, namely both PCA and cluster analysis (based on the mean intake in g/d of different food groups), followed by correlations and multiple regression analysis.

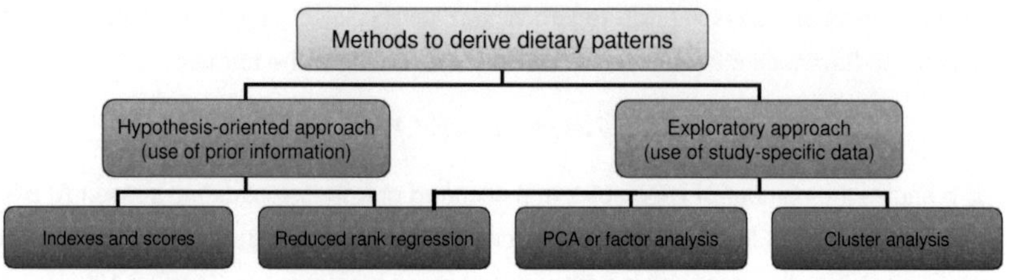

Figure 2.3   Methods of dietary pattern analysis (Schulze and Hoffmann 2006)

For the *a posteriori* approach, a factor extraction model must be selected that can be either a common factor model or a components model, such as PCA; the latter is also more widely used than common factor analysis. The purpose of a PCA is simply to reduce the number of variables by creating linear combinations that retain as much of the original measures' variance as possible. On the other hand, the goal of common factor models is to understand the latent or unobserved variables that account for relationships among measured variables (Conway and Huffcutt 2003). Still, because this technique is also called "data mining" and the process is not a statistical analysis but a data reduction, both PCA and two factor extraction models namely Principal Axis Factoring (PAF) and Maximum Likelihood Factoring (MLF) were applied to the data with three different rotation methods (varimax (orthogonal), direct oblimin and promax (both oblique)). The results in the rotated factor matrix or structure matrix were compared and, finally, the method with the highest loading factors/components and the best interpretability was chosen. Regarding the rotation method, varimax was found to be best suitable.

Similar results were obtained with all three extraction methods, i.e. always the same food groups were loading high in one component or factor, thus, all methods finally showed similar results. Compared to the two factor extraction methods, the PCA showed clearer results because, except for the food group animal products, all food groups were loading higher than 0.5 in one of the components and, consequently, this method was chosen for the present data. Results are then subject of interpretation with expert knowledge as no statistical analysis is meaningful at this point (Conway and Huffcutt 2003).

There are different options to retain the number of components. The one most widely used, Kaiser's (1956) "eigenvalue greater than one" was found to not consistently give an accurate number of factors and, at the same time, tended to produce too many factors. A combination of techniques is suggested, which should include the examination of multiple solutions with different numbers of factors that could be an advantage for interpretability (Conway and Huffcutt, 2003). With PCA and varimax rotation, the optimal number of components to be extracted was tested, namely three, four or five components (Table A1). Eigenvalues were greater than one only for the first three components. Still, the total variance explained, gained substantially when one or two more components were extracted (50.8% for three, 59.1% for four, and 67.0% for five components).

Only with five components the food group animal products loaded highly and, consequently, could be interpreted better than when only three or four components were extracted. Also, with the food group vegetables loading positively high in one component and negatively high in another when five components were extracted, more interpretation of the data was possible. Finally, the combination of extraction method and rotation method chosen for the present data was PCA with varimax rotation and the extraction of five components.

For analysing dietary patterns, factor scores for each participant and component (dietary pattern) were generated by SPSS. These values have the status of z-scores and usually range between -3 and +3 (Bühl 2006). A high positive factor score stands for a high affirmation of the respective food pattern, a high negative factor score, in turn, stands for a low or even no affirmation (Bühl 2006). Consequently, the correlation between food patterns and further variables could be tested and factor scores were used in multiple regression analysis. Furthermore, factor scores and, with them, participants were divided into quintiles for each food pattern. Especially for the highest and lowest quintile, meaning either total or no food consumption according to the respective food pattern, sample characteristics such as mean BMI or Hb, or participant's share of a certain education or origin were calculated.

PCA was initially performed with different sets of food groups, i.e. 42 different food groups (determined by *Nutrisurvey*), 30 or 13 only (determined by logical considerations). The total variance explained reached only 26% (42 food groups), 33% (30 food groups) or 55% (13 food groups) with four components. In a similar study on dietary patterns, the first and main component only explained 14.6% of total variance when using 22 food groups (Bamia et al. 2007), in another study with 52 food groups, the first seven components explained only 29.5% of total variance (Arkkola et al. 2007). Therefore, it appeared that the present data was sufficient to gain reasonable results through pattern analysis.

Among the 42 food groups, too many were not normally distributed; also from 30 food groups, only three were normally distributed. Usually, a multivariate normal distribution is necessary for confirmatory factor analysis (Überla 1972; Backhaus 1994). However, a non-normally distributed variable will have a smaller factor loading than a normally distributed one – which will simply add to the fact that is volitional anyway. The number of cases should be greater or equal three times the number of variables (Überla 1972; Backhaus 1994), which is the case in the present study if 30 food groups are chosen and 252 data sets of participants are available (ratio 1:8.4) and, in any event, if less food groups are selected.

In the analysis of 30 food groups, only few food groups were loading high and, therefore, were characteristic for dietary patterns; most food groups had factor loadings below 0.5 or even below 0.1. Even with 13 food groups the factor loading was not yet clear.

Especially the very low factor loadings below 0.2 for the food groups meat and milk led to the decision to group meat (from cow, goat, sheep), poultry, milk and eggs in one group named animal products. Thus, to find dietary patterns among the participants in this study, a data set with twelve food groups was chosen for PCA. These twelve food groups were arranged as shown in Table 2.3.2.

Table 2.3.2   The twelve food groupings used in dietary pattern analysis

No	Food Group	Foods within this group
1	Cereals	Rice; *ugali* or *uji* from maize, sorghum, millet
2	Bread/cakes	White wheat bread; dough made from wheat or rice fried or baked in oil, e.g. scones, donuts, halfcake, *vitumbua, chapati, mandazi*; noodles
3	Fruits	Mango, avocado, banana, guava, jackfruit, passion fruit, pawpaw, pineapple, watermelon, tangerine, orange, lemon, pear, wild fruits
4	Vegetables	Leafy vegetables, cabbages, tomato, onion, eggplant, okra, pumpkin, sweet pepper, carrot, peas, mushrooms
5	Nuts	Coconut, groundnut
6	Pulses	Kidney beans, chickpea, cowpea, bambara groundnut
7	Starchy plants	Sweet and Irish potatoes, cassava, taro, plantain
8	Tea	Black tea
9	Oil/fat	Sunflower and palm oil, vegetable/cooking oil, margarine, butter
10	Sugar	Sugar; sugarcane; soft drinks (10% sugar)
11	Fish	Anchovy-type, cod-type, salmon-type
12	Animal products	Goat, beef, pork, chicken, duck; fresh and curdled milk; eggs

Also cluster analysis can be used to generate dietary patterns. Thereby, mutually exclusive clusters of individuals are defined based on distance measures between observations of individuals (Schulze and Hoffmann 2006). In comparison to PCA or factor analysis, cluster analysis usually does not focus on the variables (here: food groups) but on the cases (here: study participants) and intends to group/cluster them according to the variables (Brosius 2006; Bühl 2006; Field 2000).

Unlike the results of the PCA, where participants get a factor score for each dietary pattern, with cluster analysis each participant belongs exactly to one pattern. On the one hand, this is less accurate as most participants will not consume food only according to one pattern. Moreover, the patterns generated from cluster analysis are less well-defined (for each food group within a cluster factor loadings exist, but no factor scores for participants), thus, interpretation is more difficult. On the other hand, with every participant being clearly a "member" of one cluster, further analysis and characterisation of the cluster groups is more clear-cut and straight forward. Like with the data on vegetable production and consumption, a hierarchical cluster analysis was performed with SPSS Version 16.0 using the Squared Euclidian Distance as the measure of distance and Linkage Between Groups as the method of fusion. Because the variables consumption of different food groups in g/d did not have the same valuation (range of values), a z-transformation had to be applied first. To visualise the cluster, a dendrogram presents graphically which cases (food groups, vegetable types or villages) are grouped together at certain levels of similarity or dissimilarity. The length of the horizontal lines give visual clues about the strength of the clustering, whereby long lines indicate a more distinct separation between cases.

Besides using clustering for dietary pattern analysis, this method was also used for grouping study participants in terms of vegetable production and consumption (chapter 4.1). For visualisation, a dendrogram with 252 participants (cases) was rather confusing and, therefore, the clusters were both calculated and depicted with MarVis (Kaever et al. 2009). Columns were normalised with Euclidian normalisation, which sums up all values squared and divides them by the root of the sum; after grouping participants into both ten and five clusters, it was decided to work with the result of five clusters, both for dietary patterns and vegetable consumption/production. This was mainly because of the uneven group sizes becoming too low with ten clusters. Using less than five clusters was not an option because e.g. participants' vegetable consumption patterns seemed to be too diverse and too much information would have been lost.

<u>Nutritional knowledge, attitudes and physical activity</u>
Participants were asked if they had heard about vitamin A and iron as nutrients (yes/no), their function in the body and if they knew foods which contained them. Functions and foods named by participants were usually more than one. The wrong answers were subtracted from the correct answers and when the result was either -1, 0 or 1 it counted as "indifferent". If the result was -2 or below, the answer was rated as "not correct", while if the result was 2 or higher it was rated as "correct".

To check the participants´ attitudes towards vegetable consumption enumerators read out three statements to them and the women had to decide whether they did strongly agree, agree, disagree, strongly disagree or were not sure about the statement. Frequencies were counted for analysis.

The attitude towards overweight was acquired through asking the participants to name both positive and negative characteristics which they associated with a corpulent person. Next to the characteristics named, the simple count of positive and negative features was analysed. The number of positive and negative characteristics regarding a corpulent person was also grouped into three categories (positive characteristics: 0 answer; 1 answer; 2 or more answers; negative characteristics: 0-2 answers; 3 answers; 4 or more answers). Positive and negative characteristics taken together were also grouped (more negative than positive; indifferent; more positive than negative).

Physical activity (PA) was measured with a visual analogue scale (VAS). Participants had to rate their own physical activity on one average day on a scale between 0 meaning no physical activity (sleeping) and 10 meaning extremely strenuous physical activity. To better visualise these two extremes, pictures were shown to participants (see questionnaire in appendix). As the result is rather subjective, it cannot be concluded on the real physical activity and workload of one person. It also is of less value for comparing across a group of individuals at one time point (Gould et al. 2001). Yet, the data can show changes within individuals between two dates (here Nov/Dec (SR) and Mar/Apr (LR)), which is, however, only shown in the appendix (Table A2 and Figure A1).

2.3.2   Health status

In general, all health data was first checked for normal distribution, then descriptive statistics were calculated with the appropriate tools. BMI was calculated as stated above and women were grouped into four categories, namely underweight (BMI < 18.5), normal weight (BMI = 18.5 – 24.9), overweight (BMI = 25 – 29.9) and obesity (BMI ≥ 30) (WHO 2008). Yet, the BMI was mainly used as a continuous variable similar to the Hb values and all values gained through DBS. The Hb of participants was also grouped into four categories of severe anaemia (Hb<7 g/dl), moderate anaemia (Hb 7.0-9.9 g/dl), mild anaemia (Hb 10.0-11.9 g/dl), and normal iron status (Hb ≥12 g/dl). The analysis of CIC was unfortunately not possible; further details are given in Focus C.

2.3.3   Correlation and multiple regression analysis

Both categorical and continuous variables were measured and, therefore, the relationship between two variables had to be described both with correlation (for continuous variables) and association (for categorical variables). Relationships between categorical variables, such as that between ethnic group (nominal) and BMI categories (ordinal) was assessed with the chi-square (χ^2) test as a test of association. Relationships between continuous variables, for example, the relationship between age and weight, were assessed with Spearman's rank-order coefficient, rho (ρ), which is a measure of correlation for non-parametric data. If the relationship between two ordinal or one ordinal and one continuous variable should be explored, Kendall's rank-correlation coefficient, tau (τ) was applied (Lang and Secic 2006).

Bivariate correlation results always have to be handled with care, and it has to be kept in mind that correlation is not causation. Even if highly associated variables are found this might have been caused by a third variable (Lang and Secic 2006). Classification into categories, e.g. below, within and above the recommended dietary allowance of certain nutrients or low, medium and high wealth status, was usually done not for comparison of different variables but to get a better picture of each single variable.

Besides considering associations and correlations between only two variables, it was of interest to check for the relationship between more than two variables, i.e. to predict the value of a response or dependent variable from the known values of more than one explanatory or independent variable. In this study, both continuous (e.g. BMI) and categorical (e.g. wealth status) explanatory variables were included and, thus, the appropriate analysis is either multiple regression or analysis of covariance (ANCOVA) (Lang and Secic 2006), whereby here only multiple regression analysis was applied. In Table 2.3.3 all dependent variables used for regression analysis are shown as well as all independent variables that could be considered, yet, from which usually only few met all requirements for the analysis.

The data used for multiple regression analysis has to meet certain assumptions that need to be tested beforehand. Besides others, it has to be checked for a normal distribution of residuals, for possible heteroscedasticity and collinearity/multicollinearity. To avoid the latter, a variance inflation factor (VIF) was calculated for each predictor variable and, if it was above ten, the variable was excluded from the model (Chen et al. 2003). Furthermore, the summarised condition index should be below 25 to assure that no multicollinearity exists. While the variables themselves do not have to be normally distributed the residuals have. If residuals were not normally distributed, either the dependent or the predictor variables were transformed by log transformation (Chen et al. 2003). Only variables with $p \leq 0.1$ found within bivariate correlation were chosen as predictor variables.

When two predictor variables were highly correlated with each other ($\rho = 0.694$, $p < 0.001$), it was not possible to include both of them at the same time into the regression model. Consequently, to prevent suppressor effects, the variable showing lower correlation to the dependent variable was excluded from the regression.

A multiple regression with nominal or categorical predictors is possible, however, the variable, for instance the three districts in the present study, has to be adjusted, i.e. a dummy variable for each district had to be created. To avoid the effect of collinearity, only two of the three dummy variables were used in the model. Here, the largest variable with the most cases, namely Singida, was excluded. The remaining two variables indicated if and how much the dependent variable, e.g. vegetable consumption, changed in either Kongwa or Muheza compared to Singida.

Table 2.3.3   Variables for multiple regression analyses

Dependent variables	Independent variables
Vegetable Production ● number and types of vegetables (diversity) cultivated and collected; ● challenges in production; ● sales; ● purchase;	Socio-economic parameters ● Continuous: age, hh size, distance from village to town in km and min ● Ordinal: wealth, education ● Nominal: ethnic group, religion, status in hh, occupation, marital status Knowledge/attitude ● Reasons for vegetable consumption ● Attitude towards vegetable consumption
Food and nutrient consumption ● Food and food group intake ● Food pattern ● Nutrient intake ● DDS, FVS, VDS	Socio-economic parameters ● Continuous: age, hh size, distance from village to town in km and min ● Ordinal: wealth, education ● Nominal: ethnic group, religion, status in hh, occupation, marital status Knowledge/attitude ● Reasons for vegetable consumption ● Attitude towards vegetable consumption ● Nutritional knowledge (vitamin A and iron) Vegetable production ● number and types of vegetables (diversity) cultivated and collected;
Nutritional health ● Vitamin A deficiency ● Iron deficiency ● Overweight/obesity	Socio-economic parameters ● Continuous: age, hh size, distance from village to town in km and min ● Ordinal: wealth, education ● Nominal: ethnic group, religion, status in hh, occupation, marital status Knowledge/attitude ● Reasons for vegetable consumption ● Attitude towards vegetable consumption ● Nutritional knowledge (vitamin A and iron) ● Attitude towards overweight ● Physical activity Vegetable production ● number and types of vegetables (diversity) cultivated and collected; Food and nutrient consumption ● Food and food group intake ● Food pattern ● Nutrient intake ● DDS, FVS, VDS

2.4   Constraints

Data obtained through interviews always must be handled with care as there are some disadvantages, which can not be suppressed. The dietary recall, for instance, relies on the respondent's memory as well as on her ability to estimate portion sizes (Patterson and Pietinen 2004). In the present study, this was less a problem for the amount of staples or vegetables consumed because three containers were shown for estimation; yet, foods consumed by piece such as meat, fish or fruit pieces were often not accurately described in size and, thus, the amount of these foods could only be roughly estimated. Further, as four different interviewers conducted the survey, it is possible that, to a certain extent, differences between single interviews occurred. However, as the survey was interviewer-administered and participants did not have to do the food record by themselves, the data was collected in a consistent manner from all respondents (Patterson and Pietinen 2004).

For the DBS analysis in some cases not enough blood was soaked into the filter paper so that analysis was not possible. From the 299 blood samples, 240 were analysable for RBP, CRP and AGP, and 220 for TfR.

3   Results

3.1   Socio-economic status

Among all parameters, only age (Kolmogorov-Smirnov (K-S) test statistic=0.068; Shapiro-Wilk (S-W) test statistic=0.977) and household size (K-S test statistic=0.098; S-W test statistic=0.969) were normally distributed, while all others had skewed distributions. All socio-economic factors except education differed significantly between the three districts either at p≤0.05 (status within household and household size) or at p≤0.01 (all other factors).

3.1.1   Age, ethnicity, religion and family situation

As one criteria for women to participate in this study was that they were in their reproductive age, the age range was 17 to 45 years for all districts. The mean age was 33 years overall, 31 (± 7.1) in Kongwa and 34 in both Muheza (± 7.5) and Singida (± 5.7) districts (Table A3). When participants were grouped into three age groups Kongwa district had the lowest share of women in the highest age group (25.4%), while Singida (43.3%) and especially Muheza (50.0%) had a much higher share and participants were older in these districts (Figure 3.1.1). In general and regarding all districts, while it was aimed at having similar numbers of women from each age group, only about 13.9% were aged under 26, 45.8% had a medium age and 40.2% were aged above 35 years (Table 3.1.1).

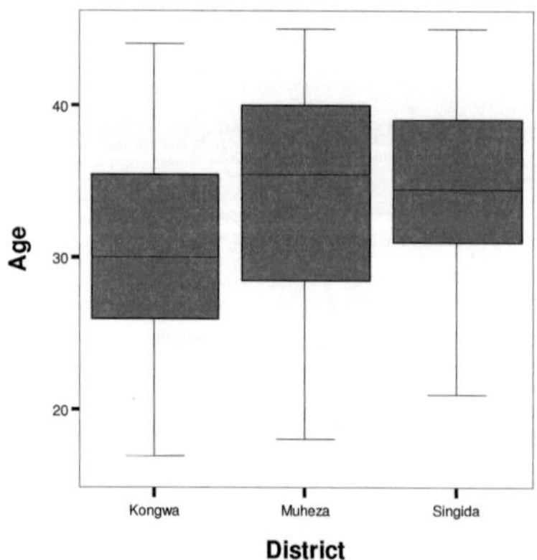

Figure 3.1.1
Age of women within three districts of Tanzania

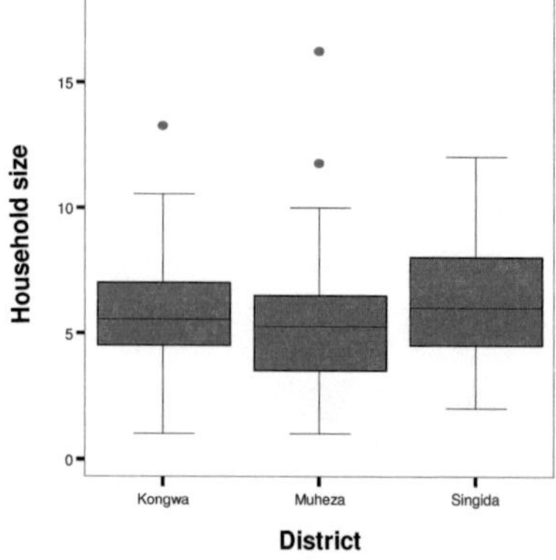

Figure 3.1.2
Household size of women within three districts of Tanzania

In Singida, only one ethnic group dominated, namely the Nyaturu, who are agriculturalists already for a long time and primarily practice agriculture with cattle integrated into the farming system (Koponen, 1988). In Kongwa, two ethnic groups were prevalent, namely Gogo and Kaguru, who used to be a semi-pastoral society, but nowadays are described as agro-pastoralists with agriculture and cattle being both essential but not closely integrated (Rigby, 1969; Koponen, 1988). In contrast, in Muheza district about 42% of all participants belonged to several different ethnic groups (Table 3.1.1). The two main ethnic groups were Shambaa, characterised as agriculturalists or agro-pastoralists with their main income from farming (Anonymous, 2003), and Bondei who are mainly grain-growers and consider hunting as a very common occupation (Koponen, 1988). The high diversity of ethnic groups in Muheza district reflects a high diversity of cultures, traditions and believes in terms of horticultural production and consumption.

While in Kongwa district the majority of participants were Christians (91.7%) and only few were Muslims (8.3%), in Muheza and Singida district only slightly more participants were Christians than Muslims (Table 3.1.1). Most participants were married (72.6% for all districts), especially in Singida district, while only about half of the women were married in Muheza district. Similarly, the status of the respondents within their household (hh) was mainly wife of head of hh (74.6%) and this was, again, especially so in Singida district. In contrary, in Muheza district about 15% of women were either mother or daughter of head of hh, and 13.2% were head of hh themselves, while only 8.3% and 5.8% were head of hh in Kongwa and Singida district, respectively (Table 3.1.1).

Household size was calculated by summing up the number of adults and children older than ten years who lived in the hh of the respondent. Thereby, children under ten years of age were counted as 0.5 only because they consume less food on average. Household size in this study ranged from 1 to 16 household members with a mean of 5.8 (± 2.3) for all districts. Mean hh size was similar in Kongwa district with 5.7 (± 2.4), smaller for Muheza with 5.3 (± 2.4) and greatest, namely 6.2 (± 2.2) in Singida district (Figure 3.1.2; Table A3). Participating women were classified into three groups with either low, medium or high hh size. Thereby, a low hh size was characterised by a household with less than five members, a hh of medium size comprised five to seven members while a large hh size meant eight or more hh members. The latter hh size was represented by only 23.6% and 14.5% of parcitpants in Kongwa and Muheza districts, respectively, while in Singida district more than one third of participants lived in a hh of large size (Table 3.1.1).

Table 3.1.1   Socio-economic characteristics of women in Tanzania for the whole study cohort and by district (share of participants)

	All districts	Kongwa	Muheza	Singida
n	252	72	76	104
Age				
Low (16-25) (%)	13.9	22.5	14.5	7.7
Medium (26-35) (%)	45.8	52.1	35.5	49.0
High (36-45) (%)	40.2	25.4	50.0	43.3
Ethnic group				
Bondei (%)	7.5	0.0	25.0	0.0
Gogo (%)	14.3	50.0	0.0	0.0
Kaguru (%)	9.5	33.3	0.0	0.0
Nyaturu (%)	41.3	2.8	1.3	97.1
Shambaa (%)	9.5	0.0	31.6	0.0
Other (%)	17.9	13.9	42.1	2.9
Religion				
Christian (%)	63.9	91.7	53.9	51.9
Muslim (%)	36.1	8.3	46.1	48.1
Marital status				
Single (%)	15.9	22.2	28.9	1.9
Married (%)	72.6	69.4	53.9	88.5
Widowed (%)	3.6	2.8	2.6	4.8
Divorced (%)	3.2	5.6	5.3	0.0
Separated (%)	4.8	0.0	9.2	4.8
Status within household				
Head of hh (%)	8.7	8.3	13.2	5.8
Wife of head of hh (%)	74.6	69.4	57.9	90.4
Mother of head of hh (%)	8.7	12.5	14.5	1.9
Daughter of head of hh (%)	7.9	9.7	14.5	1.9
Household size				
Low hh size (<5) (%)	35.3	37.5	42.1	28.8
Medium hh size (5-7) (%)	39.7	38.9	43.4	37.5
Large hh size (>7) (%)	25.0	23.6	14.5	33.7

3.1.2   Education, occupation and village location

Education of participants was the only parameter, which showed no significant differences between the three districts. Across all districts, only about 8% of women had no or only few years of education at primary school, while nearly 90% had completed primary school. Higher education, i.e. few years or completed secondary school or, in one case, even college, was a privilege for only about 3% of participants with a higher share in Kongwa and a slightly lower share in Singida district (Table 3.1.2).

The typical occupation of women was mixed crop and livestock farmer, while in Kongwa nearly 20% were only crop farmers. In contrary, in Muheza district nearly 20% of women were farmers and additionally had a small business, did some casual labour service or were handcrafter. This share of participants with another income source next to farming was smaller in Singida (13.5%) and especially small in Kongwa district (2.8%) (Table 3.1.2).

The distance in km from each village to the town center was estimated as well as the distance in minutes that one approximately would need to travel from the village to town centre (personal communication with Agricultural District Officers of respective districts). These distances were classified into the categories short, medium and long. It was noticeable that villages in Singida, classified as peri-urban, were located rather far away from town, both in kilometres and minutes travelling time. Yet, it should be noted that Singida town is larger than Kongwa and Muheza and, thus, will influence the surrounding villages differently. Unlike Singida, in Kongwa 50% of women lived in villages with short distance to the town centre (both km and min). In Muheza, even more than 50% of women had only a short distance in km from their village to town, however, not in minutes travelling time. Consequently, not only the actual distance is decisive but how long it takes to cover this distance which, in turn, depends on road conditions and transport supply (Figure 3.1.3).

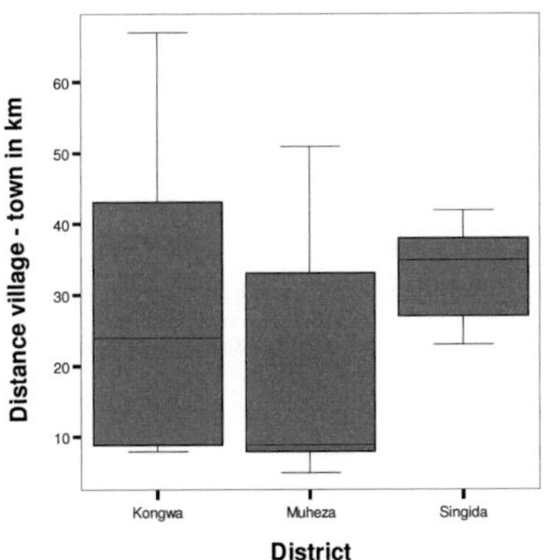

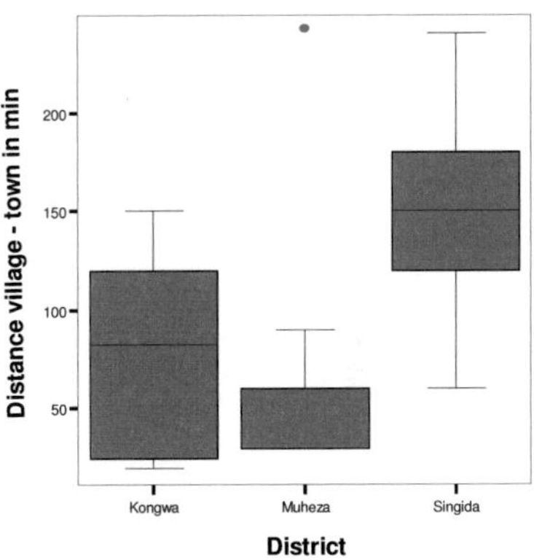

Figure 3.1.3   Distance between villages of women and next town in km and in minutes travelling time within three districts of Tanzania

Table 3.1.2   Socio-economic characteristics of women in Tanzania for the whole study cohort and by
               district (share of participants)

	All districts	Kongwa	Muheza	Singida
n	252	72	76	104
Education				
No or few years education (%)	7.9	9.7	9.2	5.8
Primary school completed (%)	89.3	86.1	86.8	93.3
Higher education (%)	2.8	4.2	3.9	1.0
Occupation				
Crop farmer (%)	10.7	19.4	7.9	6.7
Crop and livestock farmer (%)	77.0	77.8	72.4	79.8
Farmer + business/service (%)	12.3	2.8	19.7	13.5
Distance from village to town in km				
Short distance (5-20 km) (%)	33.7	50.0	64.5	0.0
Medium distance (21-34 km) (%)	30.2	13.9	23.7	46.2
Long distance (35-67 km) (%)	36.1	36.1	11.8	53.8
Distance from village to town in min				
Short distance (20-45 min) (%)	24.2	50.0	32.9	0.0
Medium distance (60-90 min) (%)	21.8	0.0	55.3	12.5
Long distance (120-240 min) (%)	54.0	50.0	11.8	87.5

3.1.3   Wealth parameters

According to the number of possessions, the setting of the house, number of livestock (types, not
number of each type), type of occupation and if vegetables were sold, the status of each woman
was classified into either low or high wealth for each parameter (Table A4). The status according to
these five parameters combined resulted in a wealth status, which was either low (number of "high
wealth" parameters=0), medium (1) or high (2-5). Differences occured between the districts, with
an especially large share of women with high wealth in Singida district (57.7%) and a rather equal
distribution between wealth groups in Kongwa and Muheza districts (Table 3.1.3).

Table 3.1.3   Share of women (%) within three wealth categories in three districts of Tanzania

	All districts	Kongwa	Muheza	Singida
n	252	72	76	104
Low wealth	24.2	27.8	31.6	16.3
Medium wealth	29.8	36.1	28.9	26.0
High wealth	46.0	36.1	39.5	57.7

3.2   Vegetable production

3.2.1   Vegetable types produced

Vegetable types cultivated in farmers' fields or gardens and collected from the wild or fallow land varied highly both between districts and seasons. Therefore, the data is in most cases not summarised across districts and seasons. The five most important indigenous and exotic vegetables in each district and season and the share of women who stated to cultivate or collect them are shown in tables A5-A7. The most popular vegetable, which was important in all three districts and during all seasons, cultivated as well as collected, was amaranth. At least eight different amaranth species were identified to be used by the study participants. A detailed list with all vegetable species cultivated and gathered is shown in the appendix (Table A8). In general, vegetable types collected from wild and fallow land differed much more between the districts than cultivated types. Exotic vegetables were cropped by only few farmers at all, especially in Kongwa district, with not much differences between the seasons.

Regarding vegetable diversity, in Kongwa district many more different vegetable types were collected than cultivated (Figure 3.2.1), while it was the other way around in Singida district. In Muheza, the diversity of both collected and cultivated vegetables together was higher during Jun/Jul (DS) and Nov/Dec (SR) than in the other two districts, however, not during Mar/Apr (LR). In terms of traditional and exotic vegetable types, during all seasons and in all districts, more different traditional vegetable types were cultivated and collected by participants than exotic vegetables (Table 3.2.1).

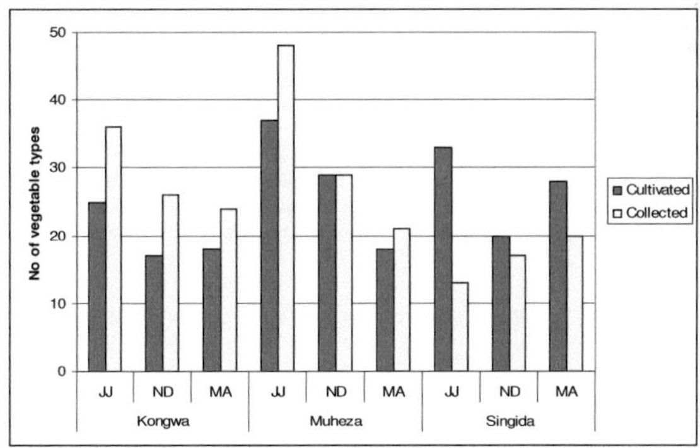

Figure 3.2.1
Overall number of vegetable types cultivated and collected by women during three different seasons in three districts of Tanzania

Table 3.2.1   Number of different traditional (TV) and exotic vegetables (EV) cultivated/collected by women in three districts of Tanzania during three different seasons

	Jun/Jul (DS)		Nov/Dec (SR)		Mar/Apr (LR)	
	TV	EV	TV	EV	TV	EV
Kongwa	45	6	40	3	38	4
Muheza	65	10	51	7	36	3
Singida	37	9	31	6	38	10

Cluster analysis was applied to group participants based on similarities regarding vegetable production (Figure 3.2.2) or consumption (see below). Table 3.2.2 shows the order in which vegetable types appear within Figure 3.2.2 (y-axis). The score for vegetables in each cluster shows the grade of affiliation to the particular cluster. In Table 3.2.3 clusters are described in showing the vegetables which were typically produced by participants belonging to each cluster Because many different vegetables had a high score for one cluster only those with a value of 0.30 or more are shown.

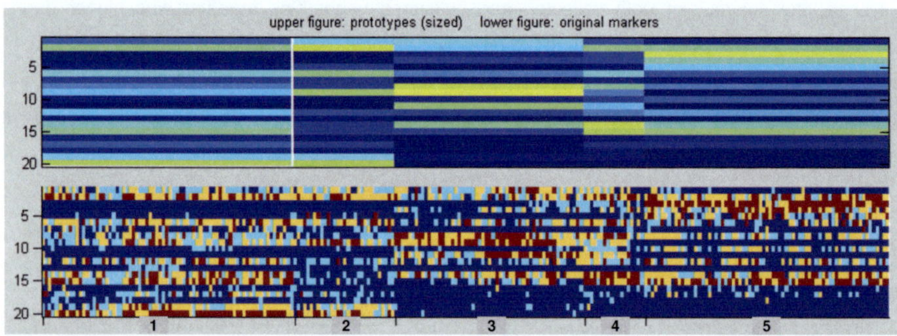

Figure 3.2.2
Five clusters showing the typical production of 20 vegetables (y-axis) by 252 women (x- axis); colours show the number of seasons per year that a vegetable was cultivated by each woman: dark red=highest value, via orange, yellow, green, to dark blue=lowest value/no production; lower figure: data for each single woman; upper figure: women's affiliation to each of the five clusters, clusters being numbered from left to right

Table 3.2.2   List of 20 vegetables produced by women as they appear in the cluster (Figure 3.2.2)

No. within cluster	Vegetable	No. within cluster	Vegetable
1	African spiderplant	11	Mhilile
2	Amaranth lvs	12	Okra
3	Bitter lettuce	13	Onion
4	Black jack	14	Pumpkin lvs
5	Cassava lvs	15	Sweet potato lvs
6	Tree cassava lvs	16	Swiss chard
7	Chinese cabbage	17	Tomato
8	Cowpea lvs	18	White cabbage
9	False sesame	19	Wild cucumber
10	Jute mallow	20	Wild simsim

Table 3.2.3   Vegetables which characterise five clusters generated from number of seasons per year each vegetable was cultivated/collected by women in Tanzania

Cluster 1 n=75		Cluster 2 n=30		Cluster 3 n=56		Cluster 4 n=18		Cluster 5 n=73	
Vegetable	Score	Vegetable	Score	Vegetable	Score	Vegetable	Score	Vegetable	Score
Wild simsim	0.39	Amaranth lvs	0.41	False sesame	0.43	Pumpkin lvs	0.42	Bitter lettuce	0.43
Amaranth lvs	0.35	Wild simsim	0.39	Cowpea lvs	0.41	Sweet potato lvs	0.39	Sweet potato lvs	0.36
Sweet potato lvs	0.34	False sesame	0.37	Mhilile	0.35	Amaranth lvs	0.34	Amaranth lvs	0.33
		Tree cassava lvs	0.34	Pumpkin lvs	0.34	Cowpea lvs	0.32	Black jack	0.32
								Pumpkin lvs	0.30

3.2.2   Number of vegetables produced per person

In this chapter the number of both indigenous and exotic vegetables cultivated and collected by each farmer will be examined. According to K-S- and S-W tests this data was not normally distributed except for data on vegetable production during Mar/Apr (LR) (K-S test statistics=0.087; S-W test statistics=0.977) and combined cultivation and collection during Mar/Apr (LR) (K-S test statistics=0.072; S-W test statistics=0.991).

<u>Cultivation vs. collection</u>

The data summarised for all districts (mean value) differed significantly between the seasons (p<0.001 for all variables). When the three seasons were observed individually, the data on vegetable cultivation and collection also differed significantly between the districts. However, when the data from all seasons was combined (mean value) only vegetable cultivation and cultivation and collection combined differed significantly between the districts (p<0.001), while vegetable collection did not vary (p=0.066).

In fact, the median number of vegetables cultivated was three during Nov/Dec (SR), four in Jun/Jul (DS) and even five during Mar/Apr (LR). The median number of vegetables collected was also highest during Mar/Apr (LR) while it was lowest for Jun/Jul (DS) (Table 3.2.4).

Table 3.2.4   Number of vegetable types cultivated and collected per woman during three different seasons and in three districts (mean across three seasons) of Tanzania

	Cultivation		Collection		Cult + Coll	
	Mean ± SD	Median (range)	Mean ± SD	Median (range)	Mean ± SD	Median (range)
All year	3.8 ± 1.7	4 (0-9)	3.6 ± 1.4	3 (0-9)	7.5 ± 2.4	7 (1-18)
Jun/Jul (DS)	3.8 ± 2.6	4 (0-12)	2.9 ± 2.2	2 (0-10)	6.7 ± 4.0	6 (0-22)
Nov/Dec (SR)	3.1 ± 2.3	3 (0-10)	3.6 ± 2.1	3 (0-11)	6.7 ± 3.5	6 (0-19)
Mar/Apr (LR)	4.7 ± 2.3	5 (0-12)	4.4 ± 2.1	5 (0-10)	9.1 ± 3.7	9 (0-19)
Kongwa	3.2±1.3	3 (1-6)	3.5±1.2	3 (1-8)	6.7±2.0	7 (2-11)
Muheza	4.4±1.5	4 (1-9)	4.1±1.8	4 (1-9)	8.5±2.7	8 (3-18)
Singida	3.9±1.9	4 (0-9)	3.4±1.0	3 (0-6)	7.3±2.2	7 (1-14)

Note: all species of a vegetable, e.g. Amaranth spp., counted as one

The median number of vegetables cultivated per person was higher in Muheza and Singida districts than in Kongwa, while the maximum number of vegetables both produced and collected was especially high for Muheza, less for Singida and smallest for Kongwa (Table 3.2.4). Further, it was noticeable that the minimum number of vegetables both cropped and collected was zero in Singida district, meaning that some farmers did not crop or collect any vegetables at all during a certain period while this was never the case in Kongwa or Muheza.

In Figure 3.2.3 the share of participants who cultivated or collected a certain number of vegetables during different seasons is considered. During Mar/Apr (LR) more farmers cultivated and collected more different vegetables than during the other two seasons. While the share of participants who cultivated different numbers of vegetables was rather evenly distributed, data on vegetable collection was clearly different between the districts. The distribution for Jun/Jul (DS) is skewed to the left, showing that less vegetable diversity is collected during that season; more women collected a higher number of vegetables during Nov/Dec (SR) and even more during Mar/Apr (LR). The ratio of cultivated to collected vegetables during Jun/Jul (DS) was 2.0, i.e. twice as many vegetable types were cultivated than collected during that time. A ratio of 1.0 during the other two seasons showed that collected vegetable types were equal in number to cultivated ones.

If the three districts were compared, it is obvious that in Muheza district more farmers cultivated and collected more different vegetables than in Singida and especially than in Kongwa (Figure 3.2.4). Differences between districts are also visible when cultivation and collection of vegetables are considered separately (but significant only for cultivation). In Singida district, the majority of participants collected either three or four different vegetables, while the number of vegetables cultivated was more evenly distributed. The ratio of cultivated and collected vegetables was 1.0 for both Kongwa and Muheza districts, while it was 1.3 for Singida district, suggesting that more different vegetable types were cultivated in Singida than collected from the wild.

Traditional vs. exotic vegetables

The mean number of vegetables cultivated and/or collected by women showed clearly that more traditional than exotic vegetables were used. Thereby, slightly more exotic vegetables were cultivated during Jun/Jul (DS) than during the other seasons and in Singida more than in the other districts (Figure 3.2.5). In terms of traditional vegetables the mean number cultivated/collected was highest for Muheza during Jun/Jul (DS) while it was highest for Singida districts during Mar/Apr (LR) (Figure 3.2.6).

Additionally, the ratio between the median number of traditional vegetables used by participants and the median number of exotic vegetables was calculated. In relation to traditional, more exotic vegetables were used in Singida while they were rather unimportant in Kongwa (Table 3.2.5).

The number of farmers who cultivated/collected traditional or exotic vegetables also differed. While nearly all participants in all districts and at all times cultivated or gathered some traditional vegetables, only few were involved in exotic vegetable cropping (Table 3.2.6). Overall only 30% of participating women cultivated exotic vegetables and especially few during Nov/Dec (SR) compared to the other seasons. In Kongwa, only 11% of farmers grew exotic vegetables, while in Muheza 23% and in Singida even 49% of women had exotic vegetables in their gardens.

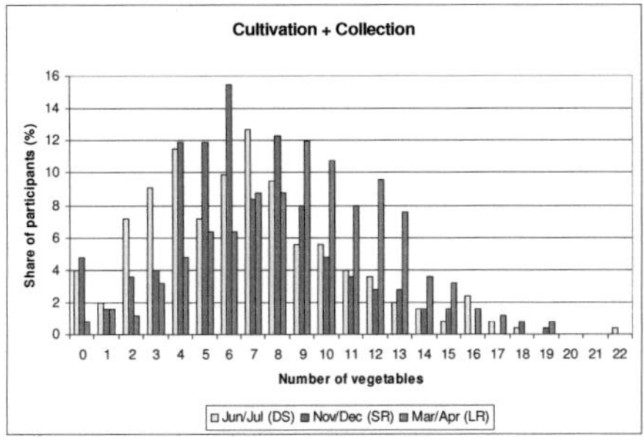

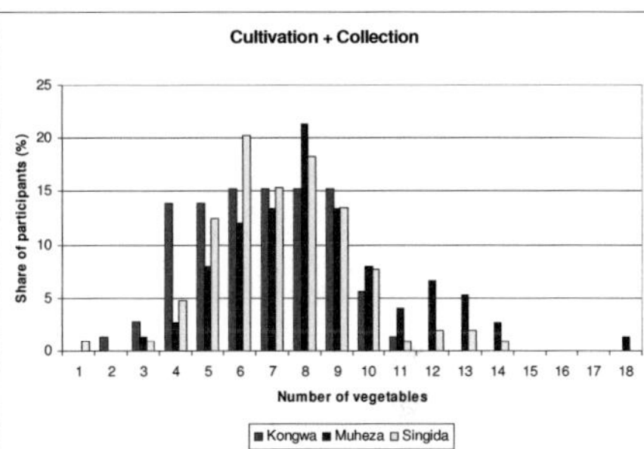

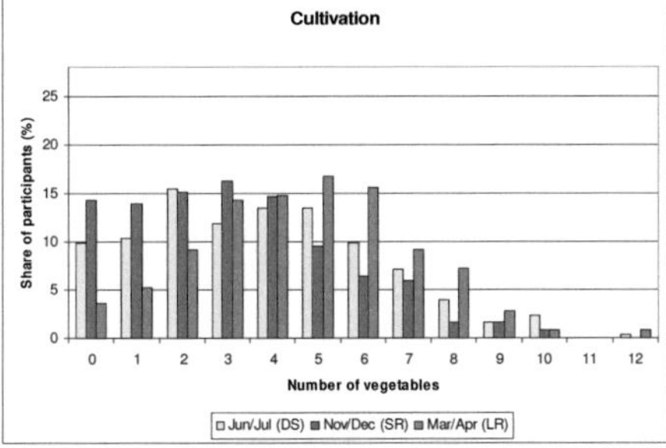

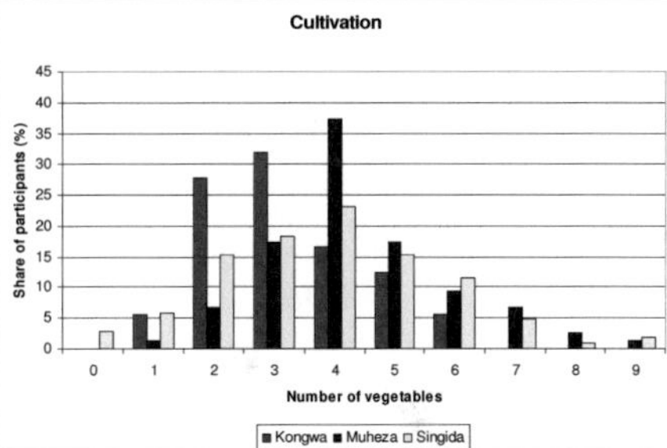

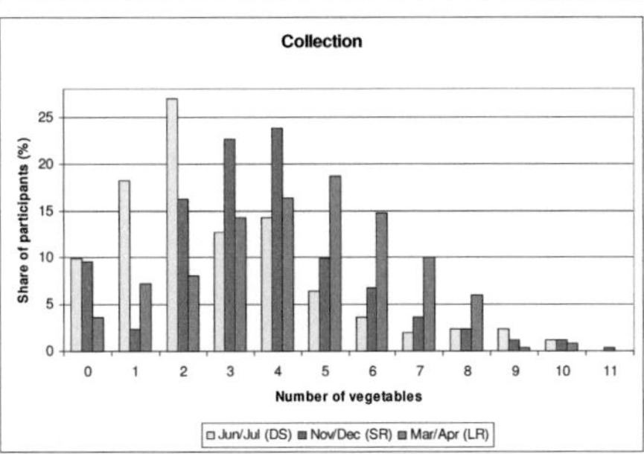

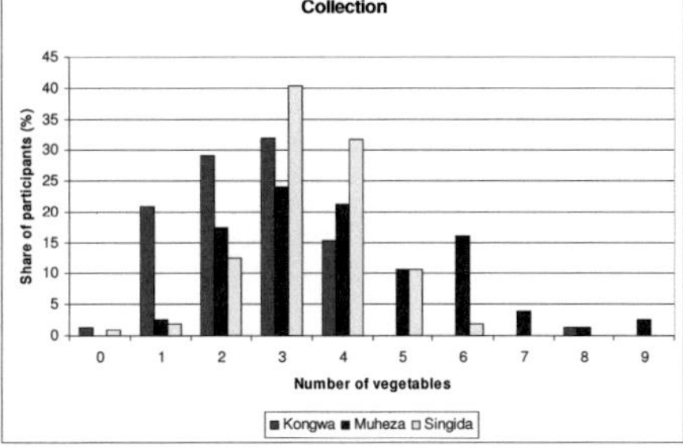

Figure 3.2.3
Share of participants that cultivated and collected a certain number of vegetable types during three different seasons

Figure 3.2.4
Share of participants that cultivated and collected a certain number of vegetable types in three districts of Tanzania (mean across three seasons)

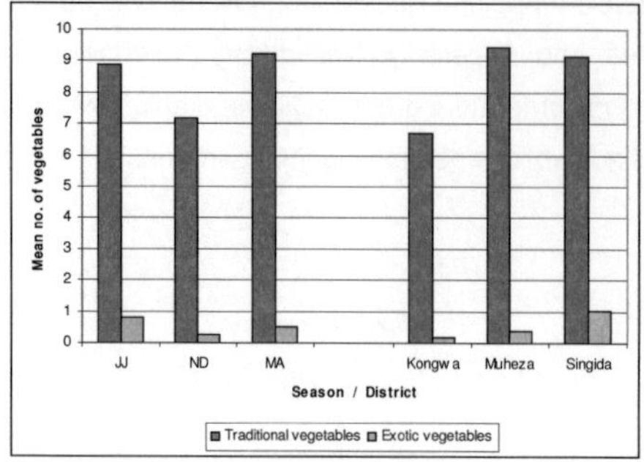

Figure 3.2.5
Mean number of traditional and exotic vegetable types cultivated/collected by women during three different seasons in three districts of Tanzania

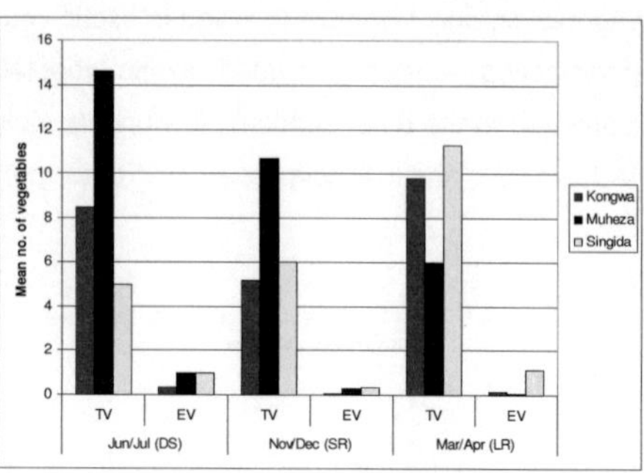

Figure 3.2.6
Mean number of traditional (TV) and exotic vegetable types (EV) cultivated/collected by women in three districts of Tanzania during three different seasons

Table 3.2.5   Ratio traditional : exotic vegetable types cultivated/collected per woman in three districts of Tanzania during three seasons

	Jun/Jul (DS)	Nov/Dec (SR)	Mar/Apr (LR)
Kongwa	25.5	74.4	70.7
Muheza	14.8	37.0	112.5
Singida	5.0	16.8	10.1

Table 3.2.6   Number and share of farmers that cultivated/collected traditional and exotic vegetables during three different seasons in three districts of Tanzania

	All district N=252		Kongwa N=72		Muheza N=76		Singida N=104	
	No	(%)	No	(%)	No	(%)	No	(%)
Traditional vegetables								
All seasons	244	96.8	70	96.8	75	99.1	99	95.2
Jun/Jul (DS)	242	96.0	70	97.2	76	100.0	96	92.3
Nov/Dec (SR)	240	95.2	67	93.1	76	100.0	97	93.3
Mar/Apr (LR)	250	99.2	72	100.0	74	97.4	104	100.0
Exotic vegetables								
All seasons	76	30.3	8	11.1	18	23.3	51	48.7
Jun/Jul (DS)	100	39.7	9	12.5	33	43.4	58	55.8
Nov/Dec (SR)	47	18.7	5	6.9	16	21.1	26	25.0
Mar/Apr (LR)	82	32.5	10	13.9	4	5.3	68	65.4

Correlations and associations

The relationship between vegetable cultivation/collection and socio-economic factors was tested as well as linear relationships between vegetable cultivation/collection and knowledge and attitudes of participants. Because these relationships were numerous, only significant correlations or associations are highlighted in the following. In order to prevent a too low number of cases when dividing participants into the three districts, for correlations it was resigned from differentiating between the three districts. However, the three investigated seasons were considered separately. Significant relationships were observed between household size and number of vegetables collected during Jun/Jul (DS) (ρ=-0.133; p=0.035) and during Mar/Apr (LR) (ρ=0.156; p=0.013), however, once negative and once positive. Only positive relationships were found between education and vegetable cultivation or collection during some seasons, while wealth was also once negatively associated to number of vegetables collected during Mar/Apr (LR). The distance from village to town, both in terms of kilometres and minutes travelling time, was correlated both positively and negatively with several cultivation/collection parameters (Table 3.2.7).

Besides the number of vegetables cultivated and collected also the relationship between the actual cultivation and collection of vegetables (yes/no) with various socio-economic parameters was tested. Significant associations are listed in Table 3.2.8 or shown as pie charts in Figures 3.2.7 and 3.2.8.

Table 3.2.7   Associations/correlations between the number of traditional (TV) and exotic vegetables (EV) cultivated/collected and certain socio-economic parameters of women in Tanzania during three different seasons

	Wealth		Education		Distance in km		Distance in min	
	Correl. coeff. tau	p	Correl. coeff. tau	p	Correl. coeff. rho	p	Correl. coeff. rho	p
Vegetable cultivation								
Jun/Jul (DS)	0.152	0.003	0.119	0.026	n.a.	n.s.	n.a.	n.s.
Nov/Dec (SR)	n.a.	n.s.	n.a.	n.s.	n.a.	n.s.	n.a.	n.s.
Mar/Apr (LR)	n.a.	n.s.	0.113	0.037	0.226	<0.001	0.306	<0.001
Vegetable collection								
Jun/Jul (DS)	n.a.	n.s.	n.a.	n.s.	-0.145	0.021	-0.253	<0.001
Nov/Dec (SR)	n.a.	n.s.	0.119	0.030	-0.251	<0.001	-0.188	3
Mar/Apr (LR)	-0.123	0.018	n.a.	n.s.	0.487	<0.001	0.563	<0.001
TV cult/coll								
Jun/Jul (DS)	n.a.	n.s	0.137	0.009	0.143	0.024	0.271	<0.001
Nov/Dec (SR)	n.a.	n.s.	0.114	0.032	-0.166	0.008	n.a.	n.s.
Mar/Apr (LR)	0.143	0.005	n.a.	n.s.	0.408	<0.001	0.503	<0,001
EV cultivation								
Jun/Jul (DS)	0.155	0.006	n.a.	n.s.	0.200	0.001	0.342	<0.001
Nov/Dec (SR)	0.191	0.001	n.a.	n.s.	n.a.	n.s.	n.a.	n.s.
Mar/Apr (LR)	n.a.	n.s.	n.a.	n.s.	0.268	<0.001	0.401	<0.001

n.a.=not applicable; n.s.=not significant

Table 3.2.8   Vegetable collection (yes/no) correlated to certain socio-economic parameters of women in Tanzania (p-values according to chi-square test) during different seasons

Season	Religion	Distance in km (grouped)
Jun/Jul (DS)	0.029	0.004
Nov/Dec (SR)	n.s.	0.006

n.s.=not significant

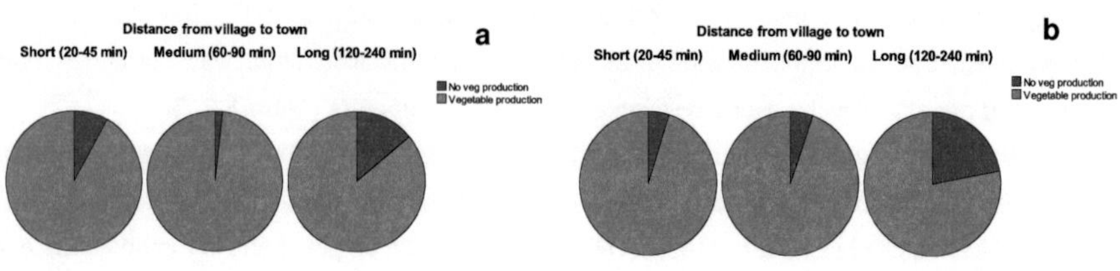

Figure 3.2.7   Association between distance from village to town (in minutes travelling time) and vegetable cultivation (yes/no) during a) Jun/Jul (DS) (p=0.034) and b) Nov/Dec (SR) (p=0.001)

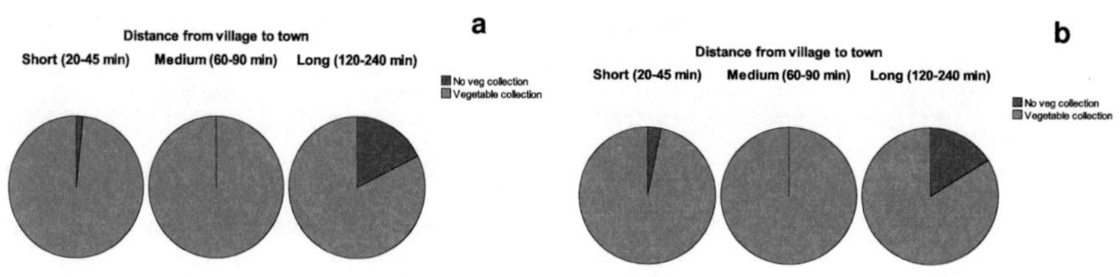

Figure 3.2.8   Association between distance from village to town (in minutes travelling time) and vegetable collection (yes/no) during a) Jun/Jul (DS) (p<0.001) and b) Nov/Dec (SR) (p<0.001)

3.2.3   Sales and purchase of vegetables

<u>Vegetable sales</u>

In general, not many study participants sold their vegetables, namely only between 5.6% and 30.3%, depending on the season and district (Figure 3.2.9). Significant differences in terms of vegetables sales (yes/no) between the three seasons occurred (p<0.001). In contrast, differences between the districts were only significant during Nov/Dec (SR) (p=0.007). Most of the women who sold their vegetables did this only during one season, only in Muheza a greater share of participants sold also during two or even three seasons (Table 3.2.9). More information on the place and frequency of vegetable sales is provided in the appendix (Tables A9 and A10; Figures A2 and A3).

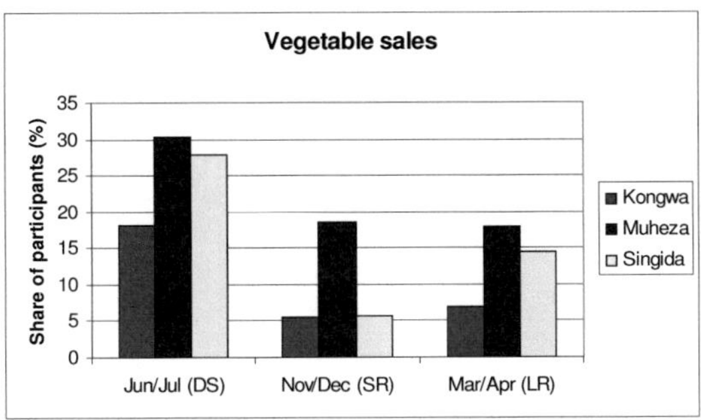

Figure 3.2.9
Share of women that sold vegetables during three different
seasons in three districts of Tanzania

Table 3.2.9   Share of women (%) that sold vegetables during a different number of seasons per year
in Tanzania

	No season	1 season	2 seasons	All seasons
Kongwa	76.4	18.1	4.2	1.4
Muheza	60.5	21.1	10.5	7.9
Sinigida	63.5	26.9	7.7	1.9

<u>Vegetable purchase</u>

Data on farmers' behaviour of purchasing additional vegetables was also allocated, however, only
during Nov/Dec (SR) and Mar/Apr (LR). Purchasing of vegetables (yes/no) differed significantly
between these two seasons (p=0.001) and within the seasons also significant differences between
the districts were found (p=0.001 and p<0.001, respectively). In fact, in Singida district always less
than 50% of participants bought vegetables, while in Muheza district even more than 60% bought
additional vegetables during both seasons. In Kongwa, interviewed women were indifferent with
75% purchasing vegetables during Nov/Dec (SR), yet, only 36% during Mar/Apr (LR) (Figure
3.2.10). Where, why, which and how often additional vegetables were purchased is shown in
Tables A11-A13 and Figures A4 and A5.

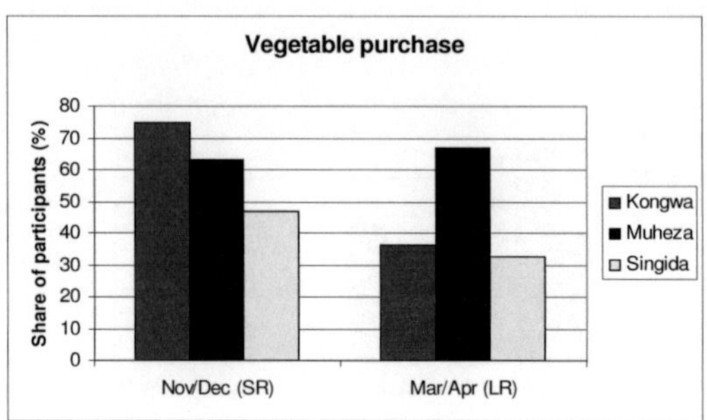

Figure 3.2.10
Share of women that bought vegetables during two different
seasons in three districts of Tanzania

It was again differentiated if traditional or exotic vegetable types were purchased and the mean number of vegetables purchased per participant calculated (Table 3.2.10). Also the ratio of traditional and exotic vegetables bought was generated and the so far existing pattern of the three districts was confirmed. In the peri-urban Singida district more exotic vegetables were purchased unlike in Muheza, where more traditional types were bought by participants. In Kongwa, nearly double the number of traditionals compared to exotic vegetables was purchased during Nov/Dec (SR).

Table 3.2.10   Mean number of traditional (TV) and exotic vegetable (EV) types purchased per woman and ratio between TV and EV purchased during two seasons in three districts of Tanzania

	Nov/Dec (SR) (n=150)			Mar/Apr (LR) (n=109)		
	Kongwa	**Muheza**	**Singida**	**Kongwa**	**Muheza**	**Singida**
TV	2.1	1.6	0.9	1.0	1.8	0.6
EV	1.2	1.1	1.3	0.6	1.2	1.5
TV+EV	3.3	2.7	2.2	1.6	3.0	2.1
Ratio TV:EV	1.8	1.5	0.7	1.5	1.5	0.4

<u>Correlations and associations</u>

While wealth was highly positively associated with vegetable sales during Jun/Jul (DS) (p<0.001), household size was positively correlated with vegetable purchase during Nov/Dec (SR) (p=0.047); the occupation of participants was positively associated with vegetable purchase during Mar/Apr (LR) (p=0.021). Further associations are listed in Table 3.2.11 or are depicted as pie charts in Figures 3.2.11–3.2.16. Seasonal differences of the association between ethnic groups (districts) and vegetable purchase (Figures 3.2.14 and 3.2.15) was already reflected in Figure 3.2.10.

Table 3.2.11   Vegetable purchase and sales (yes/no) correlated to certain socio-economic variables (p-values according to chi-square test)

	Distance in km (grouped)	Distance in min (grouped)
Vegetable sales		
Nov/Dec (SR)	n.s.	0.032
Mar/Apr (LR)	0.012	n.s.
Vegetable purchase		
Mar/Apr (LR)	<0.001	<0.001

n.s.=not significant

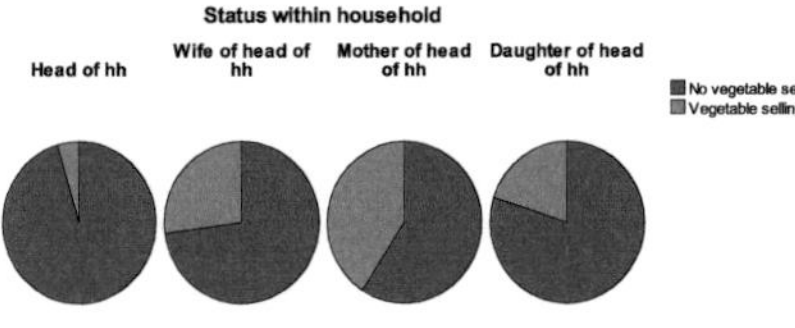

Figure 3.2.11
Association between status within household and vegetable sales (yes/no) by women in Tanzania during Jun/Jul (DS) (p=0.039)

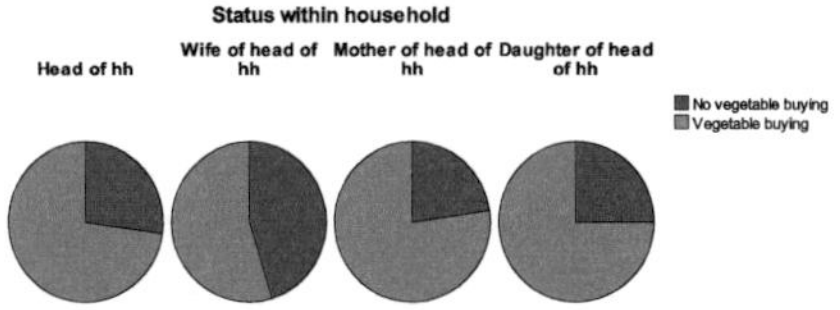

Figure 3.2.12
Association between status within household and vegetable purchase (yes/no) by women in Tanzania during Nov/Dec (SR) (p=0.042)

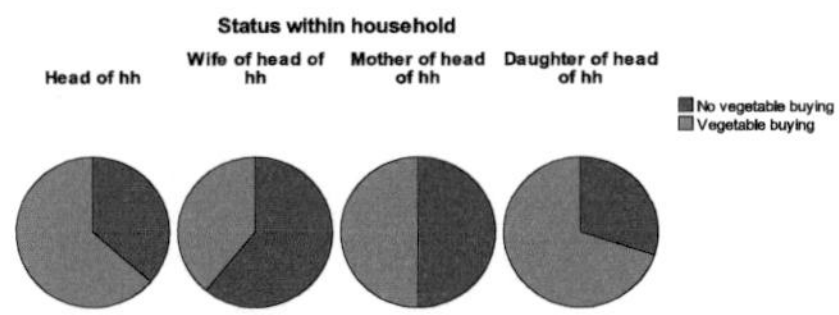

Figure 3.2.13
Association between status within household and vegetable purchase (yes/no) by women in Tanzania during Mar/Apr (LR) (p=0.012)

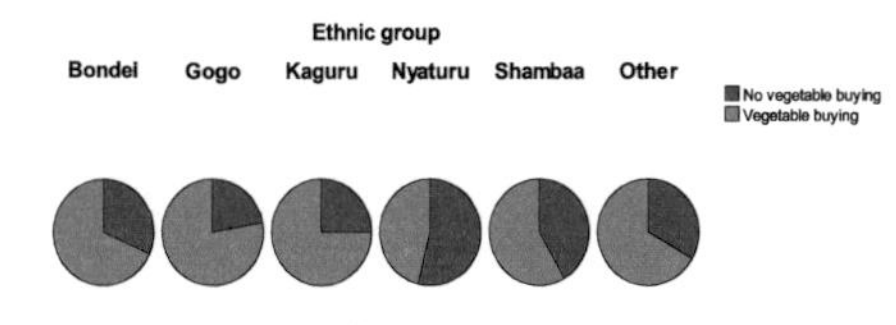

Figure 3.2.14
Association between ethnic group and vegetable purchase (yes/no) by women in Tanzania during Nov/Dec (SR) (p=0.005)

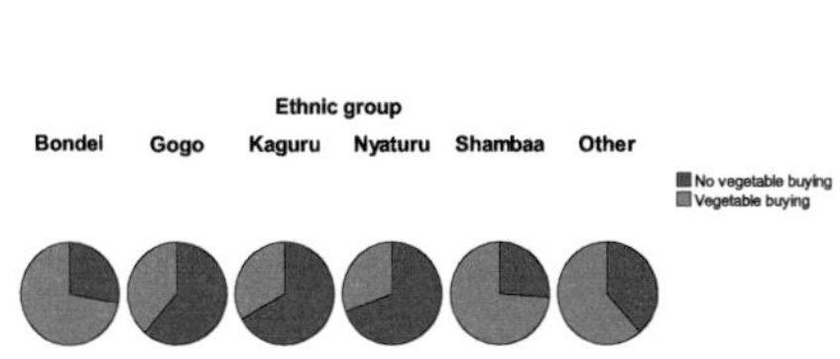

Figure 3.2.15
Association between ethnic group and vegetable purchase (yes/no) by women in Tanzania during Mar/Apr (LR) (p<0.001)

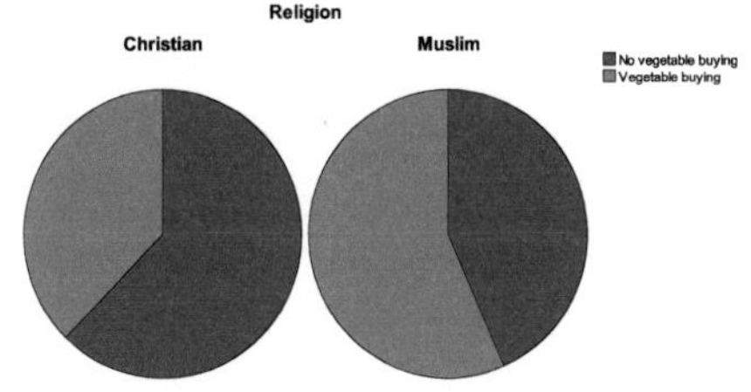

Figure 3.2.16
Association between religion and vegetable purchase (yes/no) by women in Tanzania during Mar/Apr (LR) (p=0.004)

3.3   Food consumption

3.3.1   Food intake

From the selected food groups only the data for cereals and, in some cases, vegetables were normally distributed (K-S test statistic (yearly mean): cereals=0.085, vegetables=0.093; S-W test statistic (yearly mean): cereals=0.98, vegetables=0.922). Furthermore, only these two food groups were consumed by nearly all women during all seasons (Figure 3.3.1). Seasonal differences were significant for all food groups (p≤0.002) except for cereals (p=0.071), vegetables (p=0.160), pulses (p=0.352), fish (p=0.549) and animal products (p=0.732). Especially fruits were consumed by less participants in Nov/Dec (SR), bread and cakes were more consumed during Nov/Dec (SR) and starchy plants were eaten by more women in Mar/Apr (LR) than in the other two seasons. Furthermore, the share of women that consumed a certain food group on the previous day was different for the three districts (Figures 3.3.1). It was apparent that food diversity was highest in Muheza with food groups more evenly consumed than in the other two districts. In Kongwa and Singida, clearly cereals and vegetables dominated the dietary pattern.

The mean amount of each food consumed by participants across three seasons is shown in Table 3.3.1 (data for each district see Table A16). Mean values (not median as it was zero in many cases) for certain food groups are visualised in figures 3.3.2 and 3.3.3. While in Kongwa and especially in Singida a greater amount of vegetables was consumed than in Muheza, in the latter district participants consumed much more fruits. The median amount of cereals consumed was highest in Singida, while the amount of tea, starchy plants, pulses, nuts and also breads/cakes was highest in Muheza. From the picture that resulted for Muheza, it was assumed that the general dietary diversity was higher than in the other two districts (see chapter 3.3.3). When the mean amount of food intake was displayed for each season per district, again Muheza showed the greatest differences while in Kongwa and Singida less variation between the seasons was observed (Figure A6). Food group ratios of all food groups per district and per season are shown in figures A7-A12.

To calculate the share of women that consumed foods below, within or above a recommended intake, food groups were further combined, e.g. all starchy foods, all foods from animal sources. Dietary-based guidelines are neither available for Tanzania nor internationally and, anyway, would be difficult to frame for a large group of people with different eating habits. Still, there exist some dietary suggestions for daily food intake for different countries and regions (FAO, 1997; USDHHS and USDA 2005; DGE 2004), which were combined for this study (Table 3.3.2).

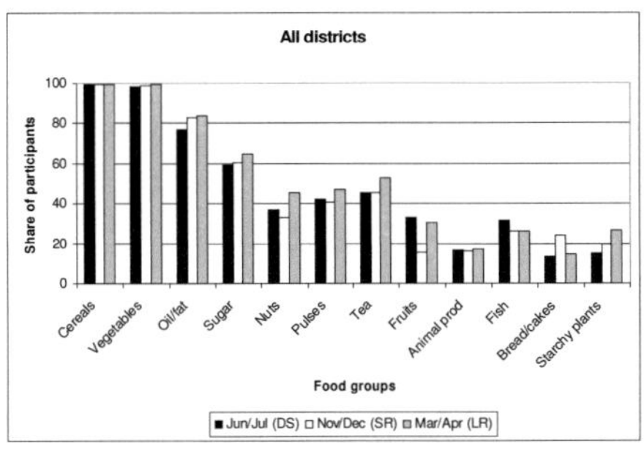

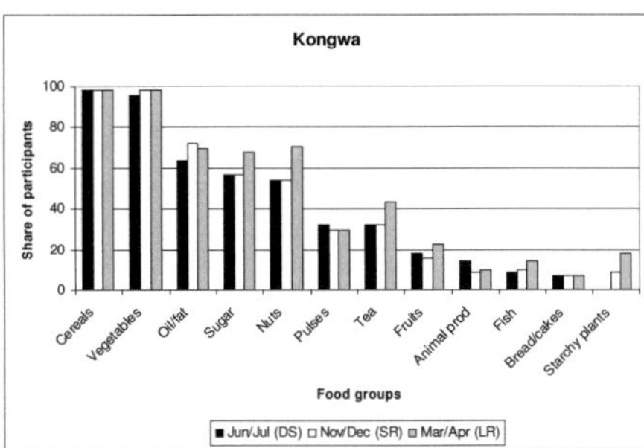

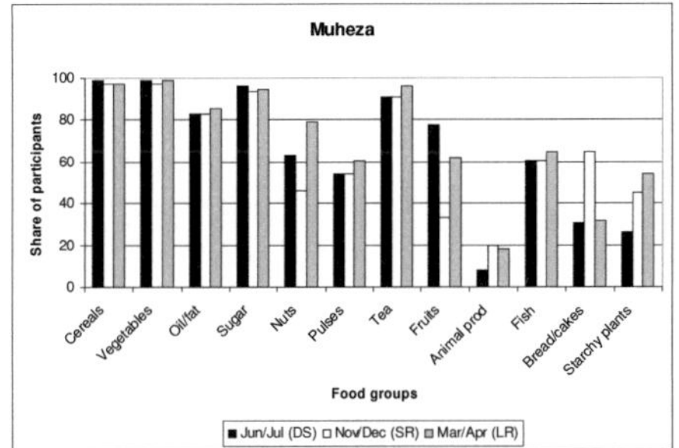

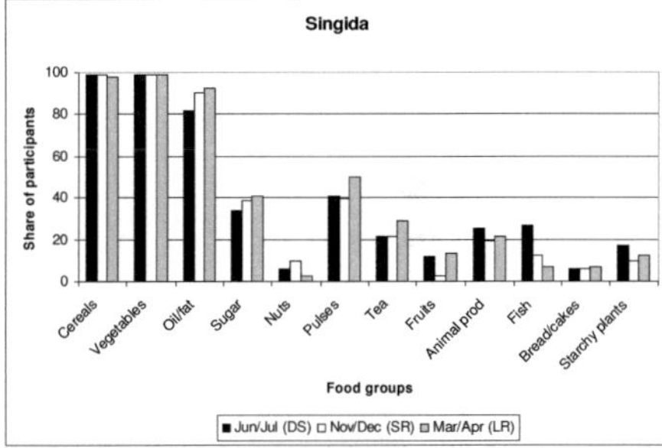

Figure 3.3.1   Share of participants (%) that consumed twelve different food groups on the previous day during three different seasons in all districts(n=252), Kongwa (n=72), Muheza (n=76) and Singida (n=104)

Table 3.3.1   Intake of selected food groups (g/day) by women of three districts in Tanzania; mean of three days during three different seasons

Food group	Mean ± SD	Median (range)
Cereals	320 ± 120	322 (0-680)
Bread/cakes	28 ± 53	0 (0-333)
Fruits	61 ± 101	0 (0-700)
Vegetables	278 ± 141	258 (0-930)
Nuts	47 ± 77	10 (0-400)
Pulses	74 ± 81	58 (0-424)
Starchy plants	55 ± 94	0 (0-326)
Tea	179 ± 190	133 (0-933)
Oil/fat	19 ± 16	16 (0-96)
Sugar	8 ± 12	5 (0-76)
Fish	29 ± 42	0 (0-230)
Animal prod (excl fish)	30 ± 54	0 (0-327)

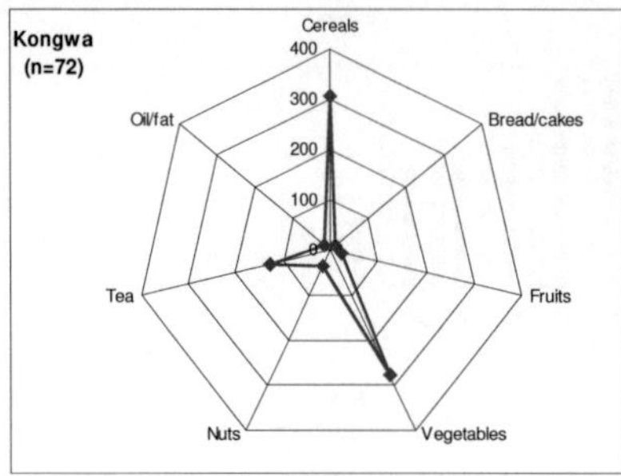

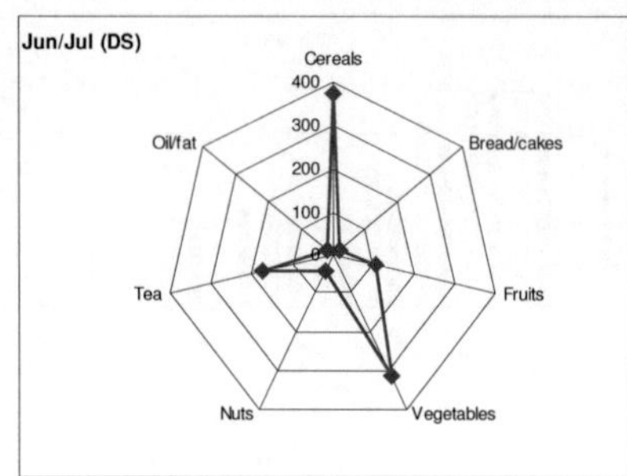

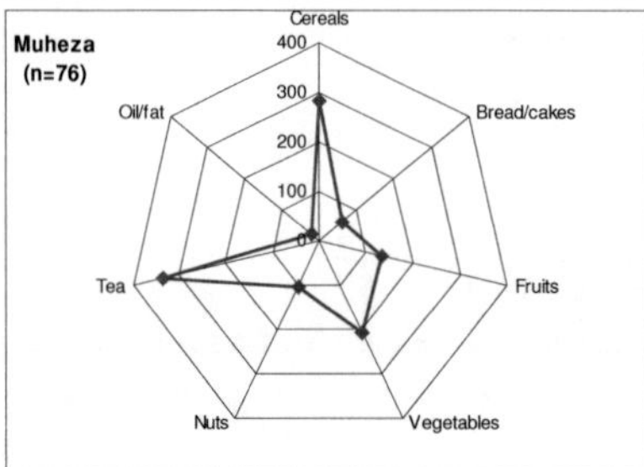

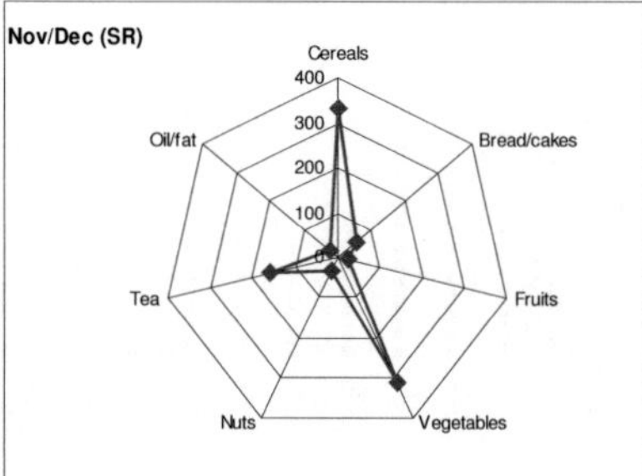

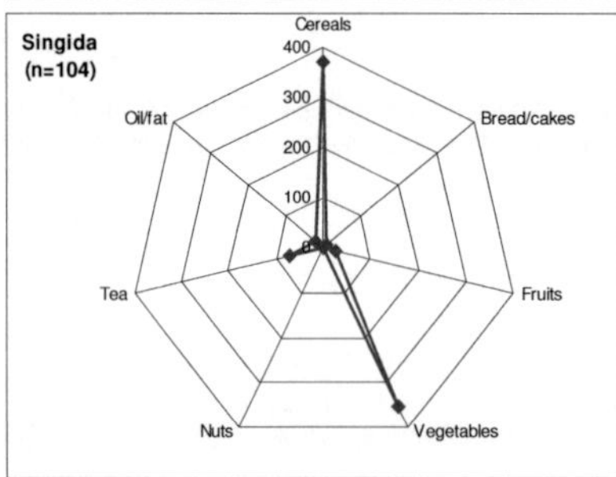

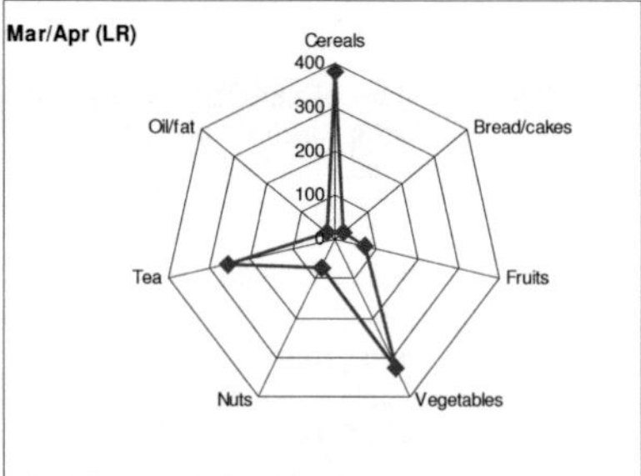

Figure 3.3.2
Mean intake of seven main food groups (g/d) by
women in three districts of Tanzania; mean
across three seasons

Figure 3.3.3
Mean intake of seven main food groups (g/d) by
women during three different seasons in
Tanzania (n=252)

Table 3.3.2   Recommended dietary allowance (RDA) for five food groups used for this study

Food Group	Recommendation in g/d (RDA)
Starchy staple food (including cereals, starchy roots, tubers, plantain etc.)	300 – 500
Fruits and vegetables	300 – 500
Fats and oils	15 – 30
Animal products (including meat, poultry, fish, eggs, milk)	50 – 150
Nuts and pulses	50 – 200

For all food groups and regarding the mean across three days, the share of women eating an amount that was within the recommendation was highest, however, only ranging between 34% for animal products and 54% for starchy staple food (Figure 3.3.4). Regarding starchy staple foods, only about 13% of participants consumed less than the recommended dietary allowance (RDA), in terms of nuts/pulses and animal products about 22% and 27%, respectively, consumed less than what is suggested; about one third of women ate less than the RDA for fruits and vegetables (31%) and fats and oils (33%) (Figure 3.3.7; Tables A17 and A18).

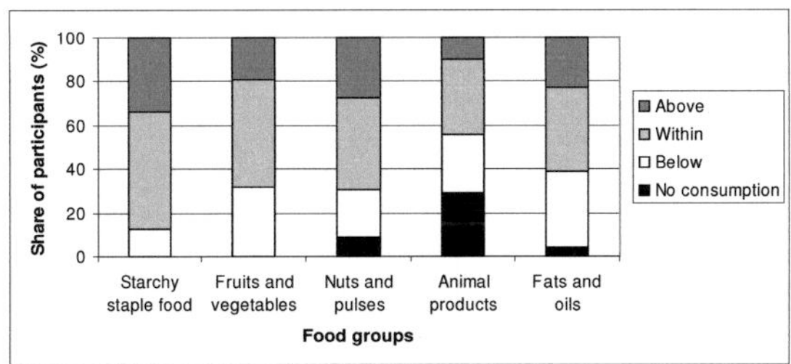

Figure 3.3.4
Share of women that consumed five food groups according to the recommended dietary allowance (RDA) set for this study; mean across three seasons

The intake of these food groups by study participants differed significantly among the districts at $p \leq 0.05$ (oil/fat, fruits and vegetables) or at $p \leq 0.01$ (all other food groups). Between seasons, food intake differed only for the food groups bread/cakes, fruits, nuts, starchy plants, tea (all with $p < 0.001$) and oil/fat ($p = 0.002$).

For each season the data of only one 24h-recall was available and, consequently, the share of women that did not consume a certain food group, e.g. animal products, was much higher than when considering the mean across three seasons. When the three seasons were compared, it was noticeable that fruit and vegetable consumption were lower during Nov/Dec (SR) with nearly 50% of participants consuming less than the RDA. Yet, during the other two seasons a share of about 40% of women did also not reach the RDA of fruits and vegetables (Figure 3.3.5). Regarding districts, Muheza was quite different from the others. There, nearly all participants consumed all five food groups and, consequently, dietary diversity was higher. Also, the share of women who ate a greater amount than the RDA was much higher in Muheza than in the other districts (Figure 3.3.6).

When considering only intake values and share of participants, dietary patterns are more assumed than calculated and are, therefore, shown for the whole districts and not for single participants. A more detailed analysis of food group consumption data was performed to clearly identify possible dietary patterns within the study population (chapter 3.3.4).

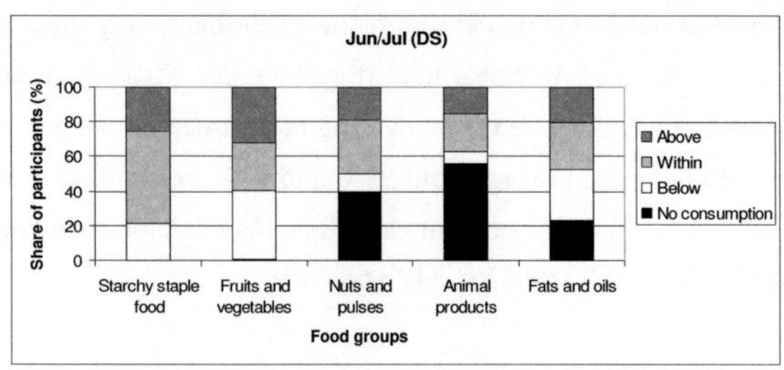

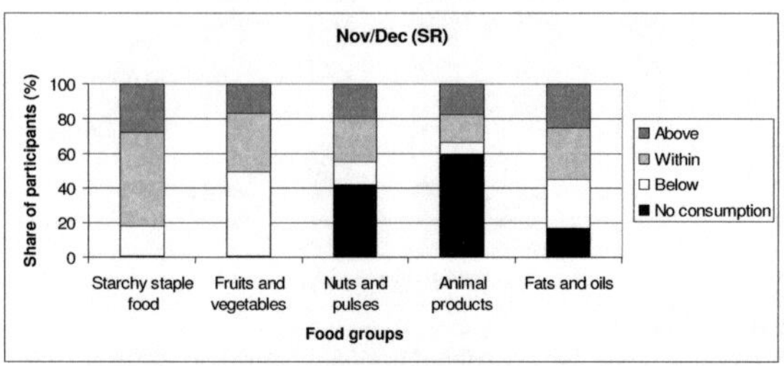

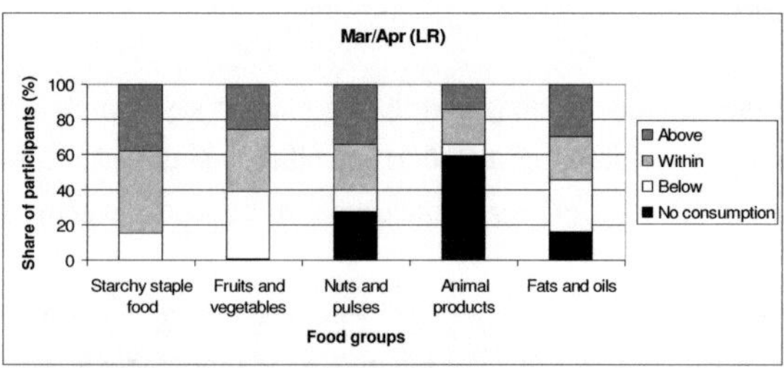

Figure 3.3.5
Share of women that consumed five food groups according to the recommended dietary allowance during three different seasons in Tanzania

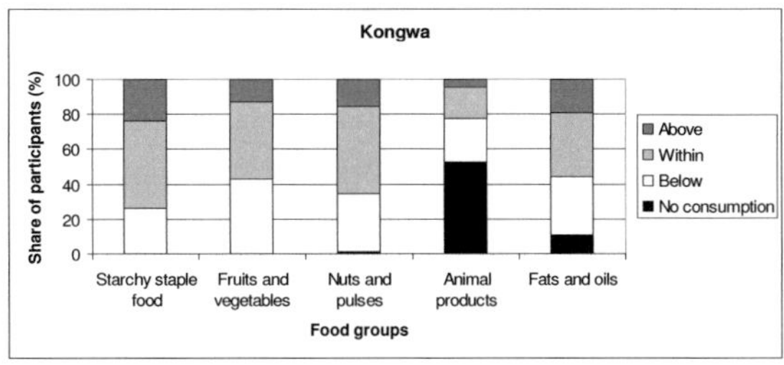

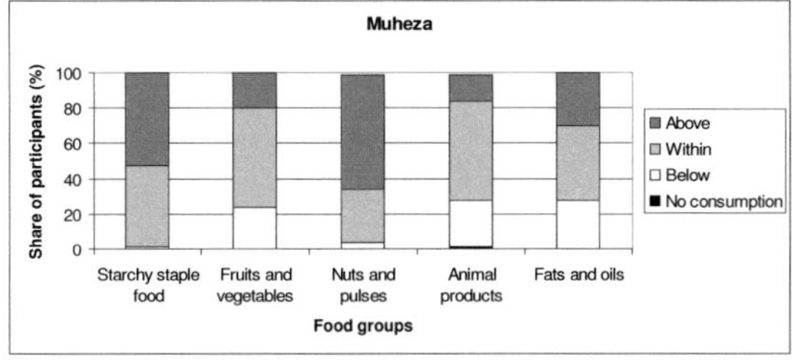

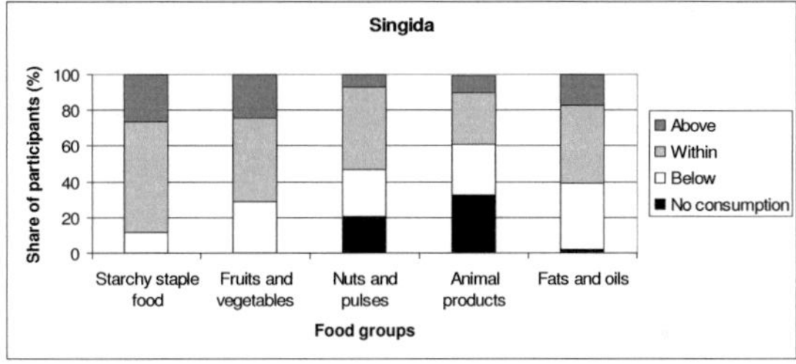

Figure 3.3.6
Share of participants that consumed five food groups according to
the recommended dietary allowance in three districts of Tanzania;
mean across three seasons

Correlations and associations

A significant and positive relationships was observed between age and tea consumption (ρ=0.156*; p=0.014). The positive association between wealth and the consumption of animal products was highly significant (τ=0.171; p=0.002), similar to the negative association between wealth and nut consumption (τ=-0.164; p=0.002). While education showed no correlation with food consumption, household size and the distance from village to town, both in terms of kilometres and minutes travelling time, were correlated with the intake of several food groups (Table 3.3.3). Household size was in all cases negatively correlated with the different food intakes. This was also true for the distances, however, not for the correlation between distance and vegetable consumption, which was highly positively correlated. In terms of nominal socio-economic variables, only few associations existed with food consumption which are illustrated in figures 3.3.7 – 3.3.10.

Table 3.3.3   Food group intake of women correlated to certain socio-economic parameters; mean across three seasons

Food group	Household size		Distance in km		Distance in min	
	Correl. coeff. rho	p	Correl. coeff. rho	p	Correl. coeff. rho	p
Bread and cakes	n.a.	n.s.	-0.214	0.001	-0.182	0.004
Fruits	-0.127	0.044	-0.190	0.002	-0.219	0.001
Vegetables	-0.142	0.024	0.233	<0.001	0.230	<0.001
Nuts	-0.193	0.002	-0.444	<0.001	-0.557	<0.001
Tea	-0.131	0.039	-0.198	0.002	-0.223	<0.001
Oils and fats	-0.312	<0.001	n.a.	n.s.	n.a.	n.s.
Sugar	-0.145	0.022	-0.153	0.015	-0.266	<0.001
Fish	-0.162	0.010	-0.173	0.006	-0.164	0.009
Animal products	n.a.	n.s.	n.a.	n.s.	0.220	<0.001

n.a.=not applicable; n.s.=not significant

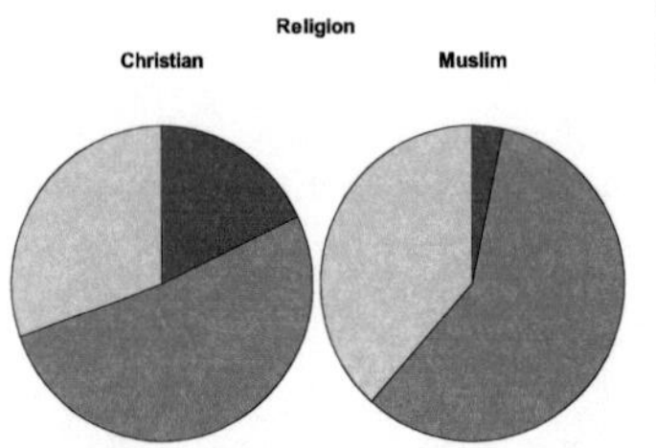

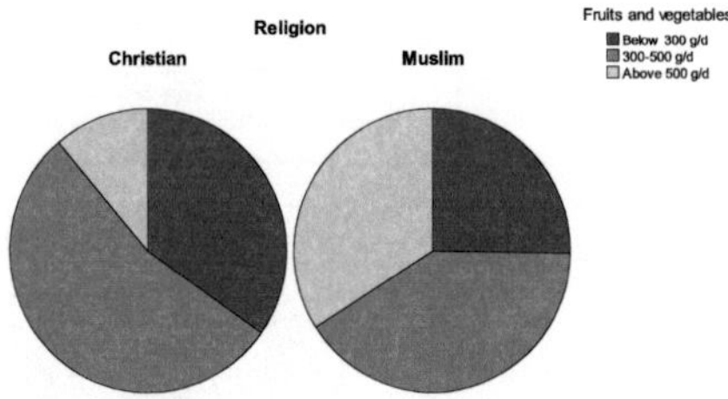

Figure 3.3.7
Association between religion and the consumption of starchy staples by women in Tanzania (mean across three seasons) (p=0.003)

Figure 3.3.8
Association between religion and the consumption of fruits and vegetables by women in Tanzania (mean across three seasons) (p<0.001)

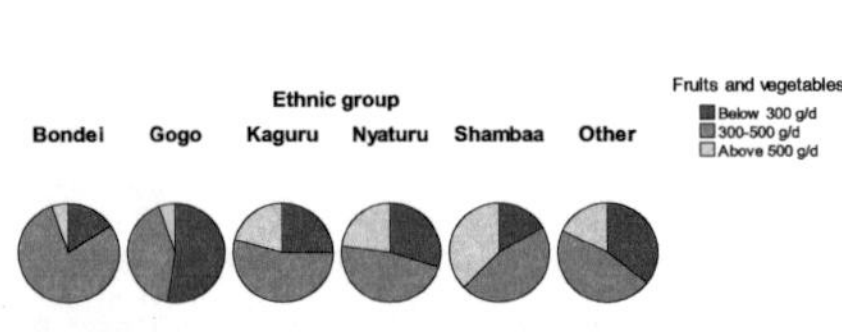

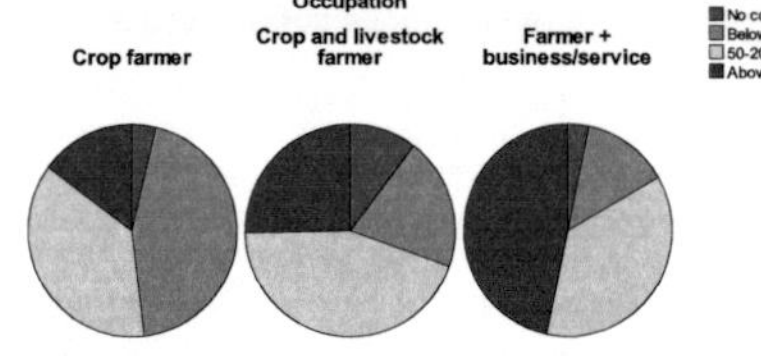

Figure 3.3.9
Association between ethnic group and the consumption of fruits and vegetables by women in Tanzania (mean across three seasons) (p=0.009)

Figure 3.3.10
Association between occupation and the consumption of nuts and pulses by women in Tanzania (mean across three seasons) (p=0.011)

The production of vegetables, namely the number of vegetables cultivated or collected by one woman or the number of traditional and exotic vegetables cultivated and collected, was also checked against the other food groups consumed. The results are presented for the overall mean of one year (Table 3.3.4), while results of single seasons are shown in Table A19. Only poor correlation was found between the vegetable production data and the consumption of animal products, pulses, oil/fat and, interestingly, vegetables for the whole year. However, there were significant correlations during Nov/Dec (SR) and, especially, during Mar/Apr (LR) between vegetable production and consumption. While the latter correlation was positive, many other food intakes were negatively correlated to vegetable production during Mar/Apr (LR).

Table 3.3.4　Correlations between vegetable cultivation/collection and food intake of eight food groups by women in Tanzania; mean across three seasons

Food group	No. of vegetables cultivated		No. of vegetables collected		No. of TVs cult/coll		No. of EVs cultivated	
	Correl. coeff. rho	p	Correl. coeff. rho	p	Correl. coeff. rho	p	Correl. coeff. rho	p
Cereals	n.a.	n.s.	n.a.	n.s.	n.a.	n.s.	0.131	0.038
Bread and cakes	n.a.	n.s.	n.a.	n.s.	0.163	0.010	n.a.	n.s.
Fruits	0.222	<0.001	0.198	0.002	0.267	<0.001	n.a.	n.s.
Nuts	0.144	0.023	0.187	0.003	-0.345	<0.001	n.a.	n.s.
Starchy plants	0.332	<0.001	n.a.	n.s.	0.221	<0.001	0.130	0.039
Tea	0.214	0.001	n.a.	n.s.	0.200	0.001	n.a.	n.s.
Sugar	0.136	0.031	0.140	0.026	0.200	0.001	n.a.	n.s.
Fish	n.a.	n.s.	0.128	0.043	n.a.	n.s.	n.a.	n.s.

n.a.=not applicable; n.s.=not significant

3.3.2　Nutrient intake

Single nutrient intake by each participant was calculated also from the 24h-recall data. While the analysis with Nutrisurvey can only give an estimate of what each participant consumed on the previous day (or on three non-consecutive days if the mean across the year is taken), it still provides a good overview on the achievement of recommended dietary intakes for the study group. The recommendations for nutrient intakes used in the following were always those for women aged between 18 and 45 years if not stated otherwise. They are compiled from Burgess and Glasauer (2004), IOM (2001) and DGE (2000) because there exist no recommendations for Tanzania.

Distribution of nutrient intake among study participants was normal only for carbohydrates (K-S test statistic=0.045; S-W test statistic=0.995) and total energy intake (K-S test statistic=0.047; S-W test statistic=0.965). The intake of these nutrients, as well as protein, differed significantly ($p \leq 0.01$) between the three districts, while they did not differ between seasons (p=0.599 for energy; p=0.144 for protein; p=0.538 for carbohydrates). In contrary, for fat intake no significant differences between the districts were observed, while differences between the seasons were significant (p<0.001).

Mean energy intake was rather low for all districts (mean across three seasons) with 1845 kcal/day only (Table 3.3.5). The highest mean energy intake was found in Muheza (2053 kcal/day), a medium intake in Kongwa (1836 kcal/day) and the lowest in Singida (1694 kcal/day) (Table A20 and Figure A13). Overall, about 44% of participants consumed the recommended dietary intake of energy (1800-2600 kcal/d), while 47% consumed less, but 9% even more than that. Though no seasonal differences became apparent across the whole study cohort, the single districts showed seasonal differences in energy intake, especially in Kongwa district (Figure 3.3.11).

The median protein intake of participants for all districts (57 g/day) was within the recommended range of 45 to 65 g/d and clearly exceeded the suggested 41 g/d by FAO (2005). In Singida district, median protein intake was lowest with 50 g/d, higher in Kongwa with 58 g/d and even bordered the RDA in Muheza with 65 g/d (Figure A15), which is also reflected in Figure 3.3.17. For all districts and all seasons, 18% of participants consumed less, and 30% more than the RDA, while 52% where within the recommended range.

The daily fat intake is recommended to be 30% of nutritional energy, which in turn depends on age, sex, bodyweight and physical activity level (PAL) of a person. Consequently, to be very precise the fat intake of each single women would have to be checked individually which, however, was not done in the present study. The range of recommended fat intake for adult women was calculated to be between 50 and 85 g/day, while 72 g/day are recommended for women aged between 18-59 years with a body weight of 55 kg (FAO 2005). Because fat consumption was not normally distributed, the median fat intake was considered, which was 35 g/day and, thus, below the suggested intake range. This represented only about 17% of the mean energy intake of study participants (instead of the recommended 30%). In fact, for all districts and all seasons, 74% of participants consumed less, and 7% more than the recommended dietary intake, while only 19% were within the recommended range (Figure 3.3.12). Results on the carbohydrate intake regarding RDA are shown in Figure A14.

Table 3.3.5   Mean intake of macronutrients per day by women in Tanzania as estimated from three non-consecutive 24-h recalls applying Nutritsurvey (n=252; outliers excluded for energy (n=236) and fat (n=244))

Nutrient	Mean ± SD	Median	(range)
Energy (kcal)	1845 ± 379	1821	(922-2811)
Protein (g)	58 ± 15	57	(21-103)
Fat (g)	38 ± 17	35	(7-109)
Carbohydrate (g)	323 ± 67	320	(142-521)
Dietary fibre (g)	37 ± 13	34	(16-92)
PUFA (g)	12 ± 8	9,5	(2-54)
Cholesterol (g)	23 ± 31	13	(0-167)

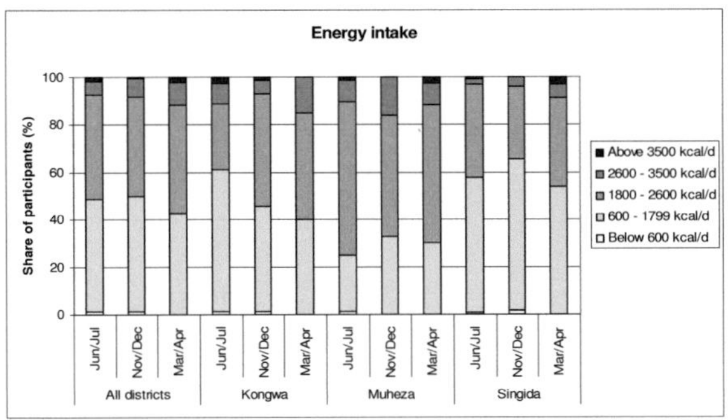

Figure 3.3.11
Share of participants with a certain energy intake during three different
seasons in three districts of Tanzania (RDA=1800-2600 kcal/d)

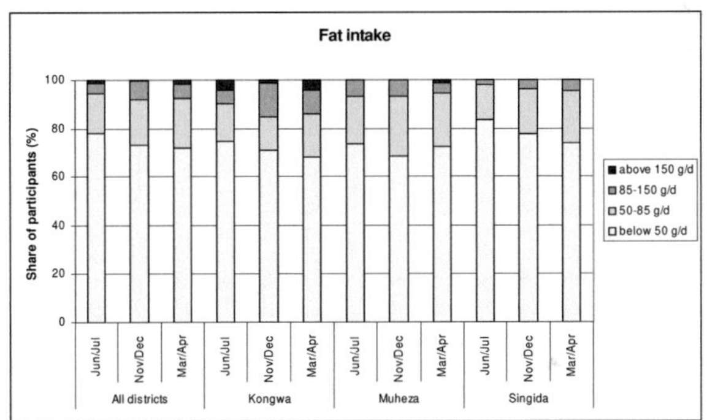

Figure 3.3.12
Share of participants with a certain fat intake during three different
seasons in three districts of Tanzania (RDA=50-85 g/d)

In terms of micronutrients, eight major vitamins and seven major minerals were analysed, whereby mainly vitamin A and iron were of interest. The intake of none of the micronutrients was normally distributed; differences between the districts were significant ($p \leq 0.01$). The median vitamin A intake (831 µg/d RE) was rather high but within the recommended daily allowance of 500-850 µg/d RE (Table 3.3.6). While the median intake in Singida district was only 381 µg/d RE, it was substantially higher in Kongwa (1034 µg/d RE) and especially in Muheza district (1315 µg/d RE; Table A22). Only few participants (16%) fell within the range of the recommended intake, but 38% of women for all districts and all seasons consumed more than 850 µg/d RE. However, about 47% of participants consumed less than 500 µg/d RE. This share was lower for Muheza, yet, noticeably higher for Singida district with as much as 74% of women not reaching the recommended intake of vitamin A (Figure 3.3.13). No significant differences between the seasons were observed for vitamin A intake (p=0.275).

Table 3.3.6   Mean intake of vitamins per day by women in Tanzania as estimated from three non-consecutive 24-h recalls applying Nutrisurvey (n=252; outliers excluded for vitamin A (n=208))

Nutrient	Mean ± SD	Median (range)
Vitamin A (µg/d)	1022.3 ± 770.2	831 (89-3546)
Carotene (mg/d)	4.3 ± 4.4	3 (0-21)
Vitamin C (mg/d)	136.8 ± 95.5	112 (29-597)
Vitamin E (mg/d)	9.0 ± 7.0	7 (0-40)
Vitamin B1 (mg/d)	1.1 ± 0.3	1 (0-4)
Vitamin B2 (mg/d)	1.0 ± 0.3	1 (0-2)
Vitamin B6 (mg/d)	1.7 ± 0.6	2 (1-3)
Folic Acid (µg/d)	391.1 ± 175.4	362 (127-1149)

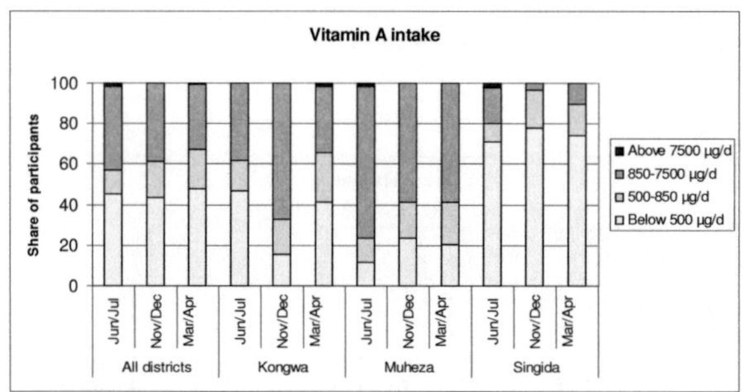

Figure 3.3.13
Share of participants with a certain vitamin A intake during three different seasons in three districts of Tanzania (RDA=500-850 µg/d)

Regarding iron intake of participants, the median of 14 mg/d was just below the RDA of 15-30 mg/d (Table 3.3.7). Districts differed significantly (p≤0.01), and iron intake was lowest for Kongwa (13 mg/d) and Muheza (14 mg/d), but highest for Singida (17 mg/d), while the widest range of the amount of iron consumed was found in Kongwa district (Figure A16 and Table A23). The share of women with an iron intake below 15 mg/d was, consequently, high with more than 40% for nearly all districts and seasons except Singida during Nov/Dec (SR) and Mar/Apr (LR) (Figure 3.3.14). Iron intake levels showed no significant differences among the seasons (p=0.102).

Regarding the relationship between nutrient intake and socio-economic variables, age was negatively correlated to vitamin A intake (ρ=-0.127; p=0.047) and carotene intake (ρ=-0.141; p=0.027). Similarly, wealth was negatively associated with vitamin A consumption (τ=-0.132; p=0.008) and carotene consumption (τ=-0.121; p=0.021), while education had no relationship to any nutrient intake at all. Household size and the distance from village to town were correlated with the consumption of several nutrients (Table 3.3.8).

Table 3.3.7 Mean intake of minerals per day by women in Tanzania as estimated from three non-consecutive 24-h recalls applying Nutrisurvey (n=252; outliers excluded for iron (n=172))

Nutrient	Mean ± SD	Median (range)
Iron (mg/d)	16.4 ± 7.1	14 (5-42)
Zinc (mg/d)	7.0 ± 2.7	7 (2-21)
Calcium (mg/d)	458.1 ± 162.1	442 (206-966)
Sodium (mg/d)	386.3 ± 283.3	300 (32-1757)
Potassium (mg/d)	2390.9 ± 850.1	2266 (814-6347)
Magnesium (mg/d)	331.4 ± 125.5	307 (147-1163)
Phosphorus (mg/d)	962.6 ± 284.2	961 (330-2366)

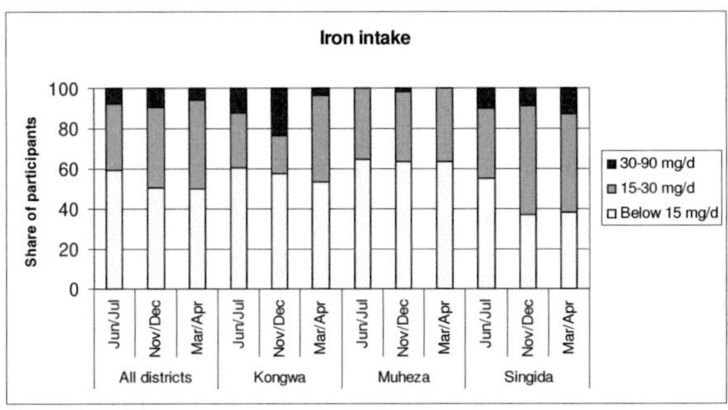

Figure 3.3.14
Share of participants with a certain iron intake during three different seasons within three districts of Tanzania (RDA=15-30 mg/d)

Table 3.3.8 Correlations between socio-economic parameters and intake of seven nutrients by women in Tanzania

Nutrient	Household size		Distance in km		Distance in min	
	Correl. coeff. rho	p	Correl. coeff. rho	p	Correl. coeff. rho	p
Energy	-0.241	<0.001	-0.167	0.006	-0.231	<0.001
Protein	-0.204	0.001	-0.285	<0.001	-0.342	<0.001
Fat	-0.317	<0.001	n.a.	n.s.	n.a.	n.s.
Vitamin A	-0.265	<0.001	n.a.	n.s.	-0.280	<0.001
Carotene	-0.314	<0.001	-0.199	0.002	-0.388	<0.001
Iron	n.a.	n.s.	-0.191	0.003	-0.199	0.002
Zinc	n.a.	n.s.	-0.204	0.001	-0.225	<0.001

n.a.=not applicable; n.s.=not significant

The relationship between vitamin A intake and religion was significant (Figure 3.3.15) as well as the relationship between iron intake and religion (Figure 3.3.16). A highly significant relationship was also detected for ethnic group and protein consumption (Figure 3.3.17) as well as for ethnic group and fat intake (p=0.003).

Furthermore, vegetable production data was checked for correlation with the nutrient intake of participants. The results were calculated for the overall mean of one year (Table A21) and for single seasons (Table A24). Only when considering the individual seasons, protein intake was significantly associated with vegetable collection (yes/no) in Jun/Jul (DS) (Figure 3.3.18). Protein intake was further associated with vegetable sales (Figure 3.3.19) and vegetable purchase (Figure

3.3.20) during Nov/Dec (SR). Vitamin A intake was also associated with vegetable purchase during Nov/Dec (SR) (Figure 3.3.21), while iron intake was significantly associated with vegetable purchase (p=0.009) during Mar/Apr (LR).

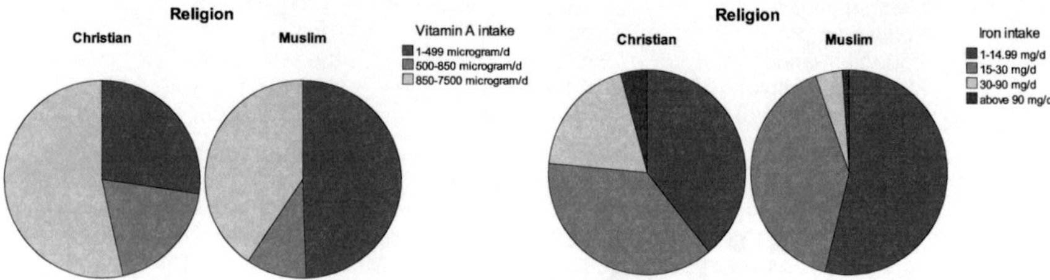

Figure 3.3.15
Association between religion and vitamin A intake by women in Tanzania (mean across three seasons) (p=0.001) (RDA = 500-850 µg/d)

Figure 3.3.16
Association between religion and iron intake by women in Tanzania (mean across three seasons) (p=0.003) (RDA = 15-30 mg/d)

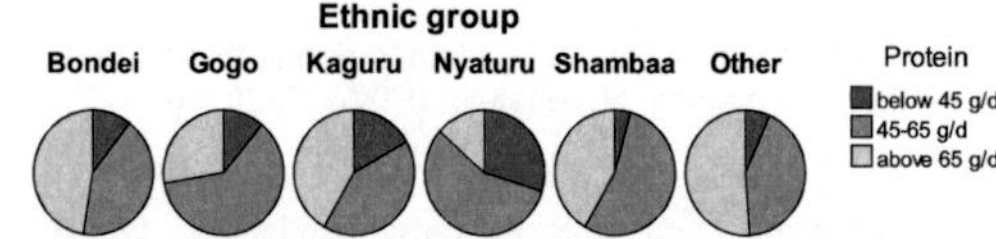

Figure 3.3.17
Association between ethnic group and protein intake by women in Tanzania (mean across three seasons) (p<0.001) (RDA = 45-65 g/d)

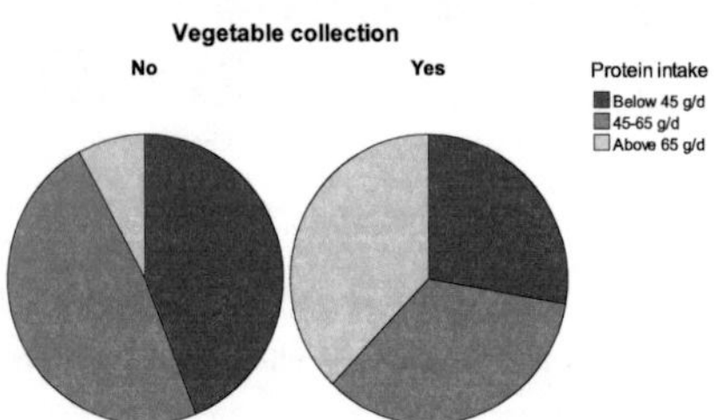

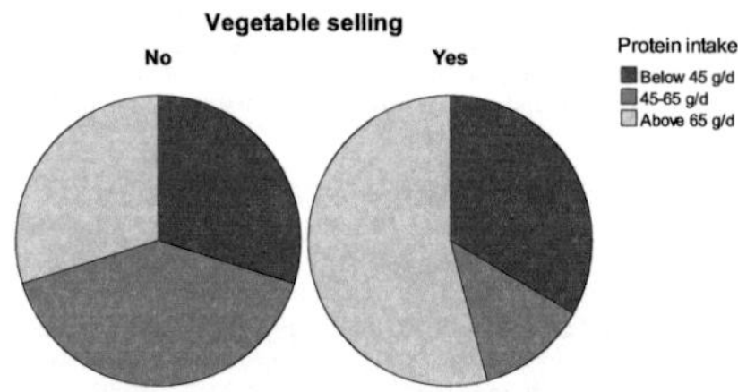

Figure 3.3.18
Association between vegetable collection (yes/no) and protein intake by women in Tanzania during Jun/Jul (DS) (p=0.011) (RDA = 45-65 g/d)

Figure 3.3.19
Association between vegetable sales (yes/no) and protein intake by women in Tanzania during Nov/Dec (SR) (p=0.014) (RDA = 45-65 g/d)

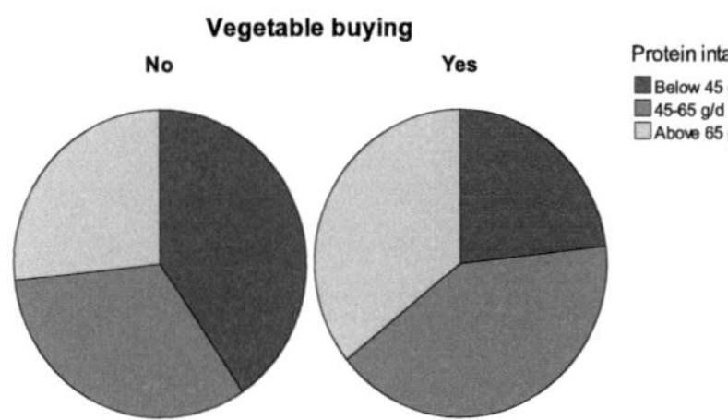

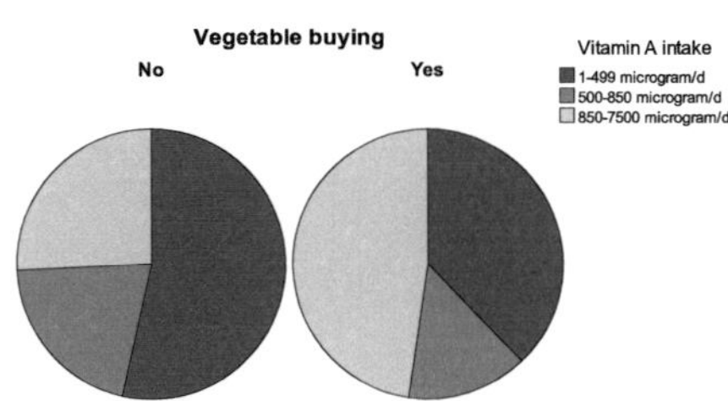

Figure 3.3.20
Association between vegetable purchase (yes/no) and protein intake by women in Tanzania during Nov/Dec (SR) (p=0.013)

Figure 3.3.21
Association between vegetable purchase (yes/no) and vitamin A intake by women in Tanzania during Nov/Dec (SR) (p=0.002)

3.3.3   Dietary Diversity Score (DDS) and Food Variety Score (FVS)

<u>Dietary Diversity Score (DDS)</u>

For the DDS which counts the food groups consumed on one day, a different classification of food groups was assumed than that applied in chapters 3.3.1 and 3.3.2. While the groups of cereals and bread/cakes were combined, the group "animal products" was split up into meat, poultry, milk and eggs, thus, resulting in 14 different food groups for DDS analysis.

Overall, participants consumed between two and eleven (two and ten for the mean across three days) different food groups on one day. Distribution was skewed and, therefore, the median instead of the mean DDS had to be considered, which was about six for all seasons and districts taken together (Table 3.3.9). The median DDS for Kongwa and Singida was only five and four, respectively, while it was eight for Muheza. In fact, not only the differences of DDS between districts were significant ($p \leq 0.01$), but also those between seasons ($p < 0.001$).

Participating women were classified into three DDS groups with either low, medium or high DDS: women within the low DDS group consumed two to four different food groups on one day, participants with medium DDS consumed five or six, while a high DDS was characterised by seven up to eleven food groups per day (Figure 3.3.22).

Table 3.3.9   Dietary diversity score (DDS) of women during three different seasons in three districts of Tanzania (median and range)

	All year	Jun/Jul (DS)	Nov/Dec (SR)	Mar/Apr (LR)
All districts	6 (2-10)	6 (2-11)	6 (2-10)	6 (2-11)
Kongwa	5 (3-8)	5 (2-10)	5 (2-10)	5 (3-10)
Muheza	8 (5-10)	8 (5-11)	8 (4-10)	8 (5-11)
Singida	4 (2-8)	4 (2-10)	4 (2-8)	4 (2-8)

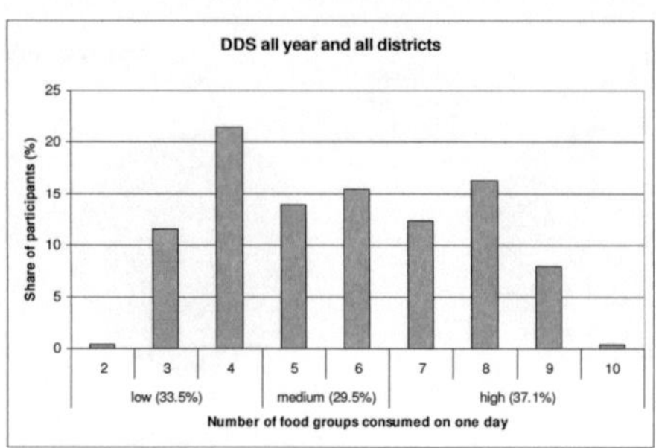

Figure 3.3.22
Share of participants (%) with a certain dietary
diversity score (DDS); mean across three seasons

When comparing districts, the share of women with a low, medium or high DDS was highly different (Figure 3.3.23): the DDS of women in Kongwa was evenly distributed without any peak. In contrary, the DDS was mainly medium or high for women in Muheza district, with an emphasis on women eating 8 different food groups per day. The DDS of women in Singida was either low or medium with a peak at four different food groups per day. Consequently, the share of women having a low, medium or high DDS was highly different between the three districts (Table A31). Seasonal differences (Figure 3.3.24) were less distinctive.

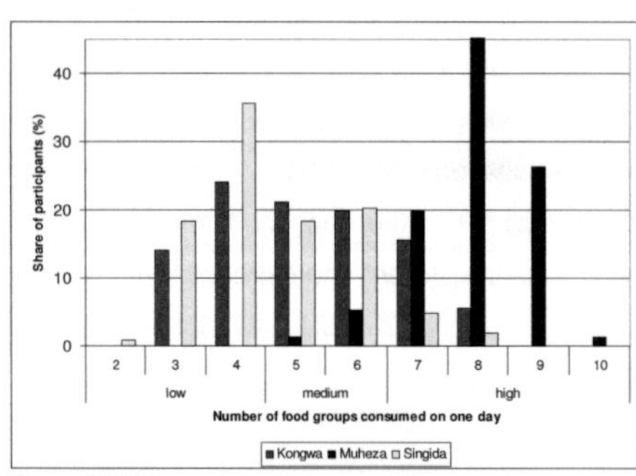

Figure 3.3.23
Share of participants (%) with a certain dietary
diversity score (DDS) in three districts of Tanzania;
mean across three seasons

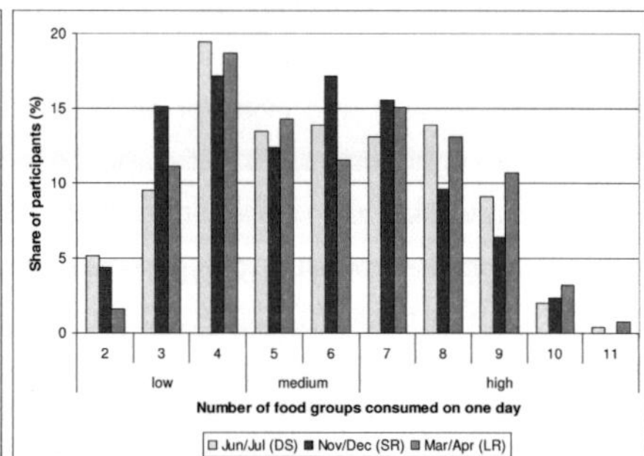

Figure 3.3.24
Share of participants (%) with a certain dietary
diversity score (DDS) during three different seasons

Food groups mainly consumed by participants with either low, medium or high DDS also differed. While cereals and vegetables were consumed by everyone, this was not so for the other twelve food groups (Figure 3.3.25). Women with a low DDS consumed, besides cereals and vegetables mainly oils/fats, and about 54% consumed also grain legumes (Figure 3.3.26). Women with a medium DDS consumed cereals, vegetables and oils/fats and most of them also sugars. About 80% consumed beverages and grain legumes, 50% nuts, and about 40% fish, starchy roots and fruits. Less often consumed were milk and especially meat, poultry and eggs (Figure 3.3.26).

The high DDS group, in which members ate seven different food groups or more, cereals, vegetables, oils/fats, beverages and sugars were consumed by everyone. About 90% also consumed nuts, grain legumes and fruits, while 83% ate fish and 67% starchy roots. Only 20% of participants or less consumed milk, meat, poultry or eggs (Figure 3.3.26). Also the intake of some nutrients was depicted as a function of the DDS categories (Figure A17), whereby the intake of all nutrients varied significantly between the three DDS categories (p<0.001).

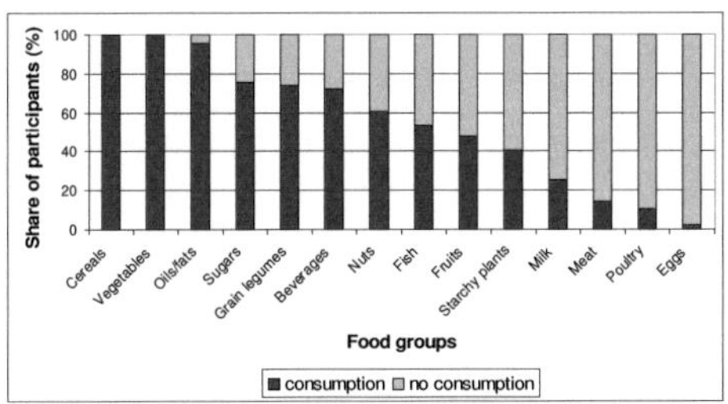

Figure 3.3.25
Share of participants (%) that consumed certain food
groups; mean across three different seasons

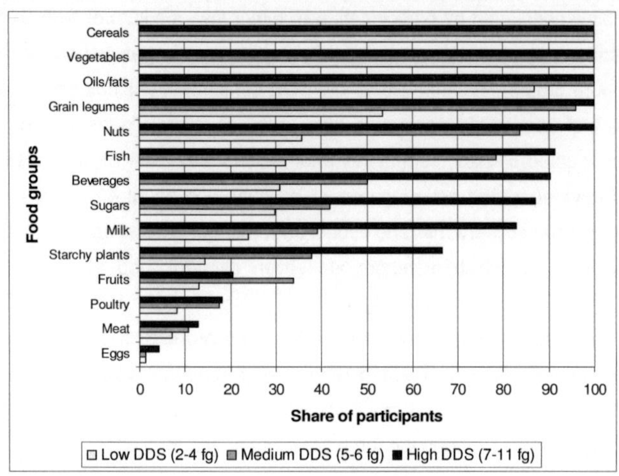

Figure 3.3.26
Share of participants (%) consuming each food group as a function of the
DDS categories; n=252 women from three districts in Tanzania interviewed
during three different seasons; DDS=dietary diversity score, fg=food groups

Food Variety Score (FVS)

The FVS, counting all different foods which were consumed by one person on one day, ranged
between two and 16 (three and 14 for the mean of three days) (Table 3.3.10). It was normally
distributed (K-S test statistic=0.096; S-W test statistic=0.978), with an overall mean FVS of eight
foods per day. Like for DDS, participants were classified into three FVS groups with either low (two
to six different foods), medium (seven to nine) or high (ten or more) FVS (Figure 3.3.27).

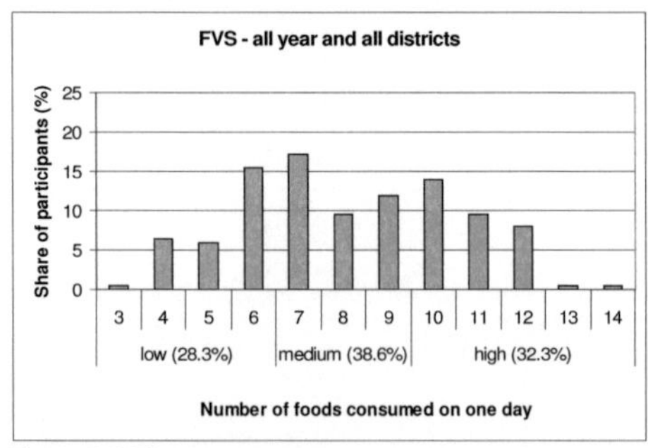

Figure 3.3.27
Share of participants (%) with a certain food variety
score (FVS); mean across three seasons

Table 3.3.10   Food variety score (FVS) of women during three different seasons in three districts of Tanzania

	All year	Jun/Jul (DS)	Nov/Dec (SR)	Mar/Apr (LR)
All districts				
Mean ± SD	8.2 ± 2.4	8.2 ± 2.9	7.8 ± 2.7	8.6 ± 2.5
Median (range)	8 (3-14)	8 (2-16)	8 (2-15)	9 (4-14)
Kongwa				
Mean ± SD	7.1 ± 2.1	6.6 ± 2.6	6.9 ± 2.9	7.8 ± 2.3
Median (range)	7 (3-12)	7 (2-15)	7 (2-14)	8 (4-12)
Muheza				
Mean ± SD	10.6 ± 1.4	10.9 ± 2.0	10.1 ± 1.9	11.0 ± 1.7
Median (range)	11 (7-14)	11 (5-16)	10 (6-15)	11 (7-14)
Singida				
Mean ± SD	7.1 ± 1.7	7.3 ± 2.2	6.7 ± 2.1	7.3 ± 1.8
Median (range)	7 (4-11)	7 (2-14)	7 (2-11)	7 (4-11)

Significant differences ($p \leq 0.01$) were found between the districts with a mean FVS of 7.1 for Kongwa and Singida and of 10.6 for Muheza district (Table 3.3.10). Like for the DDS, the FVS in Kongwa was more evenly distributed than in the other two districts with no clear focus on a particular number of foods, however, with a greater share of women consuming a low and medium food variety (Table A32, Figure 3.3.37). In Muheza, the majority of women, namely about 72%, had a high FVS with no clear favourite number of foods consumed per day. In contrast, nearly 50% of participants had a medium FVS with a main focus on seven different foods consumed per day in Singida district (Figure 3.3.28).

Differences between seasons were also significant ($p < 0.001$), with slightly higher values during Mar/Apr (LR) than during the other two seasons (Figure 3.3.29). During Mar/Apr (LR) only 23% of participants had a low FVS, while during Jun/Jul (DS) 29% and during Nov/Dec (SR) even 34% had a low FVS. In contrast, 39% of participants had a high FVS during Mar/Apr (LR), in Jun/Jul (DS) 30% had a high FVS, while during Nov/Dec (SR) FVS was lowest with only 28% having a high FVS (Figure 3.3.29).

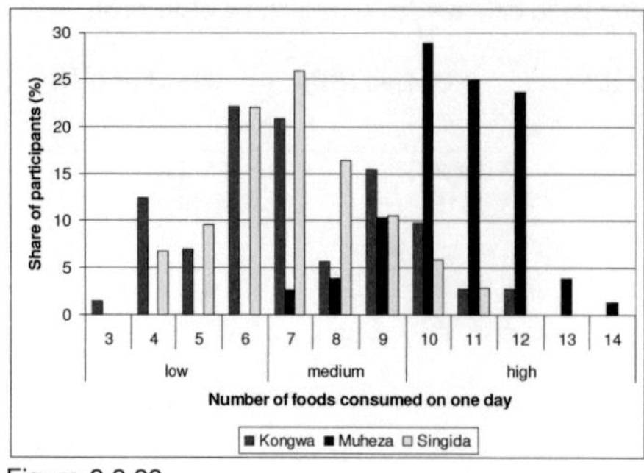

Figure 3.3.28
Share of participants (%) with a certain food variety
score (FVS) in three districts of Tanzania; mean
across three seasons

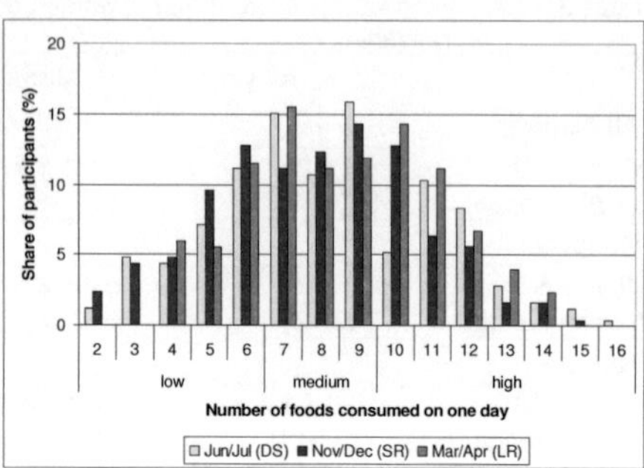

Figure 3.3.29
Share of participants (%) with a certain food variety
score (FVS) during three different seasons

<u>Correlations and associations</u>

Negative relationships between household size and distance from village to town were found with
most of the food scores (Table 3.3.11). Dietary scores and nominal socio-economic variables
showed several relationships. While ethnic group, the status within the hh and occupation were
related to both scores (Figures 3.3.30 – 3.3.32), religion was related only to FVS (Figure 3.3.33).

Table 3.3.11   Correlations between socio-economic values and food scores of women in Tanzania

Score	Household size		Distance in km		Distance in min	
	Correl. coeff. rho	p	Correl. coeff. rho	p	Correl. coeff. rho	p
DDS	-0.127	0.044	-0.236	<0.001	-0.266	<0.001
FVS	n.a.	n.s.	-0.179	0.004	-0.174	0.006

n.a.=not applicable; n.s.=not significant

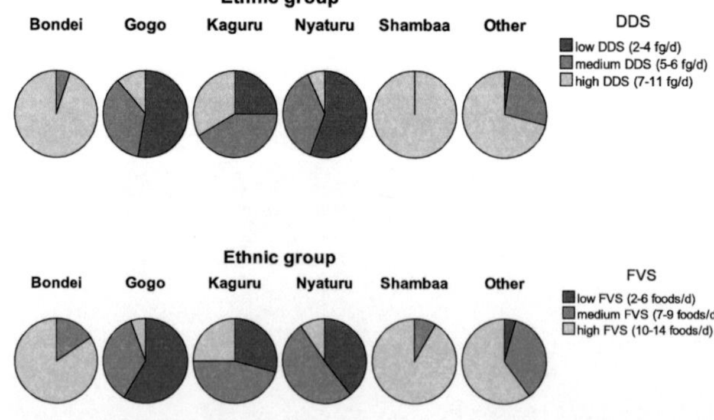

Figure 3.3.30
Association between ethnic group and dietary diversity score (DDS;
p<0.001) and food variety score (FVS; p<0.001) (mean across three
seasons)

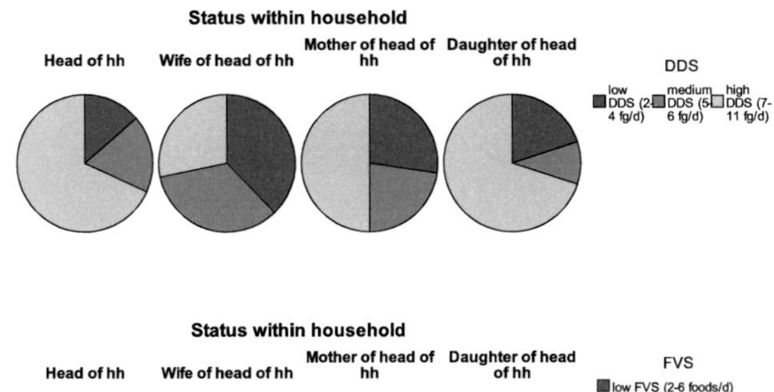

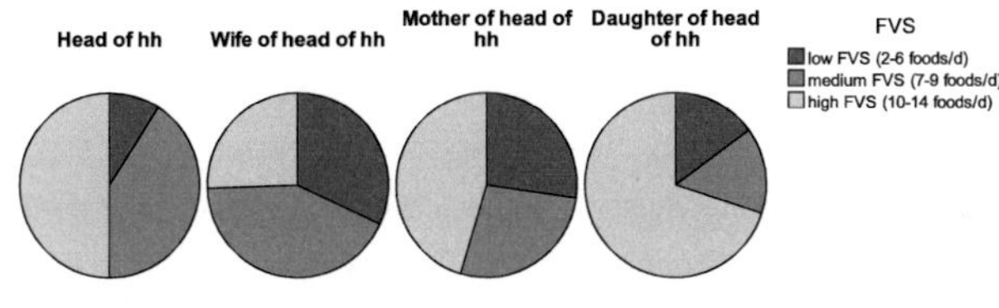

Figure 3.3.31
Association between status within household and dietary diversity score (DDS; p<0.001) and food variety score (FVS; p=0,001) (mean across three seasons)

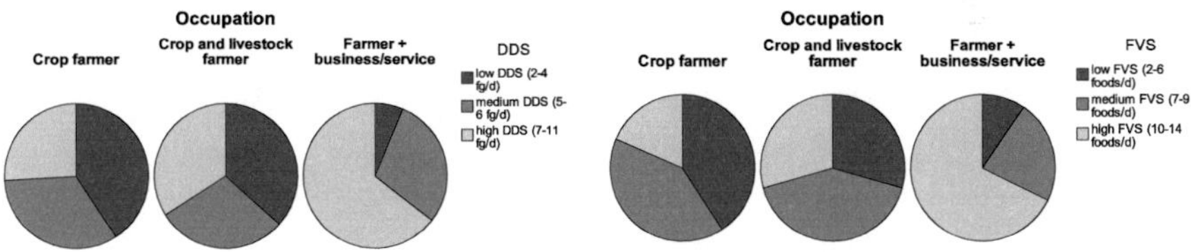

Figure 3.3.32   Association between occupation and dietary diversity score (DDS; p=0.004) and food variety score (FVS; p<0.001) (mean across three seasons)

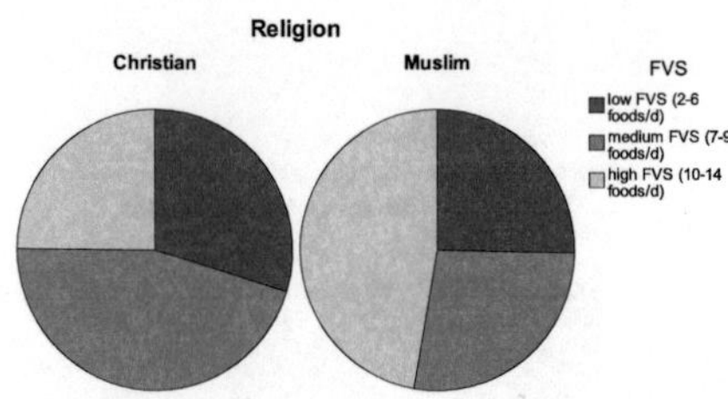

Figure 3.3.33
Association between religion and food variety score
(FVS) (mean across three seasons; p=0,001)

Both DDS and FVS were highly and positively correlated to the number of vegetables cultivated and the number of traditional vegetables cultivated/collected, when considering the whole year (Table 3.3.12). Further correlations existed, depending on the season. For example, in Mar/Apr (LR), DDS and FVS were negatively correlated to all vegetable production parameters.

The number of seasons per year vegetables were cultivated or collected, and if and how often they were bought and sold during one year, was not associated significantly with the food consumption scores. Only when analysing the individual seasons, vegetable cultivation (yes/no) was correlated to both DDS (p=0.002) and FVS (p=0.006) during Jun/Jul (DS). Also during this season vegetable collection (yes/no) was significantly correlated with the DDS (p=0.039). During Nov/Dec (SR) both DDS and FVS were associated with vegetable sales (p=0.021 and p=0.020, respectively), while FVS was also associated with vegetable purchase (p=0.034). In the Mar/Apr (LR) season, both DDS and FVS were associated with vegetable purchase (both p<0.001). Note that purchasing of vegetables was not surveyed during Jun/Jul (DS). Both food scores were significantly correlated to the intake of most food groups (p<0,001) except cereals and animal products. All correlations were positive but those with vegetable intake (Table A34).

Table 3.3.12 Correlations between cultivation/collection of vegetables and food scores of women in Tanzania

	No of vegetables cultivated		No of vegetables collected		No of TVs cult/coll		No of Evs cultivated	
	Correl. coeff. rho	p	Correl. coeff. rho	p	Correl. coeff. rho	p	Correl. coeff. rho	p
All year								
DDS	0.313	<0.001	n.a.	n.s.	0.300	<0.001	n.a.	n.s.
FVS	0.247	<0.001	n.a.	n.s.	0.241	<0.001	n.a.	n.s.
Jun/Jul (DS)								
DDS	0.307	<0.001	0.434	<0.001	0.485	<0.001	n.a.	n.s.
FVS	0.367	<0.001	0.392	<0.001	0.495	<0.001	n.a.	n.s.
Nov/Dec (SR)								
DDS	0.328	<0.001	0.135	0.032	0.301	<0.001	0.144	0.022
FVS	0.347	<0.001	n.a.	n.s.	0.286	<0.001	0.193	0.002
Mar/Apr (LR)								
DDS	-0.237	<0.001	-0.381	<0.001	-0.387	<0.001	-0.285	<0.001
FVS	-0.187	0.003	-0.321	<0.001	-0.316	<0.001	-0.224	<0.001

n.a.=not applicable; n.s.=not significant

3.3.4 Vegetable consumption and Vegetable Diversity Score (VDS)

Types of vegetables consumed

To investigate the diversity of vegetables consumed by study participants, the number of different vegetable types consumed per week was counted. Figure 3.3.34 shows differences between both the three districts and seasons, while distinguishing between traditional (TV) and exotic vegetables (EV). Figure 3.3.35, on the other hand, distinguishes between cultivated and collected vegetables. The number of vegetable types consumed overall was lower in Singida district than in the other two districts and it was also lower during Mar/Apr (LR) than during the other two seasons. Differences between the seasons for TVs, EVs and Vegetable Diversity Score (VDS) were significant as well as differences between the districts.

Besides the overall number of vegetable types consumed in a district and during a season, the number of vegetable types consumed per person and week was examined by using the vegetable diversity score (VDS), which is discussed below. The VDS, however, did not include onions and tomatoes; therefore, the number of vegetable types consumed per person and week was also calculated including these two vegetables (Figure 3.3.36).

When observing the different vegetable types most often consumed by participants (Tables A25 and A26), the great diversity of TVs was remarkable, while not much diversity among the consumed EVs was available. In terms of TVs, similar to production, some were consumed in all districts and during all seasons, such as amaranth and pumpkin leaves (except for Singida). Others were explicitly only consumed by many women in one district (e.g. bitter lettuce in Muheza), or during one season (e.g. carrot and cucumber during Mar/Apr (LR)).

Besides for vegetable production, cluster analysis was also performed to study vegetable consumption. The cluster obtained for consumption (Figure 3.3.37) looked much less diverse and less complex than that for vegetable production. The order of the 20 vegetables (y-axis) was the same as for production (Table 3.2.3), while the order of the 252 participants (x-axis) was, of course, different. From the vegetables that describe a cluster, only those with a score greater than 0.10 are shown (Table 3.3.13). Tomatoes and onions, being the two most important vegetables in every cluster, were excluded from further analysis. They scored from 0.53 to 0.65 (tomato) and from 0.42 to 0.72 (onion), respectively.

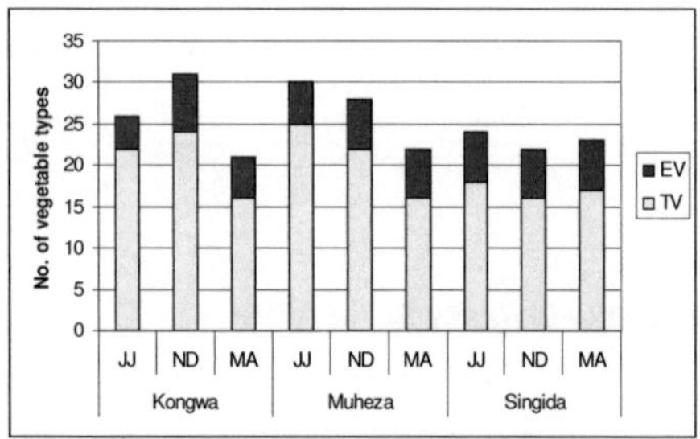

Figure 3.3.34
Mean number of traditional (TV) and exotic vegetable types (EV) consumed overall per week during three different seasons in three district of Tanzania

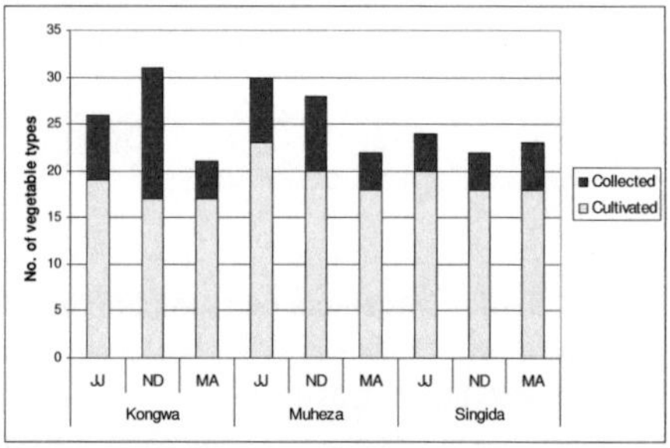

Figure 3.3.35
Mean number of cultivated and collected vegetable types consumed overall per week during three different seasons in three districts of Tanzania

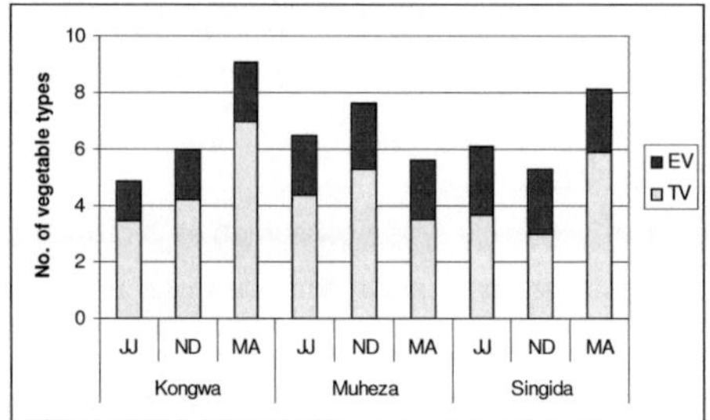

Figure 3.3.36
Mean number of different traditional (TV) and exotic vegetable types (EV) consumed per person within one week during three different seasons in three districts of Tanzania

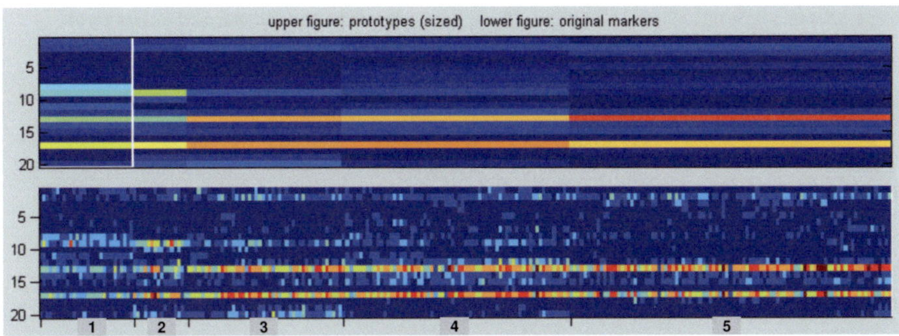

Figure 3.3.37
Five clusters showing the typical consumption of 20 vegetables (y-axis) by 252 women (x-axis); colours show the amount that was eaten of one vegetable: dark red=highest value, via orange, yellow, green, to dark blue=lowest value/no consumption; lower figure: data for each single woman; upper figure: women's affiliations to each of the five clusters, clusters being numbered from left to right

Table 3.3.13   Vegetables which characterise five clusters generated from number of days each vegetable was consumed per week by women in Tanzania

Cluster 1 n=28 Wild + cultivated; high diversity		Cluster 2 n=16 Mainly wild; high diversity		Cluster 3 n=46 Only wild; medium diversity		Cluster 4 n=67 Only cultivated; medium diversity		Cluster 5 n=95 Only cultivated; low diversity	
Vegetable	Score	Vegetable	Score	Vegetable	Score	Vegetable	Score	Vegetable	Score
False sesame	0.36	False sesame	0.48	Wild simsim	0.20	Okra	0.17	Amaranth lvs	0.16
Cowpea lvs	0.32	Wild simsim	0.17	False sesame	0.18	Sweet potato lvs	0.15	Sweet potato lvs	0.11
Mhilile	0.19	Jute mallow	0.15	Amaranth lvs	0.18	Pumpkin lvs	0.14		
Pumpkin lvs	0.16	Amaranth lvs	0.12			Amaranth lvs	0.14		
African spiderplant	0.12	Pumpkin lvs	0.11						
		Sweet potato lvs	0.11						

Additionally, factor analysis was applied to detect vegetable consumption patterns, whereby, different from cluster analysis, variables are grouped together instead of cases/participants or villages/districts. Each variable (here vegetable) scored high in only one component or pattern (comparable to a cluster) and can, therefore, be allocated quite precisely. In Table 3.3.14 the five different components that accounted for 51% of variance and factor loadings of vegetables are shown. The first component accounting for 14% of variation was difficult to explain, with cowpea leaves and Mhilile, a *Cleome* species collected from the wild, loading high; the second component was characterised by wild vegetables, which usually occured in humid areas, and one exotic vegetable (Chinese cabbage) loading negatively. In the third component wild vegetables from dry areas loaded high, while the fourth component was characterised by one exotic vegetable (Swiss chard) and no consumption of either wild cucumber or wild simsim (negative loading), but rather other exotic vegetables (Chinese cabbage, white cabbage). The fifth component includes cultivated traditional vegetables.

Table 3.3.14 Rotated component matrix of 18 vegetables (times per week consumed by women in Tanzania) and total variance explained; extraction method: principal component analysis; rotation method: Varimax with Kaiser normalisation

		Component				
		1	**2**	**3**	**4**	**5**
		Cowpea + Mhilile	Wild; humid climate	Wild; dry climate	Exotic	Cultivated; traditional
Initial Eigen-values	Total	2.552	2.394	1.628	1.331	1.196
	Percent of Variance	14.179	13.301	9.042	7.395	6.646
	Cumulative percent	14.179	27.480	36.521	43.916	50.562
Vegetables						
Mhilile		**0.816**	-0.144		-0.160	
Cowpea lvs		**0.771**	0.100			
African spiderplant		0.415	-0.208	0.329	-0.133	0.102
Bitter lettuce		-0.110	**0.729**	-0.253	0.161	
Black jack			**0.690**		0.143	0.118
Chinese cabbage		-0.189	**-0.600**	-0.186	0.397	0.151
Jute mallow		-0.130	0.115	**0.778**		
False sesame		0.174	-0.289	**0.725**	-0.134	0.103
Wild cucumber		-0.439	-0.138	0.231	**-0.566**	
Swiss chard		-0.111			**0.529**	
Wild simsim		-0.374	-0.274	0.353	**-0.504**	
White cabbage		-0.125	0.167	-0.220	0.337	0.114
Cassava lvs		-0.141	0.221		0.335	
Pumpkin lvs		0.136		0.153	0.178	**0.643**
Tree cassava lvs		0.276		-0.314	-0.148	**0.558**
Amaranth lvs		-0.476			-0.116	**0.526**
Okra		-0.154	-0.144	0.305	0.447	**0.503**
Sweet potato lvs		-0.223	0.227			0.497

<u>Vegetable consumption quantity, criteria to chose a vegetable and sources of vegetables</u>
Vegetable intake of participants (amount in g/d) was only normally distributed during Mar/Apr (LR) (K-S test statistic=0.079; S-W test statistic=0.951), yet, not during the other two seasons. If vegetable intake was compared among the three districts, significant differences existed, but not so among the three seasons. The median showed that the highest amount of vegetables per day was consumed in Singida district, followed by Kongwa and finally Muheza (Table 3.3.15). The amount of vegetables was not calculated for TVs and EVs individually because only few participants consumed exotic vegetables, e.g. during Mar/Apr (LR) only seven women.

Table 3.3.15   Amount of vegetables consumed (g/d) by women in Tanzania according to a 24h-recall conducted once during each season

	Jun/Jul (DS)	Nov/Dec (SR)	Mar/Apr (LR)
All districts			
Mean ± SD	305 ± 198	310 ± 192	321 ± 191
Median (range)	264 (0-1254)	274 (0-1160)	298 (0-1085)
Kongwa			
Mean ± SD	294 ± 206	315 ± 195	337 ± 188
Median (range)	273 (0-894)	280 (0-1160)	312 (45-934)
Muheza			
Mean ± SD	261 ± 147	235 ± 161	209 ± 123
Median (range)	231 (30-581)	210 (0-886)	195 (0-570)
Singida			
Mean ± SD	344 ± 217	361 ± 196	391 ± 198
Median (range)	292 (0-1254)	321 (56-1103)	358 (53-1085)

The most important criteria to chose a vegetable was significantly different between districts only during Jun/Jul (DS), while the most important criteria to chose a great diversity of vegetables differed significantly between districts during Jun/Jul (DS) and Mar/Apr (LR). The most important criteria to chose a vegetable was taste for all districts, yet, not during all seasons (Figures 3.3.38). Similarly, the most important criteria to consume a high diversity of vegetables was different tastes for all districts with the criteria health being more important during Jun/Jul (DS) (Figure 3.3.39).

The sources of vegetables consumed by participants (purchased, cultivated, collected or gift) are shown by district and season for the ten most important TVs and six most important EVs in tables A27-A30. Thereby, apparently the greatest share of women produced the TVs that they consumed, while the consumed EVs were mostly purchased.

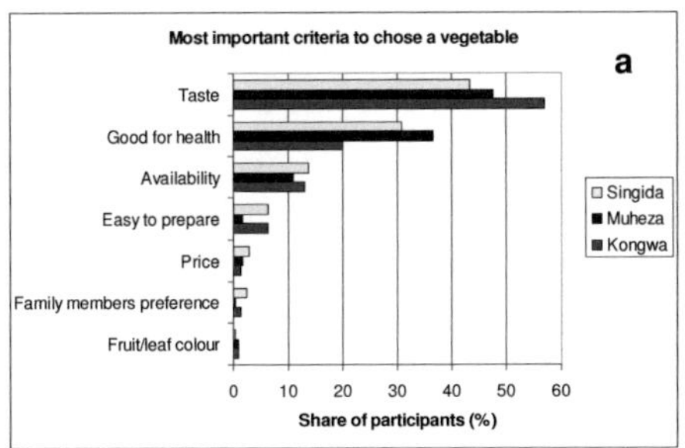

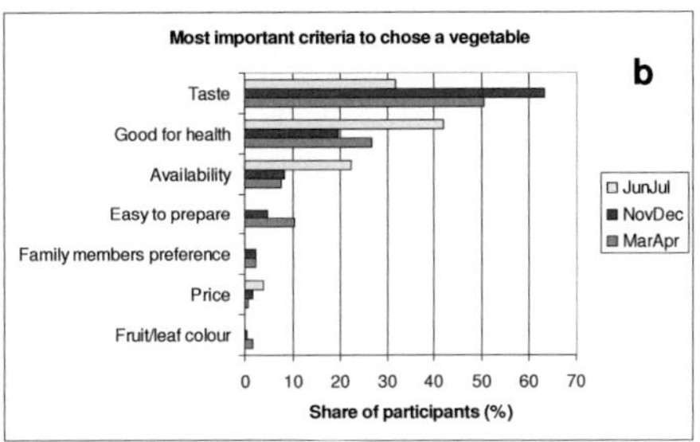

Figure 3.3.38   Share of participants that named certain criteria to chose a vegetable a) in three districts of Tanzania and b) during three different seasons (answers were not predetermined)

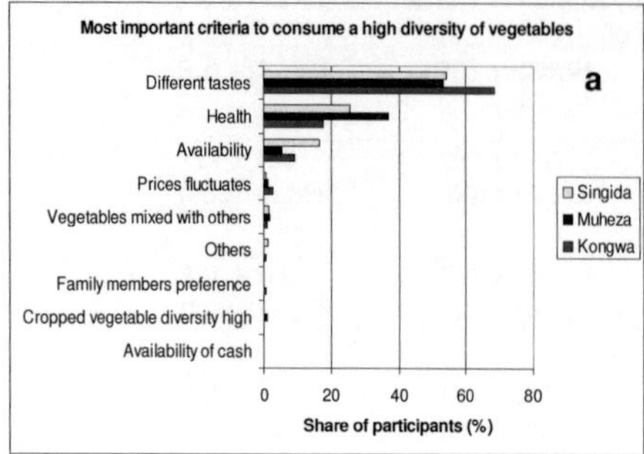

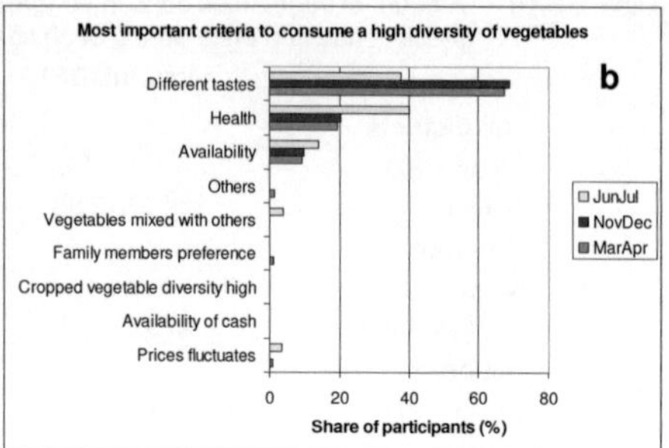

Figure 3.3.39   Share of participants that named certain criteria to consume a high diversity of vegetables a) in three districts of Tanzania and b) during three different seasons (answers were not predetermined)

Vegetable Diversity Score (VDS)

Similar to the dietary diversity score (FAO 2007), a vegetable consumption diversity score (VDS) was calculated, however, summing up all different vegetables that were consumed not during one day, but during one week, without accounting for amount or frequency. The exotic vegetables tomato and onion were not counted as they were usually the basic ingredient for every sauce, vegetable or meat dish, thus, eaten every day but normally used in very small quantities and rather considered as condiments or spices (see Focus A).

The VDS varied greatly between participants overall and ranged from one to ten different vegetables consumed per week on average, one to 15 when different seasons and districts were analysed. Distribution was skewed and, therefore, the median VDS had to be used, which was five for all districts and seasons (Table 3.3.16). The picture for the traditional vegetable diversity score (TVDS) was very similar to that of the VDS, as most of the consumed vegetables were traditional. In contrast, the exotic vegetable diversity score (EVDS) was made up of 128 data sets only while 124 participants did not consume exotic vegetables during the surveyed week. Of these 128 participants, most (97%) consumed only one exotic vegetable per week, while only 3% consumed two different ones. The EVDS was only interesting in one point, namely the number of women per district who consumed exotic vegetables at all. While in Kongwa only 23 women stated to consume exotic vegetables, there were nearly twice as many in Singida (44), and even nearly thrice as many in Muheza district (61) consuming exotic vegetables during the surveyed week. In the following only the overall VDS was studied in further detail.

Table 3.3.16   Mean and median values of the vegetable diversity score (VDS) of women in Tanzania

	n	Mean ± SD	Median (range)
VDS	252	4.7 ± 1.4	5 (1-10)
TVDS	252	4.5 ± 1.3	4 (1-9)
EVDS	128	1.0 ± 0.2	1 (1-2)

Unlike for DDS and FVS, differences among the three districts were not significant neither for the overall VDS nor for the VDS during Jun/Jul (DS). Only during the other two seasons, the VDS was significantly different (p≤0.01) among the three districts (Table 3.3.17). The VDS was also grouped into low (one to three vegetables per week), medium (four to five), and high (six to 15) VDS with the cut-off points chosen in order to achieve an equal distribution between these groups for the combined data of all districts and seasons (Figure 3.3.40).

The share of participants with a medium VDS was highest for all districts (Figure 3.3.41). This was especially true for both Kongwa district with 11%, 54% and 35% for low, medium and high VDS and Singida with 17%, 65% and 18%, respectively, while this distribution for Muheza was rather even (24%, 45% and 32%, respectively).

Major differences among the seasons were observed for VDS (p<0.001), showing that vegetable diversity consumed had seasonal fluctuations. During the Jun/Jul (DS) season clearly less different vegetables were consumed per women during one week (Table 3.3.20), and nearly 45% of women had a low VDS, while only about 18% had a high VDS. During Nov/Dec (SR) the share of participants with low, medium or high VDS was rather equal. In March/April, the situation was the other way round compared to Jun/Jul (DS), with 52% of women having a high VDS and, thus, consuming more than six different vegetables per week, while only about 20% consumed a low vegetable diversity (Figure 3.3.42).

Table 3.3.17   Vegetable diversity score (VDS) of women in three districts and during three different seasons in Tanzania

	All year	Jun/Jul (DS)	Nov/Dec (SR)	Mar/Apr (LR)
All districts				
Mean ± SD	4.7 ± 1.4	4.2 ± 2.0	4.5 ± 2.3	5.8 ± 2.5
Median (range)	5 (1-10)	4 (1-11)	4 (1-15)	6 (1-14)
Kongwa				
Mean ± SD	5.0 ± 1.2	3.6 ± 1.2	4.4 ± 1.8	7.0 ± 2.6
Median (range)	5 (3-8)	4 (1-7)	4 (1-8)	7 (2-12)
Muheza				
Mean ± SD	4.7 ± 1.8	4.7 ± 2.5	5.8 ± 3.0	4.0 ± 2.3
Median (range)	5 (1-10)	4 (1-11)	6 (1-15)	4 (1-14)
Singida				
Mean ± SD	4.6 ± 1.2	4.1 ± 2.0	3.5 ± 1.5	6.1 ± 1.7
Median (range)	4 (2-9)	4 (1-9)	3 (1-8)	6 (3-13)

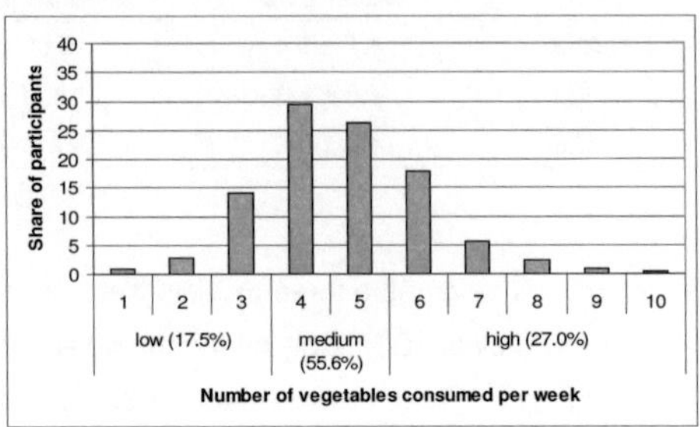

Figure 3.3.40
Share of participants (%) with a certain vegetable
diversity score (VDS); mean across three seasons

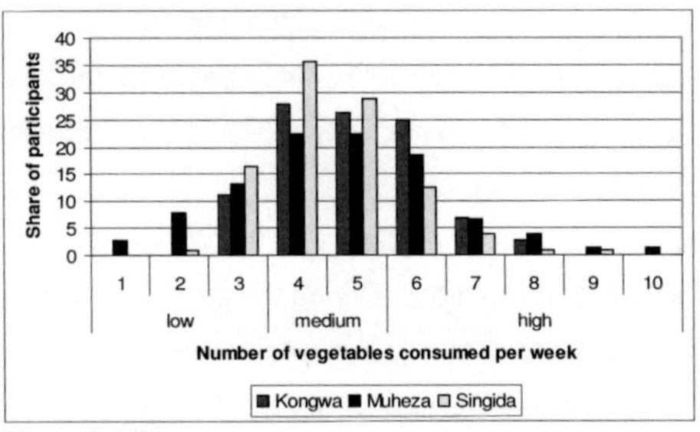

Figure 3.3.41
Share of participants (%) with a certain vegetable
diversity score (VDS) in three districts of Tanzania;
mean across three seasons

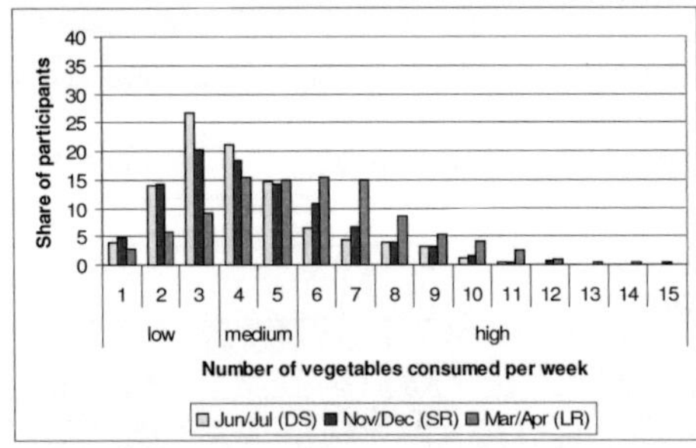

Figure 3.3.42
Share of participants (%) with a certain vegetable
diversity score (VDS) during three different seasons

The VDS (mean across three seasons) did not correlate with any socio-economic parameter but household size (ρ=0.236; p=0.016) and ethnic group (p=0.016). The VDS was further significantly correlated to different vegetable production parameters, especially with the number of vegetables cultivated and the number of traditional vegetables cultivated/collected (Table 3.3.18). Furthermore, it correlated slightly positively with fruit and sugar consumption and slightly negatively with fish consumption (Table A34).

Table 3.3.18   Correlation between vegetable diversity score (VDS) and cultivation and collection of vegetables

Season	No of vegetables cultivated		No of vegetables collected		No of TVs cult/coll		No of EVs cultivated	
	Correl. coeff. rho	p	Correl. coeff. rho	p	Correl. coeff. rho	p	Correl. coeff. rho	p
All year	0.181	0.004	0.236	<0.001	0.283	<0.001	n.a.	n.s.
Jun/Jul (DS)	0.256	<0.001	n.a.	n.s.	0.213	0.001	n.a.	n.s.
Nov/ Dec (SR)	0.257	<0.001	0.216	0.001	0.269	<0.001	n.a.	n.s.
Mar/Apr (LR)	0.365	<0.001	0.354	<0.001	0.450	<0.001	0.174	0.006

n.a.=not applicable; n.s.=not significant

The most important criteria to chose a vegetable stated by participants during June/July was significantly correlated to the VDS during that season (p<0.001). From those participants with a low VDS, 21% regarded "good for health" as the most important criteria, while this criteria was chosen to be most important by 51% of women with a high VDS. Similarly, from those participants who chose "good for health" as the most important criteria, 86% had a high VDS while only 14% had a low VDS. However, of those women who suggested the "price" to be the most important criteria to chose a vegetable, 50% each had a high and low VDS.

3.3.5   Dietary patterns

When dietary patterns were generated with PCA, based on the mean intake (g/d) of twelve main food groups during three non-consecutive days, five components and, thus, dietary patterns, were extracted explaining 67% of variance (Table 3.3.19). The first food pattern was dominated by fruits, nuts, starchy plants and fish. Fruits were mainly consumed in areas where they grew in abundance, i.e. for this study in the coastal Muheza district. Nuts could be coconuts on the one hand, which were also mainly consumed in Muheza where they were abundant, on the other hand groundnuts, which were a traditional meal ingredient in the inland Kongwa and Singida districts. Because starchy plants, namely plantains, cassava, taro and/or sweet potatoes, as well as fish were a common food in Muheza district, this pattern was called "traditional-coast".

The second food pattern was characterised by bread/cakes, sugar and tea. Bread and cakes were usually made from wheat and, in most cases, fried or baked in oil, such as donuts, *chapati*, *maandazi* or halfcake. They were sometimes made at home but often bought from small shops. Also tea leaves and sugar had to be bought by most households. Consequently, this pattern was characterised by foods usually not self-produced or home-made but bought and it was, therefore, called "purchase".

In the third pattern cereals and vegetables as well as the food group oil/fat were loading high, referring to the fact that vegetables were typically fried or cooked with oil. The food group cereals referred to the traditionally cooked rice, or *ugali/uji* made from maize, millet or sorghum flour. While in this pattern, besides the named traditional food groups, further only pulses as a traditional food loaded slightly and otherwise virtually nothing, this patter was called "traditional-inland".

Table 3.3.19   Rotated component matrix of 12 food groups (mean intake from three non-consecutive 24h-recalls of women in Tanzania) and total variance explained; extraction method: principal component analysis, rotation method: varimax with Kaiser normalization

		Component				
		1	2	3	4	5
Initial Eigen-values	Total	3.385	1.539	1.172	0.999	0.946
	Percent of Variance	28.210	12.828	9.764	8.322	7.886
	Cumulative percent	28.210	41.038	50.801	59.123	67.009
Food Group						
Fruits		**0.762**			0.196	
Nuts		**0.738**	0.244			-0.179
Starchy plants		**0.700**		-0.175		0.167
Fish		**0.531**	0.251	0.112		-0.288
Bread/cakes		0.209	**0.744**		0.156	
Sugar			**0.713**			
Tea		0.418	**0.703**		0.245	
Cereals			-0.245	**0.780**		-0.126
Oil/fat			0.453	**0.688**		0.104
Pulses		0.138	0.118	0.282	**0.827**	
Vegetables		-0.125	-0.192	**0.530**	**-0.691**	
Animal prod						**0.941**

In the fourth food pattern, only the food group pulses loaded positively high while vegetables were dominant but negatively, meaning that their consumption was minor. Pulses mainly consisted of dried and cooked common red kidney beans, but also chickpea, cowpea, mungbean, green pea (dried), pigeon pea, or soya bean. They are a traditional food but were obviously not consumed together with but rather instead of vegetables. This pattern was simply called "pulses".

The fifth pattern was characterised by the food group animal products excluding fish. Further food groups that loaded at least slightly suggested that animal products were definitely not eaten together with fish and also not with nuts or cereals. They were rather combined with oil/fat, in which meat was usually fried; also starchy plants loaded slightly, referring to one of the traditional dishes, where meat and plantain is cooked together. This dietary pattern was called "animal products".

Besides the component matrix, within each component or pattern, a specific factor score for each participant was calculated, which ranged between -2.84 and 4.62 in the present study. Besides considering each pattern at large, factor scores were also divided into percentile groups, here into quintiles. However, when observing the minimum and maximum factor values within each quintile (Table A36) it became clear that the distribution of participants was not equal. Regarding the distribution of factor scores, it was obvious that distribution was not normal and that most participants had a factor score between -1 and +1, while only a few had absolute greater values (Figure A18). Still, it was decided in favour of having equal groups in terms of participants to apply the simple ranking into five percentile groups which is a common procedure (Bühl 2006). In fact, the first quintile with values only up to -0.64 and the fifth quintile with values from at least 0.54 were found to picture the affiliation of participants with one or the other food group precisely enough. The 1st percentile in each pattern stands for "absolute no consumption according to this pattern" while the 5th percentile means "high consumption according to this pattern". The 2nd, 3rd and 4th percentile with minimum values between -0.84 and -0.64 and maximum values between 0.57 and 0.85 (Table A36) include participants being indifferent towards a pattern.

With a focus on grouping participants, based on dietary consumption similarity, the food group consumption data was clustered with MarVis (Kaever et al. 2009). After normalisation, the data was concentrated into either ten or five clusters, whereby the latter was chosen for further interpretation. Compared to ten clusters, with only five clusters more detailed information is lost, yet, dietary patterns became clearer (Figure 3.3.43, Table 3.3.25). Scores of food groups that characterised each cluster are listed in Table 3.3.26.

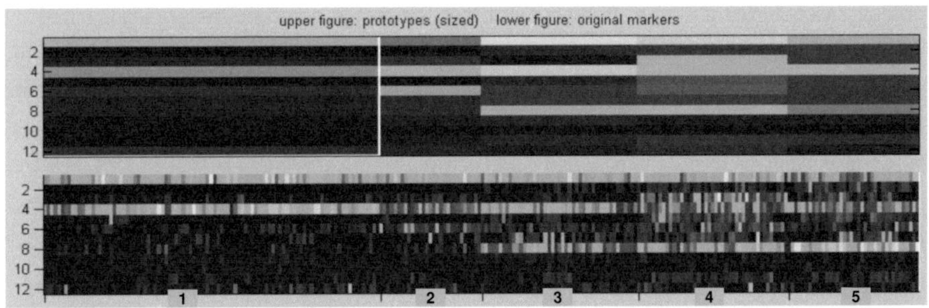

Figure 3.3.43
Five cluster/dietary patterns generated from twelve food groups (y-axis) as consumed in g/d (mean of three days) by 252 women (x-axis); colours show the amount that was eaten of one food group: dark red=highest value, via orange, yellow, green, to dark blue=lowest value/no consumption; lower figure: data for each single woman; upper figure: women's affiliation to each of the five clusters, clusters being numbered from left to right

Table 3.3.20   List of food groups as they appear in the cluster (Figure 3.3.43)

No. within cluster	Food group	No. within cluster	Food group
1	Cereals	7	Starchy plants
2	Bread/cakes	8	Tea
3	Fruits	9	Oil/fat
4	Vegetables	10	Sugar
5	Nuts	11	Fish
6	Pulses	12	Animal products

Table 3.3.21   Scores of food groups (only those ≥0.1 shown) which characterise five clusters generated from twelve food groups as consumed in g/d by women in Tanzania (mean across three days)

Cluster 1 n=97		Cluster 2 n=29		Cluster 3 n=45		Cluster 4 n=43		Cluster 5 n=38	
Food group	Score	Food group	Score	Food group	Score	Food group	Score	Food group	Score
Vegetables	0.70	Cereals	0.77	Cereals	0.62	Cereals	0.51	Tea	0.71
Cereals	0.66	Vegetables	0.45	Vegetables	0.54	Tea	0.47	Cereals	0.45
		Pulses	0.28	Tea	0.42	Vegetables	0.36	Vegetables	0.32
				Pulses	0.14	Fruits	0.34	Pulses	0.16
				Starchy plants	0.14	Nuts	0.21	Bread/cakes	0.14
						Pulses	0.19	Nuts	0.11
						Starchy plants	0.17	Starchy plants	0.11
						Fish	0.10	Fruits	0.10

Correlations and associations

Food patterns derived by PCA were significantly associated to districts. When patterns were divided into quintiles it shows in more detail that the first and second pattern were especially consumed by women in Muheza, the third and fifth pattern mainly by women in Singida, while women from Kongwa were rather indifferent (Table 3.3.22). The results of correlations/associations between further socio-economic variables and food scores are shown in Table 3.3.23; the association between food patterns 2 and 4 and occupation of participants in Table 3.3.24. For each quintile of each dietary pattern, median values (or mean if normally distributed) of continuous socio-economic parameters were calculated in order to identify typical characteristics of each pattern (Table 3.3.25). These values were tested for significant differences between the quintiles of each pattern (p^1) and for trends (p^2), i.e. if a variable increased from participants within the first quintile to participants within the fifth quintile or *vice versa*. Thereby, only data for the first and fifth quintile are shown.

Regarding vegetable production variables, food pattern 1 "traditional coast" was significantly related to the number of vegetables cultivated (ρ=0.268; p<0.001), to the number of vegetables collected (ρ=0.155; p=0.014) as well as to the number of seasons, vegetables were cultivated/collected (ρ=0.179; p=0.004). Otherwise, only pattern 5 "animal products" was slightly correlated to the number of vegetables cultivated (ρ=0.130; p=0.039).

Table 3.3.22 Distribution of participants (no., percent in brackets) within quintiles of food patterns of women in Tanzania (1st quintile: no consumption; 5th quintile: high consumption) and according to districts

	N	1st	2nd	3rd	4th	5th
Pattern 1 "Traditional-coast" (p<0.001)						
Kongwa	72	18 (25.0)	23 (31.9)	19 (26.4)	7 (9.7)	5 (6.9)
Muheza	76	4 (5.3)	3 (3.9)	3 (3.9)	25 (32.9)	41 (53.9)
Singida	104	28 (26.9)	25 (24.0)	28 (26.9)	19 (18.3)	4 (3.8)
Pattern 2 "Purchase" (p<0.001)						
Kongwa	72	10 (13.9)	17 (23.6)	19 (26.4)	15 (20.8)	11 (15.3)
Muheza	76	6 (7.9)	6 (7.9)	4 (5.3)	28 (36.8)	32 (42.1)
Singida	104	34 (32.7)	28 (26.9)	27 (26.0)	8 (7.7)	7 (6.7)
Pattern 3 "Traditional-inland" (p=0.009)						
Kongwa	72	18 (25.0)	17 (23.6)	15 (20.8)	11 (15.3)	11 (15.3)
Muheza	76	21 (27.6)	12 (15.8)	10 (13.2)	22 (28.9)	11 (14.5)
Singida	104	11 (10.6)	22 (21.2)	25 (24.0)	18 (17.3)	28 (26.9)
Pattern 4 "Pulses" (p=0.040)						
Kongwa	72	14 (19.4)	17 (23.6)	19 (26.4)	14 (19.4)	8 (11.1)
Muheza	76	9 (11.8)	11 (14.5)	15 (19.7)	20 (26.3)	21 (27.6)
Singida	104	27 (26.0)	23 (22.1)	16 (15.4)	17 (16.3)	21 (20.2)
Pattern 5 "Animal products" (p=0.001)						
Kongwa	72	11 (15.3)	20 (27.8)	17 (23.6)	16 (22.2)	8 (11.1)
Muheza	76	27 (35.5)	9 (11.8)	13 (17.1)	14 (18.4)	13 (17.1)
Singida	104	12 (11.5)	22 (21.2)	20 (19.2)	21 (20.2)	29 (27.9)

p from Pearson X^2 Test

Table 3.3.23 Bivariate correlations/associations between food patterns and socio-economic variables of women in Tanzania

		Pattern 1 "Traditional-coast"	Pattern 2 "Purchase"	Pattern 3 "Traditional-inland"	Pattern 4 "Pulses"	Pattern 5 "Animal products"
Wealth	tau*	n.a.	n.a.	n.a.	n.a.	0.171
	p	n.s.	n.s.	n.s.	n.s.	<0.001
Household size	rho**	n.a.	-0.205	-0.215	n.a.	n.a.
	p	n.s.	0.001	0.001	n.s.	n.s.
Distance in km	rho	0.183	n.a.	n.a.	-0.188	0.285
	p	0.003	n.s.	n.s.	0.003	<0.001
Distance in min	rho	n.a.	-0.193	n.a.	-0.174	0.315
	p	n.s.	0.002	n.s.	0.006	<0.001

* correl. coeff. T; ** correl. coeff. P; n.a.=not applicable; n.s.=not significant

Table 3.3.24 Occupation of participants per quintile of food patterns 2 "purchase" and 4 "pulses"; data surveyed during three different seasons in three districts of Tanzania

	N	1st	2nd	3rd	4th	5th
Pattern 2 "Purchase" (p=0.002)						
Crop farmer	27	4 (14.8)	5 (18.5)	7 (25.9)	2 (7.4)	9 (33.3)
Crop and livestock farmer	194	44 (22.7)	45 (23.2)	37 (19.1)	39 (20.1)	29 (14.9)
Farmer + business/ service	31	2 (6.5)	1 (3.2)	6 (19.4)	10 (32.3)	12 (38.7)
Pattern 4 "Pulses" (p=0.027)						
Crop farmer	27	5 (18.5)	10 (37.0)	7 (25.9)	3 (11.1)	2 (7.4)
Crop and livestock farmer	194	43 (22.2)	38 (19.6)	33 (17.0)	41 (21.1)	39 (20.1)
Farmer + business/service	31	2 (6.5)	3 (9.7)	10 (32.3)	7 (22.6)	9 (29.0)

Values expressed as n (% within occupation group); p from Pearson X^2 Test

Table 3.3.25 Socio-economic variables (continuous or categorical data) and their mean or median values of women in Tanzania within the 1st and 5th quintile of five food patterns generated by PCA from mean intake of twelve food groups during three different seasons; 1st quitile: no consumption according to the pattern; 5th quintile: high consumption according to the pattern

	Pattern 1 "Traditional-coast"				Pattern 2 "Purchase"				Pattern 3 "Traditional-inland"				Pattern 4 "Pulses"				Pattern 5 "Animal products"			
	1st	5th	p^1	p^2	1st	5th	p^1	p^2	1st	5th	p^1	p^2	1st	5th	p^1	p^2	1st	5th	p^1	p^2
Household size	6.4 ±2.2	5.3 ±2.6	n.s.	**	6.5 ±2.9	5.1 ±2.0	*	***	6.8 ±2.7	4.9 ±1.9	**	***	5.4 ±2.9	5.8 ±2.1	n.s.	n.s.	5.9 ±2.5	5.8 ±2.1	n.s.	n.s.
Distance (km)	32	18	*	**	33	23	n.s.	*	25	33	n.s.	n.s.	33	25	*	**	9	33	***	***
Distance (min)	120	60	**	*	150	60	*	***	60	120	n.s.	*	150	90	*	*	60	150	***	***

Values expressed as mean ± SD (HH size) or median (all other variables); [1]p from Kruskal-Wallis test (grouping variable was the respective food pattern); [2]p from Jonckheere-Terpstra test for trend; n.s. not significant; * $p \leq 0.05$; ** $p \leq 0.01$; ***$p \leq 0.001$

Table 3.3.26 Dietary variables of women in Tanzania within the 1st and 5th quintile of five food patterns generated by PCA from mean intake of twelve food groups during three different seasons; 1st quitile: no consumption according to the pattern; 5th quintile: high consumption according to the pattern

	Pattern 1 "Traditional-coast"				Pattern 2 "Purchase"				Pattern 3 "Traditional-inland"				Pattern 4 "Pulses"				Pattern 5 "Animal products"			
	1st	5th	p^1	p^2	1st	5th	p^1	p^2	1st	5th	p^1	p^2	1st	5th	p^1	p^2	1st	5th	p^1	p^2
FVS	7.4 ±2.0	10.6 ±1.6	***	***	6.4 ±2.0	10.1 ±1.4	***	***	7.9 ±2.8	8.6 ±2.3	n.s.	n.s.	7.4 ±2.1	9.2 ±2.1	***	***	8.4 ±2.6	8.8 ±1.8	***	*
DDS	4.7	8.0	***	***	3.8	7.3	***	***	5.8	6.2	n.s.	n.s.	4.3	6.7	***	***	6.5	6.2	***	*
VDS	4.8	5.2	n.s.	n.s.	4.3	4.5	n.s.	n.s.	5.0	4.7	n.s.	n.s.	4.7	5.0	n.s.	n.s.	4.0	4.3	**	*
Energy (kcal/d)	1672.7 ±372.1	2157.9 ± 504.2	***	***	1834.1 ±367.9	2044.4 ±397.7	***	***	1529.1 ±402.1	2259.6 ±465.7	***	***	1926.7 ±452.2	1981.7 ±374.2	**	n.s.	2000.9 ±530.1	1862.3 ±390.6	n.s.	n.s.
Protein (g/d)	51.0	65.5	***	***	59.5	61.5	***	n.s.	47.5	65.5	***	***	54.5	64.0	***	***	67.0	54.5	***	***
Fat (g/d)	34.0	37.0	n.s.	n.s.	28.5	44.0	***	***	24.0	54.5	***	***	36.0	34.5	n.s.	n.s.	35.5	40.5	n.s.	*
Vitamin A (µg/d)	353.5	1351.5	***	***	464.5	1207.0	***	***	706	569.5	n.s.	n.s.	831.5	711.0	n.s.	n.s.	863.5	706.0	n.s.	n.s.
Iron (mg/d)	18.0	14.0	**	***	21.5	13.0	***	***	13.5	17.0	n.s.	*	14.0	16.0	n.s.	n.s.	15.5	14.0	**	**

Values shown as mean ± SD (FVS and energy) and median (all others); [1]p from Kruskal-Wallis Test (grouping variable was the respective food pattern); [2]p from Jonckheere-Terpstra Test for trend; significant values highlighted; FVS=food variety score; DDS=dietary diversity score; VDS=vegetable diversity score; n.s. not significant; * $p \leq 0.05$; ** $p \leq 0.01$; ***$p \leq 0.001$

Bivariate correlations between food patterns (factor scores of women, i.e. their affiliation to a pattern) and nutrition variables are shown in Table 3.3.27. While FVS was significantly correlated to all food patterns, DDS was not correlated to pattern 3 and VDS, in turn, was only correlated to pattern 5. Highest correlations were found between pattern 2 and both FVS and DDS, suggesting that food diversity was rather high when this pattern was followed. This trend was similar for pattern 1 but less for pattern 4 and correlations between pattern 3 and 5 and the food scores were only minor significant. The nutrition variables were also used to characterise the lowest and highest quintile of each pattern (Table 3.3.26). In some cases a trend was found while there was no difference between the quintiles.

Table 3.3.27   Bivariate correlations between food patterns and dietary variables (scores and nutrient intakes) – only those parameters shown which are significantly related to at least one dietary pattern

	Pattern 1 "Traditional-coast"		Pattern 2 "Purchase"		Pattern 3 "Traditional-inland"		Pattern 4 "Pulses"		Pattern 5 "Animal products"	
	Correl. coeff. rho	p	Correl. coeff. rho	p	Correl. coeff. rho	p	Correl. coeff. rho	p	Correl. coeff. rho	p
FVS	0.498	<0.001	0.644	<0.001	0.126	0.046	0.311	<0.001	0.147	0.020
DDS	0.548	<0.001	0.637	<0.001	n.a.	n.s.	0.369	<0.001	0.127	0.045
VDS	n.a.	n.s.	n.a.	n.s.	n.a.	n.s.	n.a.	n.s.	0.161	0.011
Energy	0.422	<0.001	0.268	<0.001	0.620	<0.001	n.a.	n.s.	n.a.	n.s.
Protein	0.334	<0.001	0.188	0.003	0.428	<0.001	0.255	<0.001	-0.271	<0.001
Fat	n.a.	n.s.	0.374	<0.001	0.583	<0.001	n.a.	n.s.	n.a.	n.s.
Vitamin A	0.498	<0.001	0.265	<0.001	n.a.	n.s.	n.a.	n.s.	n.a.	n.s.
Iron	-0.245	<0.001	-0.395	<0.001	0.142	0.024	n.a.	n.s.	-0.192	0.002

n.a.=not applicable; n.s.=not significant

The same calculations were performed for each quintile with the health variables (Table 3.3.28). The cohort for the BMI calculations consisted of 210 participants only and, consequently, each quintile consisted of 42 participants. Significant correlations between dietary patterns and health variables were found between pattern 2 "purchase" and BMI ($\rho=0.192$; p=0.005) and pattern 5 "animal products" and the number of negative characteristics named for an obese person ($\rho=0.203$; p=0.003). Similarly, there was a significant difference of BMI values between the quintiles of pattern 2 as well as a significant increasing trend from the first to the fifth quintile.

The cohort for the analysis of iron parameters consisted of 185 participants, thus, each quintile of the food patterns consisted of 37 women. The Hb value of women was significantly correlated to three food patterns, namely negatively to the "traditional-coast" pattern ($\rho=-0.291$; p<0.001), again negatively to the "purchase" pattern ($\rho=-0.241$; p=0.001) as well as positively to the "animal production" pattern ($\rho=0.216$; p=0.003). Also the trend for Hb values from the first to the fifth quintile of pattern 1 and 2 was negative, while Hb was highest for women within the fifth quintile of pattern 5. Similarly, iron intake was lower for those women following pattern 2, while it was higher for those who consumed food according to pattern 3. Finally, the cohort for the vitamin A analysis consisted of 145 women and, consequently, each quintile was made up of 29 participants.

Table 3.3.28   Health variables (including iron intake) of women in Tanzania within the 1st and 5th quintile of five food patterns generated by PCA from mean intake of twelve food groups during three different seasons; 1st quitile: no consumption according to the pattern; 5th quintile: high consumption according to the pattern (n=210 for BMI and PHNE; n=185 for Hb and Iron)

	Pattern 1 "Traditional-coast"				Pattern 2 "Purchase"				Pattern 3 "Traditional-inland"				Pattern 4 "Pulses"				Pattern 5 "Animal products"			
	1st	5th	p^1	p^2	1st	5th	p^1	p^2	1st	5th	p^1	p^2	1st	5th	p^1	p^2	1st	5th	p^1	p^2
BMI	21.7	22.1	n.s.	n.s.	21.5	23.2	*	*	22.5	21.5	n.s.	n.s.	22.7	22.3	n.s.	n.s.	21.8	22.3	n.s.	n.s.
PHNE	3.0 ±1.2	3.0 ±1.8	n.s.	n.s.	3.0 ±1.3	3.0 ±1.3	n.s.	n.s.	3.0 ±1.4	3.0 ±1.4	n.s.	n.s.	3.0 ±1.2	3.0 ±1.5	n.s.	n.s.	3.0 ±1.3	4.0 ±1.2	**	**
Hb	13.1 ±1.4	11.7 ±1.3	***	***	13.2 ±1.7	12.3 ±1.3	*	**	12.8 ±1.4	13.2 ±1.3	n.s.	n.s.	12.8 ±1.3	12.6 ±1.6	n.s.	n.s.	12.1 ±1.8	13.2 ±1.5	n.s.	**
Iron (mg/d)	15.0	14.0	n.s.	*	22.0	13.0	***	***	13.0	18.0	n.s.	*	14.0	16.0	n.s.	n.s.	15.0	14.0	n.s.	n.s.

Values as mean ± SD or median; [1]p from Kruskal Wallis test (grouping variable is the respective food pattern); [2]p from Jonckheere-Terpstra test for trend; significant values highlighted; BMI=body mass index; PHNE= negative characteristics for a corpulent person named by participants; Hb=heamoglobin; n.s. not significant;
* p≤0.05; ** p≤0.01; ***p≤0.001

Multiple regression analysis was applied in order to analyse whether the affiliation to a food pattern is influenced by other parameters. All five models, one for each pattern, were controlled for age, districts and wealth, and 14 predictor variables were chosen which were already associated to one or more patterns within bivariate comparison. Because of the high correlation between DDS and FVS, the latter was not included to avoid suppressor effects. Also the number of seasons during which vegetables were cultivated/collected were excluded for the same reason and in favour of the number of vegetables cultivated/collected. While the residuals of the models with food patterns 3 and 4 as dependent variables were normally distributed, this was not the case for the residuals of the models with patterns 1, 2 and 5. Thus, factor scores of the latter food patterns had to be transformed with a logarithmic transformation. Before the logarithm was taken, the amount of 3 was added to all factor scores to avoid negative values. While after this procedure residuals of patterns 1 and 2 were normally distributed those of pattern 5 were not and, therefore, this multiple regression model was not further studied.

The five food patterns derived by cluster analysis also showed significant associations with socio-economic variables (Table 3.3.29), DDS and FVS (Table 3.3.30), nutrient intake variables (Table 3.3.31) and health variables (Table 3.3.32). Again, a district effect could be shown with women from Muheza consuming food mainly to cluster 5, while those in the other two districts rather favoured cluster 1. Furthermore, women with a low DDS/FVS mainly followed cluster 1, those with a medium DSS/FVS were concentrated in cluster 3 and women having a high DDS/FVS ate mainly food according to cluster 4. The mean value for energy intake was highest for cluster 4 as well as the median values for protein and vitamin A intake. The median value for iron intake was, however, highest for clusters 1 and 2. Also women in clusters 1 and 2 had the highest mean Hb values, while the highest median BMI was detected for women within cluster 5.

Table 3.3.29   Five food clusters characterised by different socio-economic variables; data surveyed during three different seasons in three districts of Tanzania

	n	p (χ^2)	**Cluster 1** Veg/ cereals n=97 n (%)	**Cluster 2** Cereals/ veg/pulses n=29 n (%)	**Cluster 3** Cereals/ veg/tea n=45 n (%)	**Cluster 4** Cereals/tea/ veg/fruits/ nuts/pulses n=43 n (%)	**Cluster 5** Tea/cereals/ veg/pulses/ bread/cakes n=38 n (%)
District		<0.001					
Kongwa	72		**36 (50.0)**	9 (12.5)	18 (25.0)	2 (2.8)	7 (9.7)
Muheza	66		1 (1.5)	1 (1.5)	8 (12.1)	27 (40.9)	**29 (43.9)**
Singida	104		**60 (57.7)**	19 (18.3)	19 (18.3)	4 (3.8)	2 (1.9)
Religion		0.018					
Christian	161		**64 (39.8)**	24 (14.9)	29 (18.0)	19 (11.0)	25 (15.5)
Muslim	91		**33 (36.3)**	5 (5.5)	16 (17.6)	24 (26.4)	13 (14.3)
Ethnic group		<0.001					
Bondei	19		0 (0.0)	0 (0.0)	0 (0.0)	**11 (57.9)**	8 (42.1)
Gogo	36		**20 (55.6)**	8 (22.2)	3 (8.3)	1 (2.8)	4 (11.1)
Kaguru	24		**12 (50.0)**	0 (0.0)	9 (37.5)	1 (4.2)	2 (8.3)
Nyaturu	104		**60 (57.7)**	20 (19.2)	19 (18.3)	2 (1.9)	3 (2.9)
Shambaa	24		0 (0.0)	0 (0.0)	1 (4.2)	**15 (62.5)**	8 (33.3)
Other	45		5 (11.1)	1 (2.2)	13 (28.9)	13 (28.9)	13 (28.9)

			Cluster 1 Veg/ cereals n=97	Cluster 2 Cereals/ veg/pulses n=29	Cluster 3 Cereals/ veg/tea n=45	Cluster 4 Cereals/tea/ veg/fruits/ nuts/pulses n=43	Cluster 5 Tea/cereals/ veg/pulses/ bread/cakes n=38
Position in hh		0.014					
Head of hh	22		5 (22.7)	1 (4.5)	3 (13.6)	5 (22.7)	**8 (36.4)**
Wife of head of hh	188		**81 (43.1)**	25 (13.3)	35 (18.6)	25 (13.3)	22 (11.7)
Mother of head of hh	22		6 (27.3)	1 (4.5)	5 (22.7)	5 (22.7)	5 (22.7)
Daughter of head of hh	20		5 (25.0)	2 (10.0)	2 (10.0)	**8 (40.0)**	3 (15.0)
Marital status		0.006					
Single	40		7 (17.5)	4 (10.0)	6 (15.0)	**12 (30.0)**	11 (27.5)
Married	138		**81 (44.3)**	24 (13.1)	33 (18.0)	25 (13.7)	20 (10.9)
Widowed	9		2 (22.2)	1 (11.1)	**4 (44.4)**	1 (11.1)	1 (11.1)
Divorced	8		3 (37.5)	0 (0.0)	2 (25.0)	2 (25.0)	1 (12.5)
Separated	12		4 (33.3)	0 (0.0)	0 (0.0)	3 (25.0)	5 (41.7)
Occupation		0.003					
Crop farmer	27		**16 (59.3)**	1 (3.43.7)	3 (11.1)	3 (11.1)	4 (14.8)
Crop and livestock farmer	194		**77 (39.7)**	27 (13.9)	34 (17.5)	32 (16.5)	24 (12.4)
Farmer + business/service	31		4 (12.9)	1 (3.2)	8 (25.8)	8 (25.8)	**10 (32.3)**
Distance village-town (km)		<0.001					
Short (5-20 km)	85		**24 (28.2)**	5 (5.9)	12 (14.1)	23 (27.1)	21 (24.7)
Medium (21-34 km)	76		**30 (39.5)**	7 (9.2)	15 (19.7)	15 (19.7)	9 (11.8)
Long (35-67 km)	91		**43 (47.3)**	17 (18.7)	18 (19.8)	5 (5.5)	8 (8.8)
Distance village-town (min)		<0.001					
Short (20-45 min)	61		**24 (39.3)**	5 (8.2)	8 (13.1)	11 (18.0)	13 (21.3)
Medium (60-90 min)	55		11 (20.0)	1 (1.8)	6 (10.9)	**25 (45.5)**	12 (21.8)
Long (120-240 min)	136		**62 (45.6)**	23 (16.9)	31 (22.8)	7 (5.1)	13 (9.6)

Table 3.3.30   Five food clusters characterised by dietary diversity score (DDS) and food variety score (FVS); data surveyed during three different seasons in three districts of Tanzania

			Cluster 1 Veg/ cereals n=97 n (%)**	Cluster 2 Cereals/ veg/pulses n=29 n (%)	Cluster 3 Cereals/ veg/tea n=45 n (%)	Cluster 4 Cereals/tea/ veg/fruits/ nuts/pulses n=43 n (%)	Cluster 5 Tea/cereals/ veg/pulses/ bread/cakes n=38 n (%)
	n	p (χ^2)*					
DDS		<0.001					
Low (2-4 fg)			**74 (76.3)**	10 (34.5)	0 (0.0)	0 (0.0)	0 (0.0)
Medium (5-6 fg)			20 (20.6)	16 (55.2)	**32 (71.1)**	1 (2.3)	6 (15.8)
High (7-11 fg)			3 (3.1)	3 (10.3)	13 (28.9)	**42 (97.7)**	32 (84.2)
Median		<0.001	4,0	4,7	6,3	8,0	7,7
FVS		<0.001					
Low (2-6 foods)			**59 (60.8)**	10 (34.5)	2 (4.4)	0 (0.0)	0 (0.0)
Medium (7-9 foods)			37 (38.1)	17 (58.6)	**28 (62.2)**	6 (14.0)	10 (26.3)
High (10-14 foods)			1 (1.0)	2 (6.9)	15 (33.3)	**37 (86.0)**	28 (73.7)
Mean (±sd)		<0.001	6.2 ±1.4	7.1 ±1.7	8.9 ±1.3	**10.9** **±1.4**	10.2 ±1.3

* p for differences between food score groups and for differences between mean/median values of cluster groups according to K-W test if applicable
** absolute number of participants and percent within cluster, if not mean/median

Table 3.3.31   Five food clusters characterised by different nutrient intake variables; data surveyed during three different seasons in three districts of Tanzania

	n	p (χ^2)*	Cluster 1 Veg/ cereals n=97 n (%)**	Cluster 2 Cereals/ veg/pulses n=29 N (%)	Cluster 3 Cereals/ veg/tea n=45 n (%)	Cluster 4 Cereals/tea/ veg/fruits/ nuts/pulses n=43 n (%)	Cluster 5 Tea/cereals/ veg/pulses/ bread/cakes n=38 n (%)
Energy		0.003					
600-1799 kcal/d	121		**58 (59.8)**	**21 (72.4)**	18 (40.0)	12 (27.9)	12 (31.6)
1800-2599 kcal/d	122		37 (38.1)	7 (24.1)	**25 (55.6)**	**28 (65.1)**	**25 (65.8)**
2600-3500 kcal/d	7		1 (1.0)	1 (3.4)	1 (2.2)	3 (7.0)	1 (2.6)
Above 3500 kcal/d	2		1 (1.0)	0 (0.0)	1 (2.2)	0 (0.0)	0 (0.0)
Mean (±sd)		<0.001	1708.4 ±452.7	1753.1 ±329.8	1992.3 ±481.9	**2101.7 ±387.5**	1932.3 ±360.2
Protein							
Below 45 g/d	45	0.001	28 (28.9)	4 (13.8)	7 (15.6)	3 (7.0)	3 (7.9)
45-65 g/d	131		**52 (53.6)**	**19 (65.5)**	**21 (46.7)**	**20 (46.5)**	**19 (50.0)**
Above 65 g/d	76		17 (17.5)	6 (20.7)	17 (37.8)	**20 (46.5)**	16 (42.1)
Median		<0.001	51,0	56,0	58,0	**65,0**	63,5
Fat		0.087					
Median		0.008	34,0	33,0	**39,0**	36,0	37,5
Vitamin A		<0.001					
1-499 µg/d	89		**52 (53.6)**	**16 (55.2)**	13 (28.9)	4 (9.3)	4 (10.5)
500-849 µg/d	40		19 (19.6)	2 (6.9)	7 (15.6)	7 (16.3)	5 (13.2)
850 – 7500 µg/d	123		26 (26.8)	11 (37.9)	**25 (55.6)**	**32 (74.4)**	**29 (76.3)**
Median		<0.001	467	363	1125	**1303**	1198
Iron		<0.001					
1-14.99 mg/d	112		29 (29.9)	5 (17.2)	**24 (53.3)**	**26 (60.5)**	**28 (73.7)**
15-29.99 mg/d	97		**43 (44.3)**	**16 (55.2)**	15 (33.3)	17 (39.5)	6 (15.8)
30-90 mg/d	35		22 (22.7)	6 (20.7)	4 (8.9)	0 (0.0)	3 (7.9)
Above 90 mg/d	8		3 (3.1)	2 (6.9)	2 (4.4)	0 (0.0)	1 (2.6)
Median		<0.001	**19**	**19**	14	14	13

* p for differences between nutrient groups and for differences between mean/median values of cluster groups according to K-W test if applicable
** absolute number of participants and percent within cluster, if not mean/median

Table 3.3.32   Five food clusters characterised by body mass index (BMI) and haemoglogin (Hb) level; data surveyed during three different seasons in three districts of Tanzania

	N	P (χ^2)*	Cluster 1 Veg/ cereals N (%)**	Cluster 2 Cereals/ veg/pulses N (%)	Cluster 3 Cereals/ veg/tea N (%)	Cluster 4 Cereals/tea/ veg/fruits/ nuts/pulses N (%)	Cluster 5 Tea/cereals/ veg/pulses/ bread/cakes N (%)
BMI		0,073	n=72	n=26	n=39	n=39	n=34
Underweight	15		**5 (6.9)**	1 (3.8)	3 (7.7)	3 (7.7)	3 (8.8)
Normal weight	149		57 (79.2)	21 (80.8)	28 (71.8)	26 (66.7)	17 (50.0)
Overweight	33		9 (12.5)	4 (15.4)	4 (10.3)	5 (12.8)	**11 (32.4)**
Obese	13		1 (1.4)	0 (0)	4 (10.3)	**5 (12.8)**	3 (8.8)
Median		<0.001	20,7	21,9	21,7	22,2	**24,4**
Hb		<0.001	n=71	n=21	n=33	n=30	n=30
Severe anaemia			0 (0)	1 (4.8)	0 (0)	0 (0)	0 (0)
Moderate anaemia			0 (0)	0 (0)	0 (0)	4 (13.3)	0 (0)
Mild anaemia			10 (14.1)	1 (4.8)	12 (36.4)	10 (33.3)	8 (26.7)
Normal iron status			61 (85.9)	19 (90.5)	21 (63.6)	16 (53.4)	22 (73.3)
Mean (±sd)		<0.001	**13.2 ±1.2**	**13.2 ±1.9**	12.7 ±1.3	11.8 ±1.7	12.4 ±1.1

* p for differences between BMI and Hb groups and for differences between mean/median values of cluster groups according to K-W test if applicable
** absolute number of participants and percent within cluster if not mean/median

3.4   Nutritional health

3.4.1   Vitamin A status

All participants with an infection according to c-reactive protein (CRP) and acid glucoprotein (AGP) values, all women who stated to have taken vitamin A supplements during the last three month, as well as all who were identified to most likely have HIV/Aids, were excluded from the analysis. Therefore, only a total of 145 participants were included for vitamin A status examinations.

The retinol binding protein (RBP) values were normally distributed (K-S test statistic=0.082; S-W test statistic=0.969), however, regarding the three districts individually, RBP values only showed normal distribution in Muheza (K-S test statistic=0.084; S-W test statistic=0.978). No significant differences were found among the three districts (p=0.590) in terms of vitamin A status of participants. The mean RBP value for all districts (1.05 µmol/l ± 0.29) was above the vitamin A deficiency (VAD) cut-off of 0.7 µmol/l, while the range was between 0.54 µmol/l and 1.89 µmol/l RBP, and both mean and range were similar for all districts (Table 3.4.1).

Table 3.4.1   Retinol binding protein (RBP) values of participants (n=145) during Mar/Apr (LR) in three different district of Tanzania

	All districts	Kongwa	Muehza	Singida
n	145	48	38	59
Mean ± SD	1.05 ± 0.29	1.03 ± 0.35	1.03 ± 0.24	1.08 ± 0.27
Median (range)	1.07 (0.54-1.89)	0.93 (0.54-1.89)	1.01 (0.59-1.66)	1.14 (0.56-1.67)

Most participants with vitamin A deficiency (VAD), according to RBP value, were found in Kongwa (15%), less in Singida (10%) and fewest in Muheza district (5%). Still, a great share of participants was classified to have marginal vitamin A status (Kongwa 44%, Muheza 47%, Singida 29%). Consequently, the greatest share of women without VAD was found in Singida district (61%), followed by Muheza (47%) and Kongwa (41%) (Figure 3.4.1).

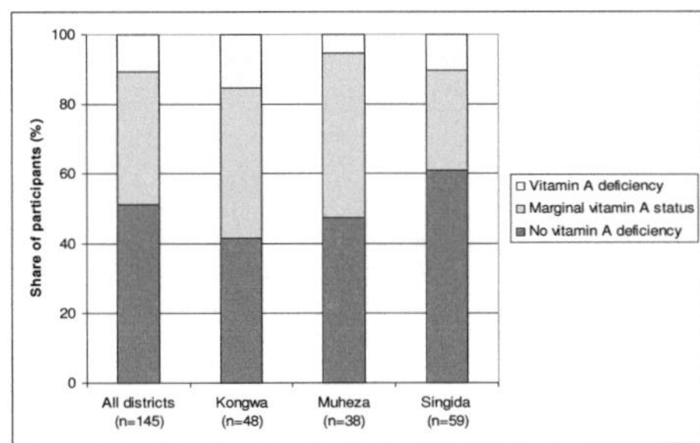

Figure 3.4.1
Share of participants (n=145) with either vitamin A deficiency (RBP <0.70 µmol/l), marginal   vitamin A status (RBP <1.05 µmol/l) or no vitamin A deficiency (RBP≥1.05 µmol/l) in three   different districts of Tanzania

As one parameter alone can not show a complete picture of a possible deficiency, more than one factor was checked for the vitamin A status of a person. Besides RBP in dried blood spots (biochemical indicator), during all seasons the eyes of each participant were investigated for Bitot spots and signs of night blindness (functional indicator). Additionally, during Nov/Dec (SR) and Mar/Apr (LR) CIC with transfer was carried out (histological indicator). As data from the latter approach was not analysable to date, only the functional indicators were taken into account. These also showed that the prevalence of severe vitamin A deficiency among women was rather marginal in the three examined districts, confirming the biochemical indicators.

Out of the 145 participants, none had Bitot spots in Jun/Jul (DS), only one from Kongwa district in Nov/Dec (SR) and again the same participant in Mar/Apr (LR). In terms of self-reported night blindness, two participants were identified during Jun/Jul (DS) (one from Muheza, one from Singida), none reported to have night blindness during Nov/Dec (SR) and two participants during Mar/Apr (LR) (one from Kongwa and one from Singida).

<u>Correlations and associations</u>

When correlated to socio-economic parameters, VAD categories showed a significant association with the wealth status (p<0.001), yet, no linear relationship (Figure 3.4.2). While all other socio-economic variables were not significantly associated to VAD, the relationships with ethnic group (Figure A19), education (Figure A20) and age (Figure A21) were interesting. In terms of food consumption data, a significant negative correlation was found between sugar consumption and vitamin A status ($\rho=-0.171$; $p=0.40$). Neither nutrient intakes nor the food scores were significantly correlated to RBP values. Still, the associations between VDS and VAD and between the intake quantity of fruit/vegetables and VAD are visualised in figures A22 and A23.

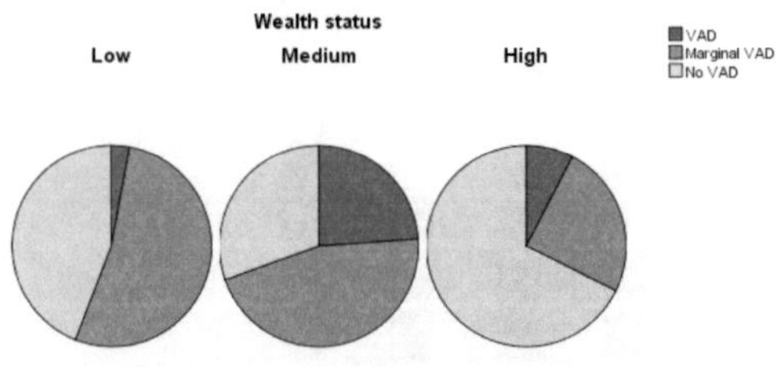

Figure 3.4.2
Association between wealth status and vitamin A deficiency of women in Tanzania (n=145; p<0.001)

3.4.2 Iron status

When analysing iron status as measured by Hb level, several women had to be excluded from the study, namely those who were in the 3rd trimester of pregnancy, women who stated to take iron supplements and those who most likely had HIV/Aids. Consequently, the data of 185 women were analysed.

Hb levels of participants were normally distributed during Jun/Jul (DS) (K-S test statistic=0.067; S-W test statistic=0.973), Nov/Dec (SR) (K-S test statistic=0.071; S-W test statistic=0.958) and when the mean of all three seasons was taken (K-S test statistic=0.055; S-W test statistic=0.969), however, not during Mar/Apr (LR). The median Hb level of all participants ranged between 11.8 and 14.3 g/dl during all seasons. Hb values were in general highest for Singida and lowest for Muheza district, and they were also higher during Jun/Jul (DS) than during Nov/Dec (SR) and Mar/Apr (LR) (Table 3.4.2).

According to ANOVA, the mean Hb values were significantly different between the three districts during Jun/Jul (DS) (p<0.001) and Nov/Dec (SR) (p<0.001) and also when the mean value of all three seasons was compared (p<0.001). Also during Mar/Apr (LR) significant differences between the three districts were observed according to K-W-test (p<0.001). Significant differences existed between the three seasons (p<0.001).

Table 3.4.2   Mean and median Hb values (g/L) of participants (n=185) in three different district of Tanzania during three different seasons

	All districts	Kongwa	Muheza	Singida
n	185	52	58	75
All year				
Mean ± SD	12.7 ± 1.5	12.7 ± 1.5	11.9 ± 1.4	13.4 ± 1.1
Median (range)	12.9 (6.9-15.8)	12.9 (6.9-15.4)	12.1 (8.1-14.1)	13.6 (10.3-15.8)
Jun/Jul (DS)				
Mean ± SD	13.3 ± 1.8	13.2 ± 1.7	12.4 ± 1.7	14.2 ± 1.6
Median (range)	13.4 (6.5-17.5)	13.4 (8.6-16.9)	12.6 (6.5-15.8)	14.3 (10.0-17.5)
Nov/Dec (SR)				
Mean ± SD	12.5 ± 1.5	12.3 ± 1.7	12.1 ± 1.4	13.1 ± 1.1
Median (range)	12.5 (5.9-15.5)	12.5 (5.9-15.2)	12.1 (7.7-14.4)	13.1 (10.8-15.5)
Mar/Apr (LR)				
Mean ± SD	12.4 ± 1.7	12.5 ± 1.7	11.4 ± 1.5	13.1 ± 1.4
Median (range)	12.5 (6.1-16.0)	12.7 (6.1-16.0)	11.8 (7.5-14.1)	13.3 (6.6-15.5)

When participants were grouped according to their Hb level (Figure 3.4.3), differences between seasons were observable with most cases of anaemia occurring during Nov/Dec (SR) and Mar/Apr (LR). Yet, more obvious differences appeared between the districts: in Singida, there was only one case of severe anaemia during Mar/Apr (LR), while otherwise 80% of participants or more had a normal Hb. While in Kongwa district only between 64 and 77% of participants, depending on the season, had a normal Hb, in Muheza district the situation was even worse: only for between 43 and 66% of women a normal Hb was measured, while between 26 and 35% had mild anaemia, in Mar/Apr (LR) 23% had even moderate anaemia.

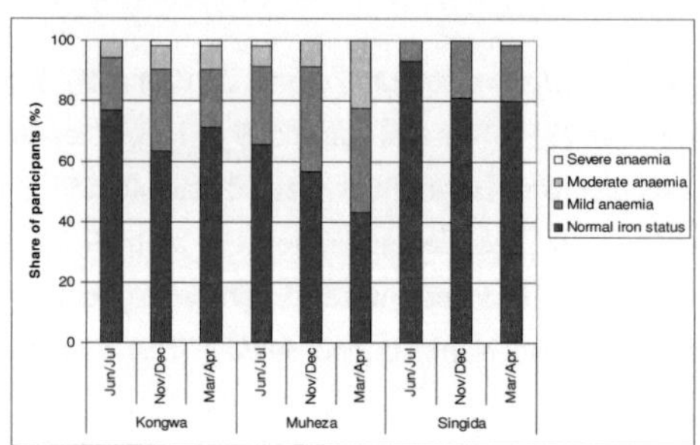

Figure 3.4.3
Iron status according to Hb level from participants
(n=185) of three districts in Tanzania during three
different seasons; normal iron status Hb≥12 g/dl; mild
anaemia 10.0-11.9 g/dl; moderate anaemia 7.0-9.9 g/dl;
severe anaemia <7 g/dl

Besides Hb, another parameter to determine the iron status was measured, namely transferrin receptor (TfR) in dried blood spots (DBS), which was only done during Mar/Apr (LR) (Table 3.4.3). As not all DBS could be analysed, data from only 131 women could be used for this analysis. This data was not normally distributed; significant differences were found among the three districts (p=0.001).

Table 3.4.3   Transferrin receptor (TfR) (g/L) values* of women (n=131) in three district of Tanzania during March/April 2007

	All districts	Kongwa	Muehza	Singida
n	131	42	40	49
Mean ± SD	7.6 ± 5.5	8.3 ± 8.1	8.8 ± 4.2	6.1 ± 2.8
Median (range)	6.1 (2.3-53.4)	5.9 (2.4-53.4)	7.5 (3.8-21.5)	5.3 (2.3-14.3)

*TfR > 8g/L means iron deficiency

The cut-off point for iron deficiency is 8 g/L, whereby participants with a TfR below have a normal iron status and those with a TfR equal or above can be classified as iron-deficient. Similar to the results of the Hb measures, TfR values showed that differences between the districts existed and that only 20% of participants had iron deficiency in Singida district, 24% in Kongwa, and that women in Muheza were worse off with 48% of participants having iron deficiency (Figure 3.4.4).

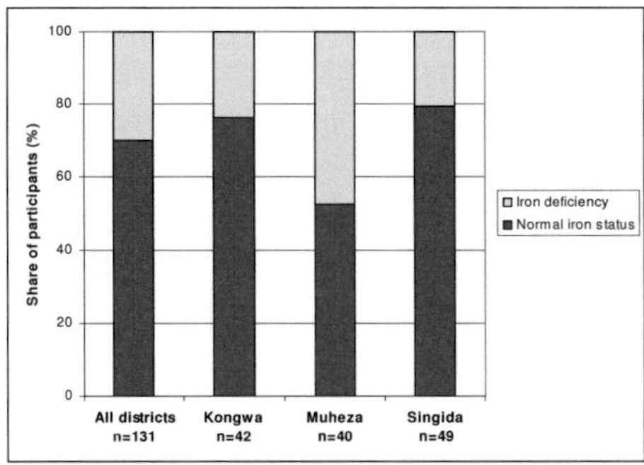

Figure 3.4.4
Share of participants (n=131) with a normal/subnormal
iron status according to TfR values in three districts of
Tanzania during March/April

When the link between Hb values from Mar/Apr (LR) and TfR was analysed, a highly significant relationship became apparent (ρ=-0.421; p<0.001). The relationship was negative, as low Hb values indicate iron deficiency while low TfR values stand for a normal iron status.

There was no significant relationship between any of the socio-economic parameters and the Hb or TfR values of participants. The only exception was TfR and wealth, which were negatively associated (τ=-0.176; p=0.009) meaning that the higher the wealth status the more likely it was for a participant not to suffer from iron deficiency.

The Hb value was both negatively and positively correlated to different vegetable production parameters, especially during Mar/Apr (LR) (Table 3.4.4). The TfR value was, however, only negatively correlated to the number of exotic vegetables cultivated (ρ=-0.186; p=0.034). No associations were found between the dichotomous vegetable production/purchase/sales parameters and Hb or TfR values.

Table 3.4.4   Correlations between Hb values and cultivation and collection of vegetables by women in Tanzania during different seasons (n=185)

Season	No of vegetables cultivated		No of vegetables collected		No of TVs cult/coll		No of EVs cultivated	
	Correl. coeff. rho	p	Correl. coeff. rho	p	Correl. coeff. rho	p	Correl. coeff. rho	p
All year	n.a.	n.s.	n.a.	n.s.	n.a.	n.s.	0.172	0.019
Jun/Jul (DS)	n.a.	n.s.	-0.201	0.006	-0.177	0.016	n.a.	n.s.
Nov/Dec (SR)	n.a.	n.s.	-0.170	0.021	-0.181	0.014	n.a.	n.s.
Mar/Apr (LR)	0.189	0.010	0.207	0.005	0.227	0.002	0.293	<0.001

n.a.=not applicable; n.s.=not significant

Further correlations of the Hb value was with other variables were calculated, especially food group and nutrient intake. No significant correlation was found between Hb values and iron intake, however, between Hb values and vitamin C intake. This correlation was always negative, but highly significant for the whole year (ρ=-0.218; p=0.003), but less so for Jun/Jul (DS) (ρ=-0.167; p=0.023) and Mar/Apr (LR) (ρ=-0.179; p=0.015). The Hb was also negatively correlated to FVS, DDS and several food groups during all year (Table 3.4.5). Bread and cakes showed a negative relationship with Hb only during Mar/Apr (LR) and with the mean across all seasons. Furthermore, there was a positive relationship between Hb and vegetable as well as animal products consumption during Jun/Jul (DS) and the mean across all seasons. Another positive relationship was found between VDS and Hb during Mar/Apr (LR).

All these correlations must be considered with caution: while correlations at the 0.01 significance level are usually considered as meaningful, the scatter plot of the correlation between Hb and DDS, as an example, clearly shows that there was only a trend of a relationship (Figure 3.4.5). While the correlations between Hb and the food scores (FVS, DDS, VDS) were calculated for the whole group of participants, correlations of Hb with specific food groups were treated with care, as usually not all participants consumed all food groups on the day of measurement.

Also the TfR values showed some positive correlations, namely with FVS and DDS, and the food groups nuts, tea and fish (Table 3.4.6). Again, because low TfR values mean a normal iron status, the correlations do coincide with those between Hb and the other parameters.

Table 3.4.5   Correlations between the Hb values and different food scores and food group intakes of women in Tanzania during different seasons (n=185)

	All year		Jun/Jul (DS)		Nov/Dec (SR)		Mar/Apr (LR)	
	Correl. coeff. rho	p	Correl. coeff. rho	p	Correl. coeff. rho	p	Correl. coeff. rho	p
FVS	-0.343	0.004	-0.211	0.004	-0.209	0.004	-0.398	0.000
DDS	-0.364	0.000	-0.289	0.000	-0.243	0.001	-0.384	0.000
VDS	n.a.	n.s.	n.a.	n.s.	n.a.	n.s.	0.192	0.009
Fruits	-0.341	0.000	-0.297	0.000	-0.163	0.027	-0.216	0.003
Vegetables	0.171	0.020	0.169	0.022	n.a.	n.s.	n.a.	n.s.
Nuts	-0.401	0.000	-0.216	0.003	-0.232	0.002	-0.361	0.000
Starchy plants	-0.213	0.004	-0.163	0.026	-0.212	0.004	-0.207	0.005
Bread/cakes	-0.253	0.001	n.a.	n.s.	n.a.	n.s.	-0.213	0.004
Tea	-0.338	0.000	-0.244	0.001	-0.241	0.001	-0.301	0.000
Sugar	-0.264	0.000	-0.207	0.005	-0.179	0.015	-0.218	0.003
Fish	-0.349	0.000	-0.203	0.006	-0.167	0.023	-0.262	0.000
Animal prod	0.206	0.005	0.199	0.007	n.a.	n.s.	n.a.	n.s.

DDS=dietary diversity score; DDS=dietary diversity score; VDS=vegetable diversity score
n.a.=not applicable; n.s.=not significant

Table 3.4.6   Correlations between TfR values and different food scores and food group intakes of women in Tanzania during Mar/Apr (LR) (n=131)

	FVS	DDS	Nuts	Tea	Fish
Correl. coeff. rho	0.198	0.231	0.249	0.190	0.269
p	0.023	0.008	0.004	0.029	0.002

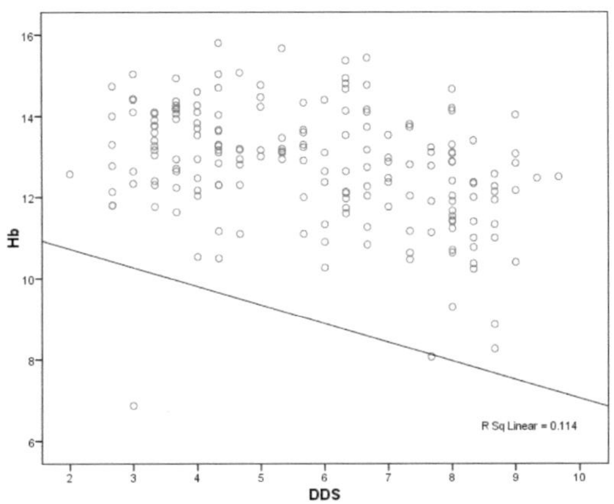

Figure 3.4.5
Correlation between Hb and dietary diversity score (DDS) of women in three districts of Tanzania; mean across three seasons (ρ=-0.346; p<0.001; n=189)

3.4.3   Body mass index (BMI)

When analysing the BMI data, those women were excluded from the study who were pregnant in the 2nd or 3rd trimester, or who most likely had TB and/or HIV/Aids. Thus, the study cohort consisted of 210 women for BMI analysis. The BMI data was not normally distributed neither for all districts together nor for each single district. Mean, median and maximum values were slightly higher in Muheza (Table 3.4.7) and during Nov/Dec (SR) (Table 3.4.8). However, differences between the three districts were not significant (K-W test; p=0.197) while there were marginally significant differences among the seasons (Friedman test; p=0.045). Figure 3.4.6 show these minor differences.

In terms of BMI categories, it was especially remarkable that less than 10% of participants, in Kongwa even less than 2%, were underweight, while in all districts more than 20% of participating women were either overweight or even obese (Figure 3.4.6 and Table A41).

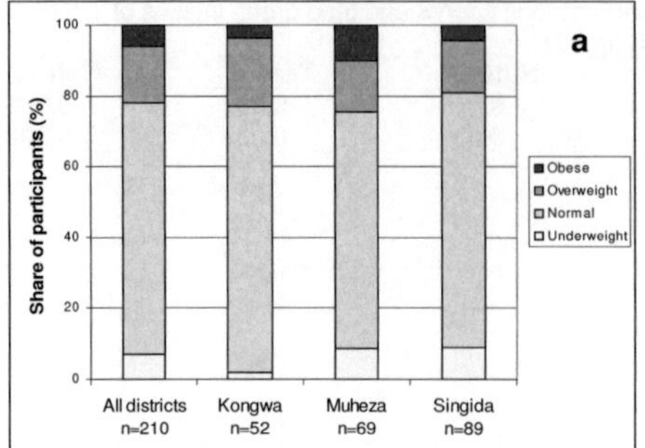
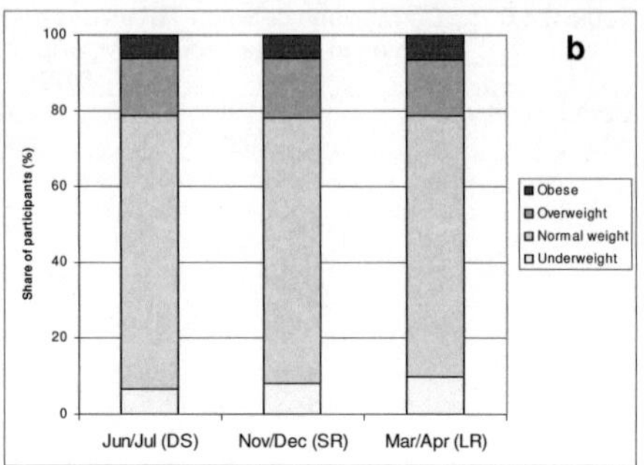

Figure 3.4.6   Share of participants within different BMI categories a) in three districts of Tanzania (mean across three seasons) and b) during three different seasons (n=210); underweight BMI<18.5 kg/m^2; normal 18.5-24.9 kg/m^2; overweight 25-29.9 kg/m^2; obese ≥30 kg/m^2

Table 3.4.7   BMI values of women in three different districts of Tanzania (mean across three seasons)

	All districts	Kongwa	Muheza	Singida
n	210	52	69	89
Mean ± SD	22.7 ± 3.9	22.7 ± 3.6	23.4 ± 4.4	22.3 ± 3.7
Median (range)	21.7 (14.9-37.7)	21.6 (17.7-34.7)	22.5 (14.9-37.7)	21.4 (16.4-35.2)

Table 3.4.8   BMI values of women in Tanzania during three different seasons (n=210)

	Jun/Jul (DS)	Nov/Dec (SR)	Mar/Apr (LR)
Mean ± SD	22.7 ± 3.9	22.8 ± 4.0	22.7 ± 4.1
Median (range)	21.7 (14.3-37.3)	21.9 (15.3-37.7)	21.7 (12.3-37.2)

No correlation was observed between BMI (mean value across three seasons) and any socio-economic parameter. However, the BMI was positively correlated with both FVS (Figure 3.4.7) and DDS (Figure 3.4.8), suggesting that the greater the diversity of foods eaten the higher the BMI and *vice versa*. The BMI was further positively correlated with some food groups, namely bread/cakes (ρ=0.240, p<0.001), sugar (ρ=0.259, p<0.001) and tea (ρ=0.216, p=0.002). There were no correlations between BMI and any of the surveyed nutrient intakes.

When BMI values were correlated to vegetable cultivation/collection, the only significant correlation was found between exotic vegetable cultivation and BMI (ρ=-0.164; p=0.017). No correlation existed between BMI and the dichotomous vegetable production, purchase and sales variables.

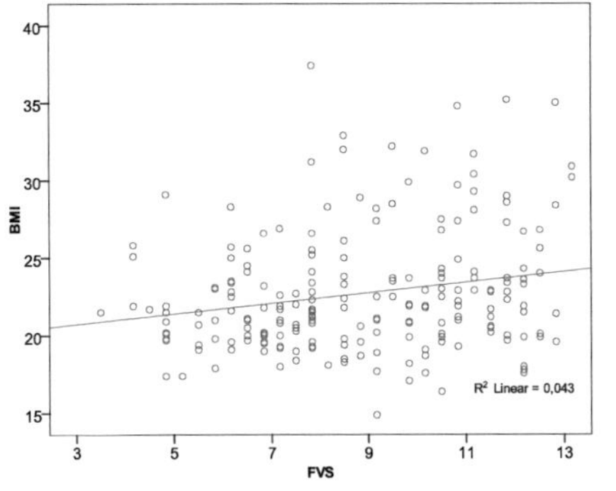

Figure 3.4.7
Association between body mass index (BMI) and
food variety score (FVS) of women in Tanzania
(n=210; ρ=0.204; p=0.003)

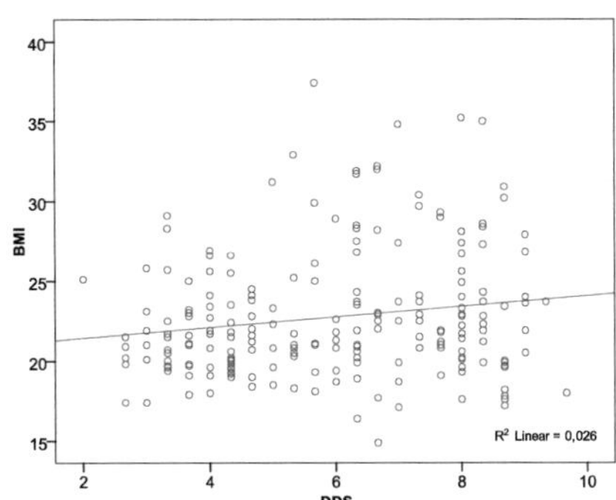

Figure 3.4.8
Association between body mass index (BMI) and
dietary diversity score (DDS) of women in Tanzania
(n=210; ρ=0.147; p=0.033)

3.4.4 Nutritional knowledge and attitudes

<u>Attitude towards vegetable consumption</u>

The answers to questions on participants' attitudes towards vegetable consumption were not
normally distributed and also showed no significant differences among the three districts. While the
first question was either strongly agreed or agreed to by all participants (Figure 3.4.9), the other
two questions were not as clearly answered, but all answering possibilities were chosen at least
once. Especially in Muheza, 8% each of participants agreed and even agreed strongly that
"vegetables are inferior foods that are good when one doesn't have much money or food at home".
Also in Kongwa and Singida, 19% and 17% of women, respectively, agreed to this question. About
4% of women in Kongwa district disagreed that "fresh vegetables are likely to contain more
nutrients than dried ones", while only about 1% did so in both Muheza and Singida district. When
attitude towards vegetable consumption was correlated to other values, no relationship could be
detected.

<u>Attitude towards overweight</u>

The number of positive and negative features for a corpulent person, named by participants
(Figure 3.4.10), showed no normal distribution. Nevertheless, both the number of positive
characteristics (p<0.001) and that of negative ones (p=0.044) showed significant differences
between the three districts.

In general, much less positive than negative characteristics were named by participants (Table
3.4.9). More than 60% of women (in Muheza only 47%) gave no example for a positive
characteristic of a corpulent person, while nearly all participants gave at least one ore more

negative examples (Figure 3.4.11). The share of participants naming more negative than positive characteristics was much greater than those naming more positive features (Figure 3.4.12).

The attitude towards overweight was significantly associated with other parameters (Table 3.4.10), whereby, mainly the number of positive characteristics was concerned. The number of negative characteristics grouped into three categories correlated significantly with the ethnic group of participants (p=0.012); the number of positive characteristics grouped correlated only slightly with this variable. Positive and negative characteristics taken together and grouped also significantly correlated with ethnic groups of participants. However, all correlation coefficients were very low and cane, therefore, only hint at certain trends.

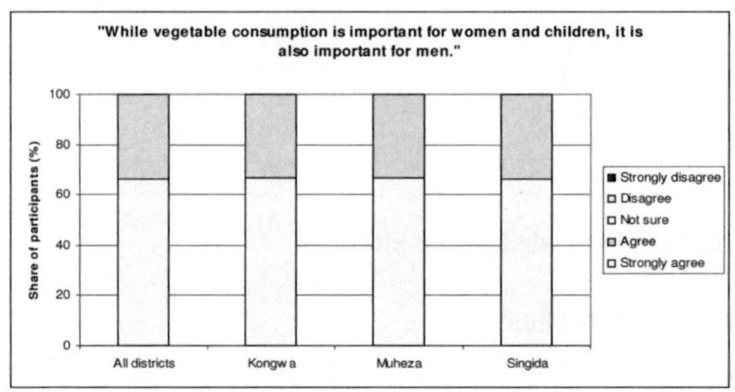

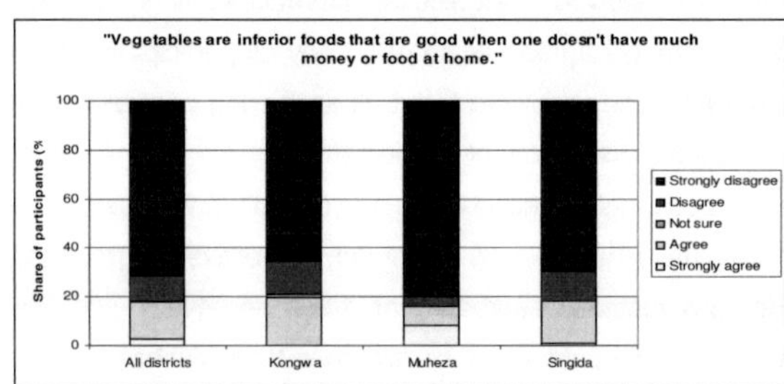

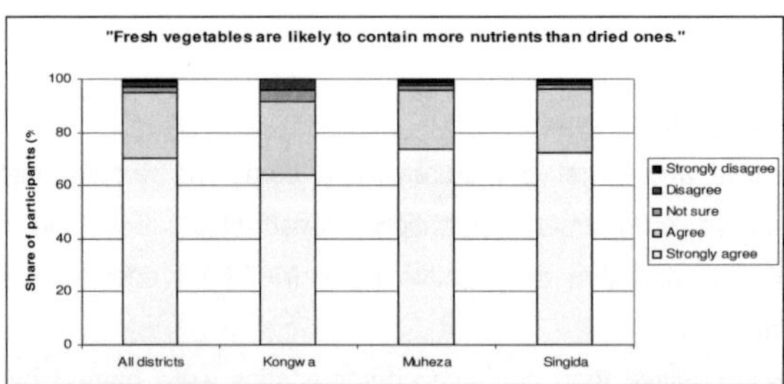

Figure 3.4.9
Share of participants from three districts in Tanzania with a certain attitude towards three different statements

Table 3.4.9 Number of positive and negative characteristics for a corpulent person named by women from three districts of Tanzania

	All districts		Kongwa		Muheza		Singida	
	Pos	Neg	Pos	Neg	Pos	Neg	Pos	Neg
n	246	249	72	72	70	73	104	104
Mean	0.7	3.1	0.5	2.7	1.3	3.3	0.4	3.1
± SD	±1.2	±1.4	±1.0	±1.2	±1.5	±1.6	±0.9	±1.3
Median	0	3	0	3	1	3	0	3
(range)	(0-6)	(0-7)	(0-4)	(0-6)	(0-6)	(0-7)	(0-4)	(1-6)

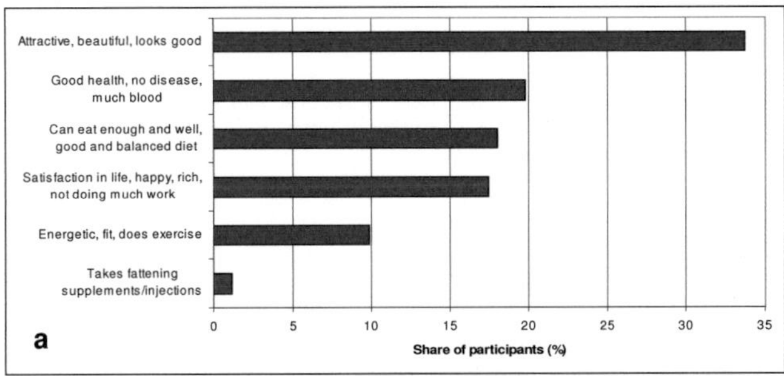

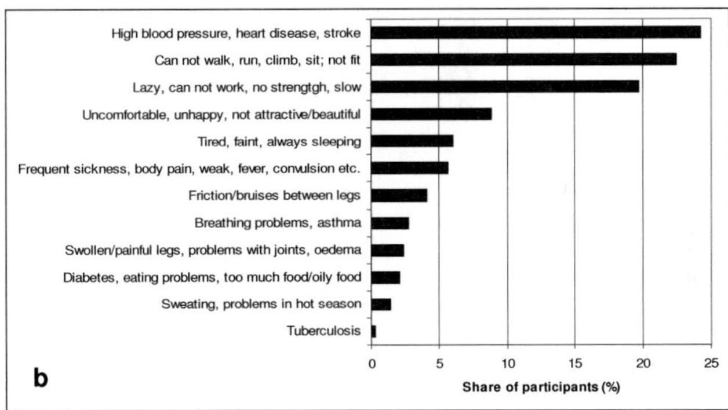

Figure 3.4.10
Share of participants that named certain a) positive and b) negative characteristics regarding a corpulent person

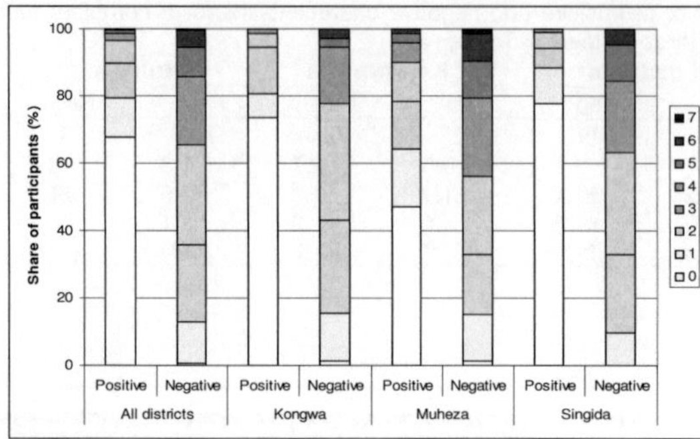

Figure 3.4.11
Share of participants from three districts in Tanzania that named
a certain number (0-7) of positive or negative characteristics
regarding a corpulent person

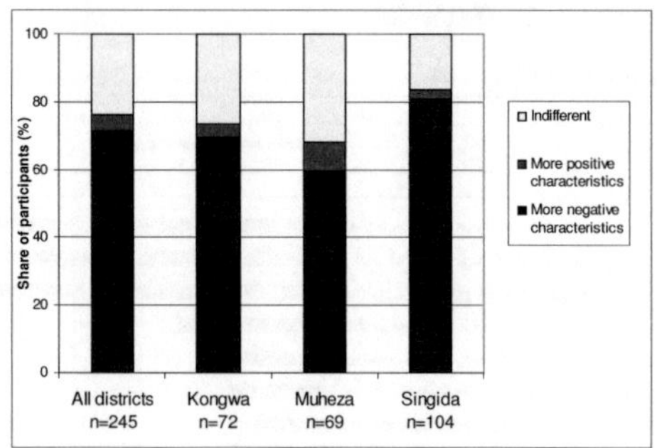

Figure 3.4.12
Share of participants of three districts in Tanzania that
named either more positive or more negative
characteristics regarding a corpulent person or were
indifferent in their answers

Knowledge about vitamin A and iron in nutrition

While nearly all interviewed women had heard about vitamin A as a nutrient, less suggested that they knew why vitamin A was important for the body (Figure 3.4.13). In Muheza, even 32% of participants stated that they did not know the function of vitamin A. However, when asked to name the function in the body, only 15% gave a wrong answer in Muheza, while in the other districts more women answered incorrectly (Figure 3.4.14). It should be pointed out that not all participants answered the question on the function of vitamin A. Yet, as this was similar for all districts, the districts are still comparable.

The functions of vitamin A named were various and most of them rather general ("protects or strengthens the body"). One of the most important functions of vitamin A, namely "good for vision and eye health" was mainly named by participants in Muheza district (Figure 3.4.16). When asked to name three to five foods that are good for sight and eye health, most of the participants answered rather randomly with both correct and incorrect answers. Only about 30% of women, in Kongwa even less than 20%, knew foods that were good for sight and eye health (Figure 3.4.15).

Table 3.4.10   Correlations between the number of positive/negative characteristics for a corpulent person named by women in Tanzania and socio-economic, vegetable production, and food/nutrient intake parameters

	No. of positive characteristics n=246		No. of negative characteristics n=249	
	Correl. coeff. rho	p	Correl. coeff. rho	p
Age	n.a.	n.s.	0.133	0.036
Distance (km)	-0.179	0.005	n.a.	n.s.
Distance (min)	-0.125	0.049	n.a.	n.s.
Hb	-0.151	0.044	n.a.	n.s.
Vegetable cultivation	n.a.	n.s.	0.145	0.022
Vegetable collection	0.191	0.003	n.a.	n.s.
TV cult/coll	0.128	0.044	0.130	0.040
Cereals	-0.165	0.009	n.a.	n.s.
Bread/cakes	0.153	0.016	n.a.	n.s.
Fruits	0.125	0.050	n.a.	n.s.
Nuts	0.174	0.006	n.a.	n.s.
Pulses	n.a.	n.s.	0.199	0.002
Tea	n.a.	n.s.	0.130	0.041
Fish	0.159	0.013	n.a.	n.s.
Dietary fibre	-0.127	0.047	n.a.	n.s.
PUFA	-0.255	0.000	n.a.	n.s.

Hb=haemoglobin; TV=traditional vegetables; PUFA=poly unsaturated fatty acids
n.a.=not applicable; n.s.=not significant

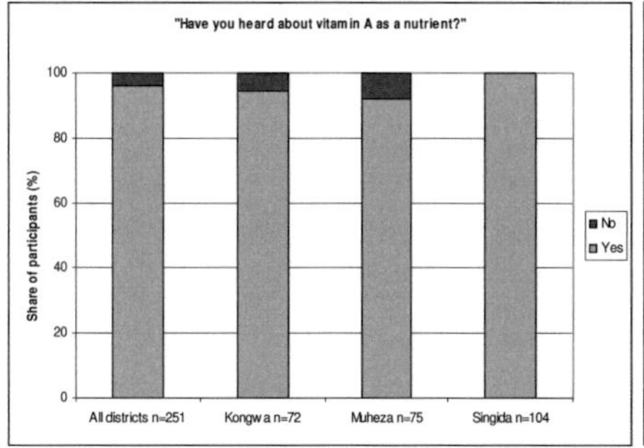

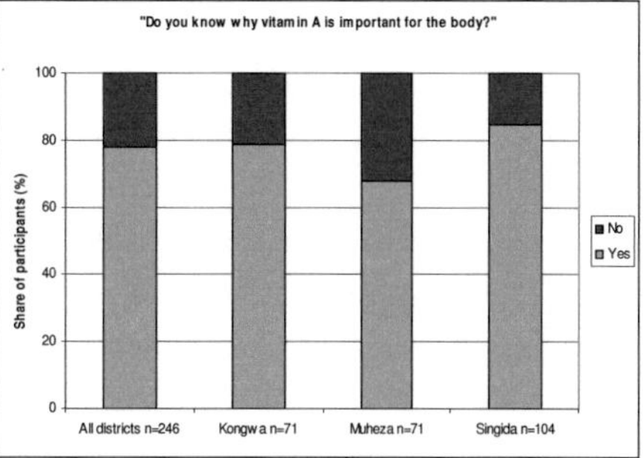

Figure 3.4.13   Share of participants that answered with "yes" or "no" to the questions above in three districts of Tanzania

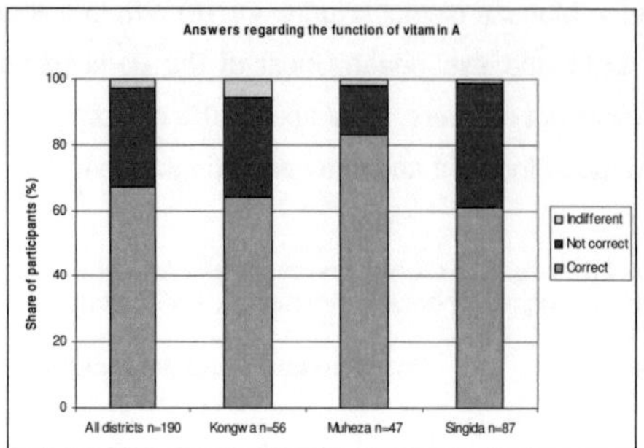

Figure 3.4.14
Knowledge of participants in three districts of
Tanzania regarding the function of vitamin A

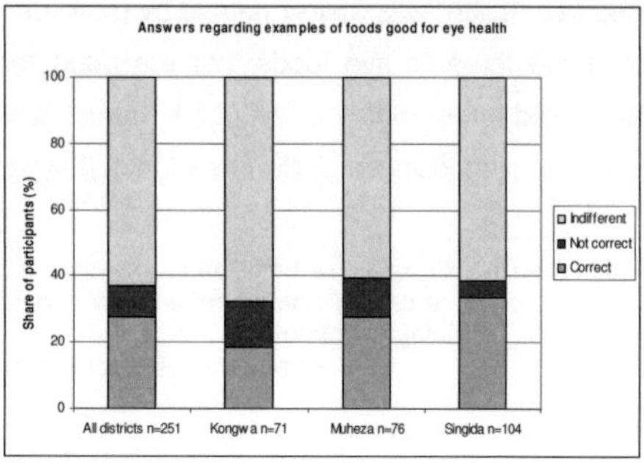

Figure 3.4.15
Knowledge of participants in three districts of
Tanzania regarding vitamin A rich foods

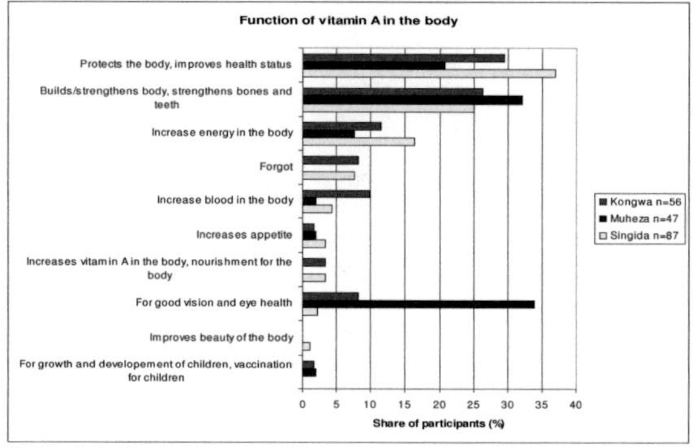

Figure 3.4.16
Functions of vitamin A in the body as stated by participants from
three districts of Tanzania (multiple answers permitted; answers
were not predetermined)

The answers to the same questions, but regarding iron, were slightly different to those of vitamin A.
Less participants had heard about iron as a nutrient and, in all districts, 50% or more stated that
they did not know why iron was important for the body (Figure 3.4.17).

Answers regarding the function of iron were analysed like those for vitamin A. However, women
were less knowledgeable about the correct function of iron in the body and, like for vitamin A, the
largest share of participants giving the correct answer were those in Muheza (Figure 3.4.18). The
answers given were, in fact, less general than those for vitamin A, but clearer like "increases blood"
or "strengthens muscles". In Singida, one participant even had heard that there was a connection
between iron and vitamin C, however, could not describe the correct interrelation (Figure 3.4.20).

Participants named less correct foods that contained iron (described to them with "make red blood
cells and allow the muscles and brain to work properly"), compared to vitamin A rich foods. In fact,
only about 20% of women and 28% in Singida, could name iron-rich foods (Figure 3.4.19).

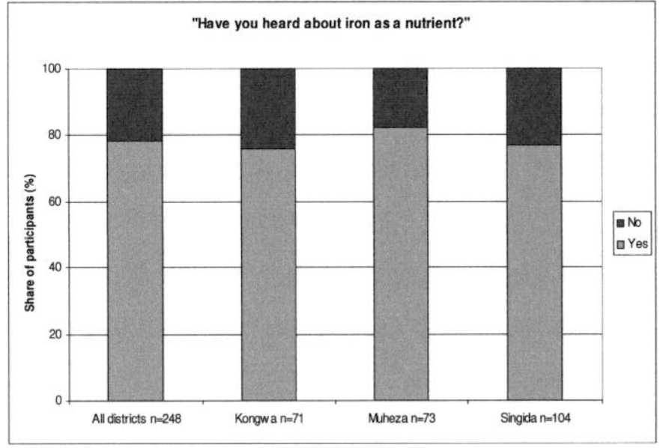

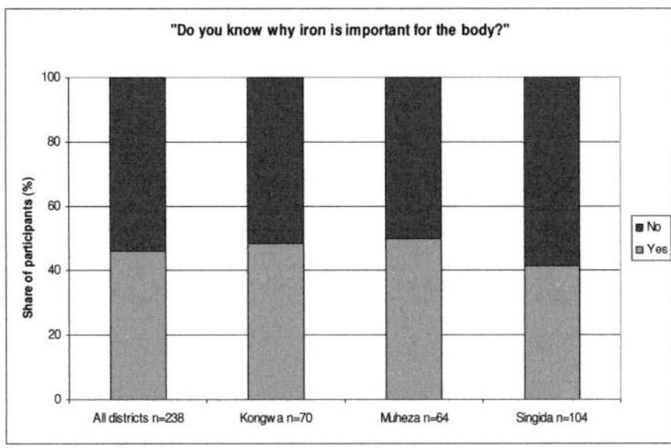

Figure 3.4.17   Share of participants that answered "yes" or "no" to the questions above in three districts of Tanzania

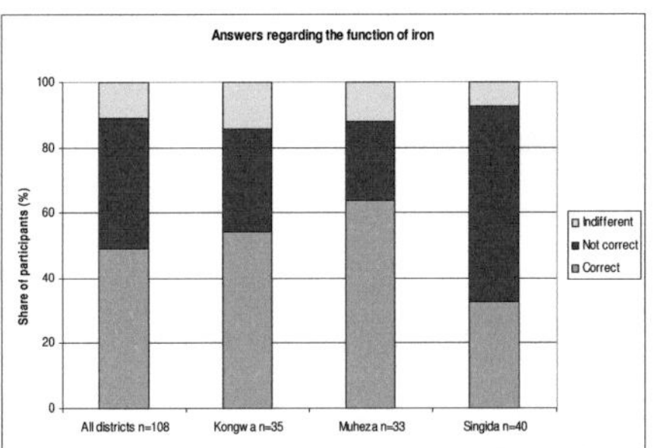

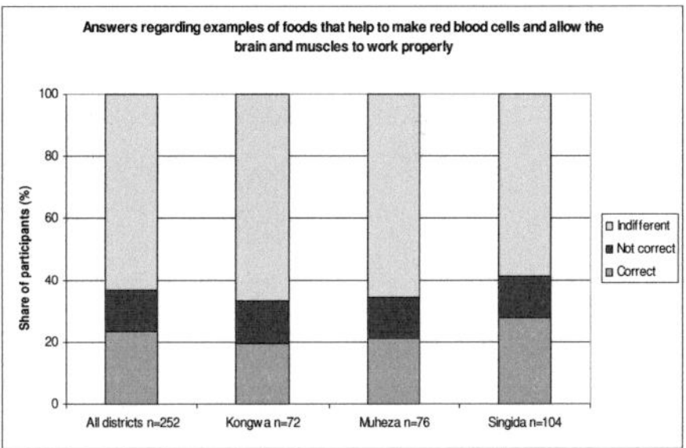

Figure 3.4.18
Knowledge of participants in three districts of Tanzania regarding the function of iron

Figure 3.4.19   Knowledge of participants in three districts of Tanzania regarding iron rich foods

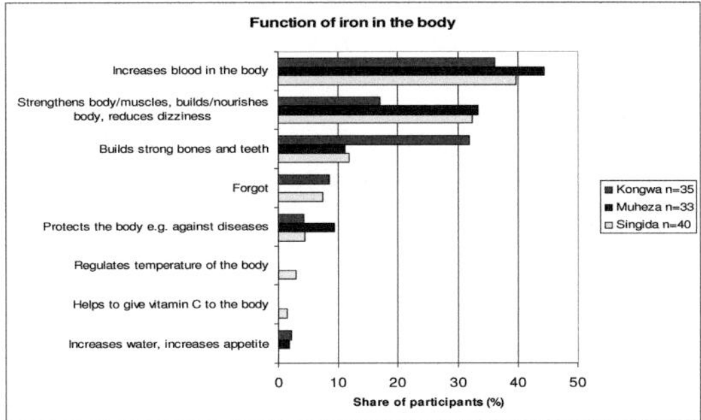

Figure 3.4.20
Functions of iron in the body as stated by participants from three districts of Tanzania (multiple answers permitted; answers were not predetermined)

The knowledge of participants on vitamin A and iron was associated with different parameters. Correct answers regarding the function of vitamin A were significantly associated with the consumption of animal products (Figure 3.4.21). If participants knew the function of iron, there was a significant correlation to the consumption of nuts (Figure 3.4.22). However, these correlations of knowledge with different parameters were not necessarily reflected in consumption patterns. For example, more women who stated that they knew why vitamin A was important for the body had a lower DDS than those who had said that they did not know it. As the relationships between the nutritional knowledge of women about vitamin A rich foods and VAD or VDS were not significant, yet, of special interest, results are depicted in tables A43 and A44 in the appendix.

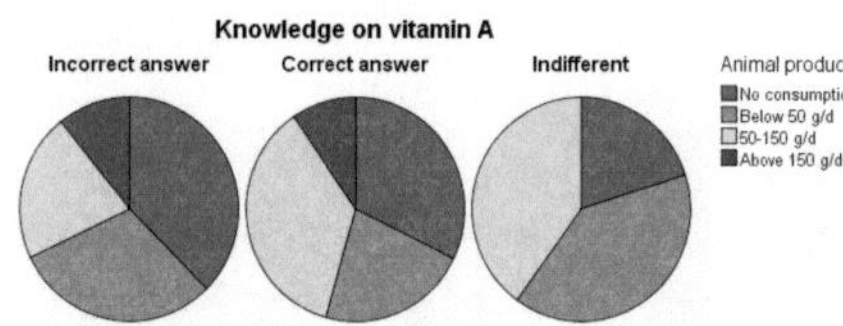

Figure 3.4.21
Association between knowledge about the function of
vitamin A and the consumption of animal products by
women in Tanzania (p=0.043)

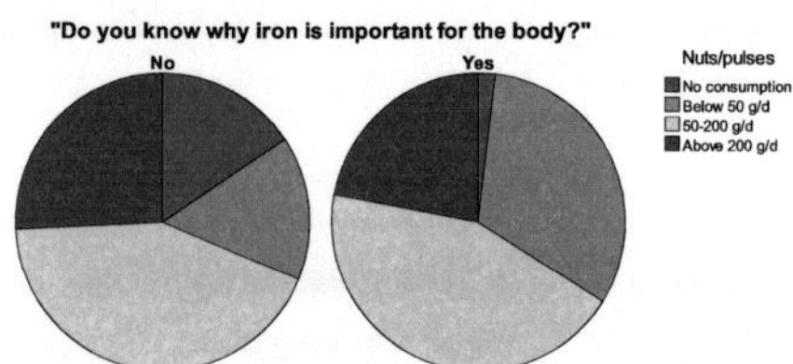

Figure 3.4.22
Association between knowledge about iron and the
consumption of nuts and pulses by women in Tanzania
(p<0.001)

4   Discussion

Within this chapter the results gained will be discussed and also will be − beyond bivariate correlations − related to each other through multiple regression, cluster or factor analysis. As the basis for a logical order in which to discuss the different subject areas, the modified model of nutrition security as explained in the introduction was applied (Figure 4.1 − see large arrow).

While the discussion chapters 4.1 up to 4.4 will focus either on single food groups, single health problems or nutrients, it is the aim of chapter 4.5 and 4.6 to take a more holistic view on dietary diversity/food variety and dietary patterns, and its relation with nutritional status and vegetable production. The discussion will start with "production" involving mainly vegetable production, its sales and purchase, and its relation to vegetable consumption, i.e. "food intake". To view this relationship in a more holistic way, also part of the further influencing data, namely those on knowledge about and attitudes towards vegetable consumption, will be discussed in this first chapter: 4.1 Linking vegetable production and consumption: "Does diversity in the field equal diversity on the plate?".

The discussion continues with three parts in the area of "nutritional health" of study participants and its relation to food intake and knowledge, socio-economic factors and attitudes: 4.2 Linking overweight/obesity, food consumption and attitudes: "how the 'nutrition transition' is on the rise in rural Tanzania"; 4.3 Linking iron status, food consumption and nutritional knowledge: "dietary diversity versus health issues"; 4.4 Linking vitamin A status, food consumption and nutritional knowledge: "food taste versus nutritional knowledge and education".

Two further chapters will concentrate on the section of "consumption" and discuss in more detail dietary diversity and food patterns: 4.5 Dietary Diversity Score (DDS) and Food Variety Score (FVS): "the measurement of diversity scores and the association between nutritional diversity and nutritional health"; 4.6 Dietary Patterns and their association with nutritional health: "the strength and weaknesses of pattern analysis".

In all discussion chapters it will not only be referred to tables and figures occurring in the discussion but also to those from chapter 3 "Results" in order to provide a better overview. At the end of each discussion chapter a conclusion will be drawn for each specific topic, while in chapter 5 an overall conclusion and an outlook with recommendations is given.

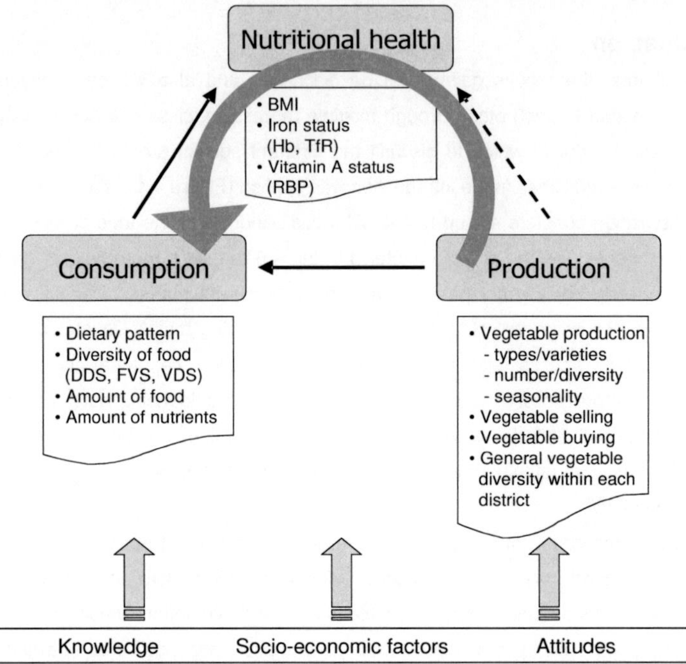

Figure 4.1   The three main areas investigated within the present study and influencing factors; large arrow indicating the order in which the three areas are discussed

4.1 Linking vegetable production and consumption: "Does diversity in the field equal diversity on the plate?"

In Tanzania, vegetables are popular everyday foods in the kitchen, and are usually consumed as relishes accompanying main starchy dishes (Lyimo et al 1991; Vainio-Mattila 2000). Traditional leafy vegetables, which were found to contain a multiple of some micronutrients compared with exotic ones (Weinberger and Msuya 2004), have been important in meeting the nutritional needs of Tanzanian people (Chweya and Eyzaguirre 1999). Generally, many traditional vegetables are available in Tanzania, with greatly varying species composition according to the country's agro-ecological zones (Ruffo et al. 2002; Keller et al. 2005). As the relationship between vegetable production and consumption is highly complex and was analysed in different ways and through different parameters, at the end of each sub-chapter results regarding this relationship are summarised.

4.1.1 Vegetable production and consumption of the study population

<u>Types of vegetables consumed and produced</u>
While there was a great diversity among the traditional vegetable types most often produced and consumed by women, mainly the exotic vegetables tomato and onion were consumed by nearly everyone, while further exotic vegetables were not used to a great extend. To compare production and consumption issues, tomatoes and onions were first included in the analysis; yet, for the calculation of the vegetable diversity score (VDS) these two vegetables were excluded, as they were usually used in very small quantities only (see also Focus A).

Vegetable types used by study participants differed both between seasons and districts, i.e. agro-ecological zones (Tables A5-A7). Differences between districts were especially pronounced for wild vegetables as they depended highly on the given eco-system and climate. Cultivated vegetables types, on the other hand, differed much less between districts which suggests that these vegetable types have spread over the country and are of similar importance in all researched districts. This is especially true for exotic vegetable types, which are only few in number but everywhere identical.

In the following, for each district three vegetables are compared exemplarily in terms of consumption and cultivation/collection. Thereby, for Kongwa three wild, for Muheza three cultivated traditional and for Singida three cultivated exotic vegetables were chosen on the basis of frequency in consumption and cultivation/collection. In Kongwa, false sesame was consumed by 72% of participants during Nov/Dec (SR), however, no one stated to collect this vegetable during this time (Figure 4.1.1a). False sesame is locally also cultivated (Bedigian and Adetula 2004) and, consequently, could be available on the market, however, it is rather unlikely that it is sold in large quantities. This discrepancy could further be due to the consumption of dried false sesame, which had been collected earlier.

Also wild simsim was collected by 49% of women during Nov/Dec (SR) but consumed during the surveyed week only by 3%, thus, differences between collection and consumption became obvious. On the one hand, wild simsim collected during this season was probably hardly consumed during the surveyed week but maybe before and after. On the other hand, in Nov/Dec (SR) being the rainy season in Kongwa district, wild and spontaneously growing vegetables were abundant in general. Yet, it is known that the leaves and young shoots of wild simsim can be dried and stored for later use (Jansen 2004), which could explain the collection without consumption. However, the question arises why wild simsim is collected by many women but not consumed, while false sesame is consumed by many but not collected? One possible reason could be that these two vegetables were usually not consumed by the same persons and were, therefore, not exchangeable. However, cluster analysis (Table 3.3.13) showed that these two vegetables occurred together in two out of five consumption clusters, meaning that the same participants preferred both vegetables, yet, concerning all three seasons (not only Nov/Dec).

The diversity of vegetables that are preserved and stored must definitely be observed for a whole year and not for one season only. In fact, when examining all three seasons, more women stated to collect false sesame during Jun/Jul (DS) and Mar/Apr (LR) than they reported to consume at those times. While from this data nothing is known about amounts of both collected and consumed vegetables it is assumed that during these two seasons a surplus of false sesame might be preserved for the Nov/Dec (SR) season. In fact, in the 7-d recall the women stated during Nov/Dec (SR) that the source of false sesame was not only collection (44%) but also cultivation (49%) and even the market (7%). It was often consumed dried (39%), thus, not directly after harvest. On the other hand, wild simsim was consumed dried only in few cases (11%) and it was more often collected (67%) than cultivated (31%) or bought from the market (2%) (data not shown in results). For wild vegetables it was obviously impossible to trace back consumption from seasonal cultivation or collection of vegetables, but diversity of consumption needs to be looked at on a long-term basis and also depends on other factors such as preservation and storage.

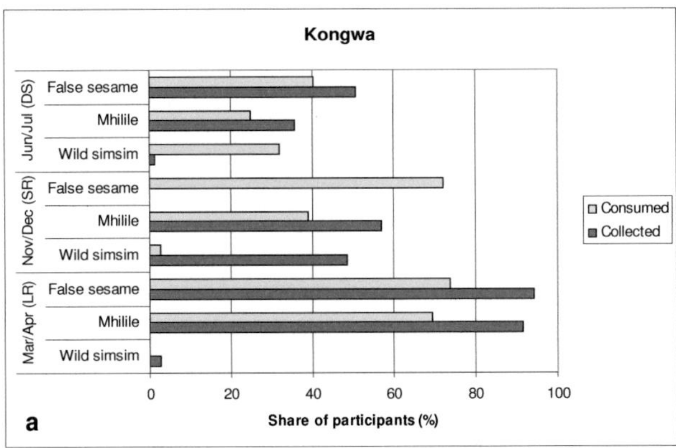

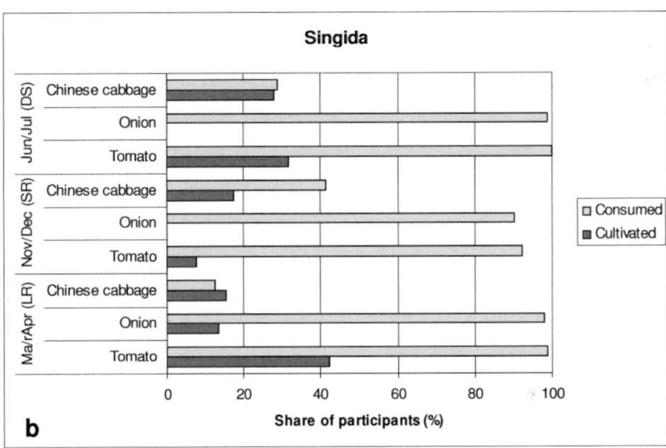

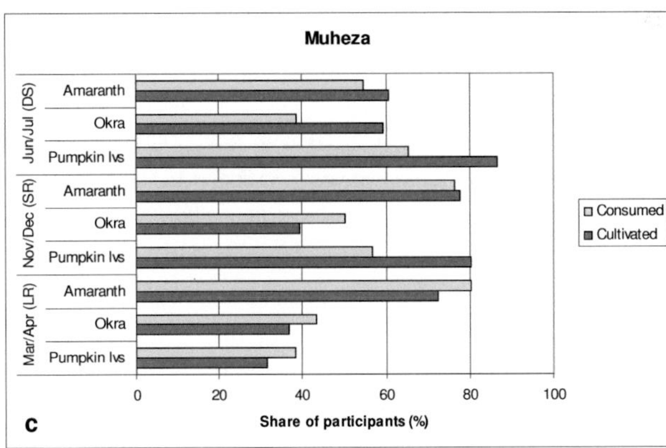

Figure 4.1.1   Share of participants who consumed and collected/cultivated three vegetable types in Tanzania, during three different seasons a) Kongwa: wild types; p=0.825 for differences between consumption and production; b) Singida: exotic types; p=0.005; c) Muheza: traditional types; p=0.596

Also in terms of exotic vegetables (Figure 4.1.1b), there was an imbalance between the share of women consuming and producing. Onions, for example, were cultivated in Singida district only during Mar/Apr (LR) and only by 14% of participants. Still, onions were consumed by more than 90% of women during all seasons. This approves the figures stated above that study participants bought this vegetable from the market as they also did with tomatoes and Chinese cabbage during Nov/Dec (SR).

The share of women who cultivated traditional vegetables, as shown in the example of Muheza district (Figure 4.1.1c), and the share of women who consumed them was rather similar. Only during Mar/Apr (LR) all listed vegetables were consumed by more participants than cultivated, while during the other seasons it was reversed but for okra. Especially during Mar/Apr (LR) purchasing traditional vegetables other than the ones produced was obviously common among study participants.

From the fact that differences between consumption and production were significant for exotic vegetables, it can be concluded that for this vegetable group there is no direct link between production and consumption, especially because most exotic vegetables were bought in the market and not homegrown. Also for wild vegetables, it is difficult to perceive a direct link between their collection and consumption, which was obviously mainly due to preservation and storage. However, for cultivated traditional vegetables, no significant differences between consumption and cultivation were found and obviously the homegrown production of these vegetables was associated with their consumption. To find a similar link for exotic and wild vegetables, more comprehensive data is needed including marketing, sales and storage.

Differences in cropping can be mainly attributed to environmental conditions, differences in consumption is usually traced back to availability – which also partly depends on environmental conditions – and especially culture and custom as well as market demand. This association was, for example, found in Kenya where vegetable consumption, species use and taste preferences reflected cultural experiences and backgrounds (Owuor and Olaimer-Anyara 2007). Thus, when culture and custom change and, for example, become more similar for different districts, this "mainstreaming" can be responsible for increasingly analogue eating habits. This would be one indication for an existing 'nutrition transition' – more or less independent from food production. In fact, in a survey from 1983 in the Morogoro region, Tanzania, it was recognised that concentrated energy sources, such as fats, oil and sugar, were scarce irrespective of the season and that the diets of all age groups were highly deficient in energy and protein during the lean season in February (Tanner and Lukmanji 1987). From the present data, however, it was learned that energy-dense foods, such as sugar and oil, were used by nearly all surveyed participants and during all seasons and that overweight and obesity was affecting double as many participants as underweight (Keding et al. 2008). The issue of 'nutrition transition' and changing food habits will be discussed in more detail in chapters 4.2 on obesity and 4.6 on dietary patterns.

In terms of vegetables, a year round availability irrespective of the season would be desirable to combat micronutrient deficiencies. Yet, consumption of similar vegetables during all seasons did not mean that also a sufficient amount of vegetables was consumed. In the following it was, therefore, investigated, whether the already found links between vegetable production and consumption regarding the share of participants consuming a certain vegetable, could be also confirmed when observing vegetable diversity and vegetable quantity.

<u>LINK between production and consumption</u>
(regarding share of particpants that prod./cons. certain vegetables):

✓ yes for production and consumption of cultivated traditional vegetables
✗ no for production and consumption of exotic and wild vegetables

<u>Number of vegetables consumed and produced, and vegetable diversity score (VDS)</u>
Vegetable diversity, i.e. the number of different vegetable types, produced and consumed by women was expected to be influenced not only by agro-ecological zones but also by seasonality, the distance of consumers to urban centres, and individual preferences, attitudes and knowledge. Vegetable diversity used over the whole year was already found to vary to a great extent among the three districts researched (in Muheza about 70 different vegetable types used, in Kongwa 35, in Singida 20) (Keller et al. 2005).

The overall number of vegetable types and, therefore, vegetable diversity used by all participants in the present study was found to be different among the seasons and also districts, which coincides with the above mentioned study. Overall, in Muheza the highest diversity of vegetables was cultivated and collected by participants with great differences between the seasons (Figure 4.1.2). Interestingly, not much seasonal differences were found in terms of vegetable types that were most important for participants in Muheza (Tables A5-A7). Consequently, in Muheza, though a great diversity of vegetables is available throughout the year, this diversity was obviously not used, but only few vegetables were favoured during all seasons. In the other two districts, however, vegetable diversity was lower and differences between the seasons, especially in Singida district, not pronounced. In both Kongwa and Muheza a substantially higher number of vegetables was noted during the dry season (Jun/Jul (DS)), cultivated as well as collected.

The share of bought vegetables was highest during Nov/Dec (SR) for all vegetables both exotic and traditional. This indicates that during this time, namely the rainy season, cropping of vegetables just started but most vegetables were not ready to be eaten as yet. A survey in the Morogoro Region, Tanzania, found that during the lean season in February, vitamin A requirements were met while during the post-harvest season, where food should be available in abundance, vitamin A deficiency was noted. This was attributed to the scarcity of leafy vegetables in the diets during August (Tanner and Lukmanji 1987), which supports the present findings in Kongwa district, which is close to Morogoro, and where the amount of vegetables consumed was also lowest during June/July and highest during Mar/Apr (LR).

Besides the produced diversity, the diversity consumed by women was of interest and if the latter coincided with the produced one. In fact, the actual number of vegetable types consumed by participants did not vary to a great extent among districts and seasons (Figure 4.1.2), yet, the number of consumed vegetable types differed between those collected and those cultivated (Figure 3.3.35) with much less collected types being consumed.

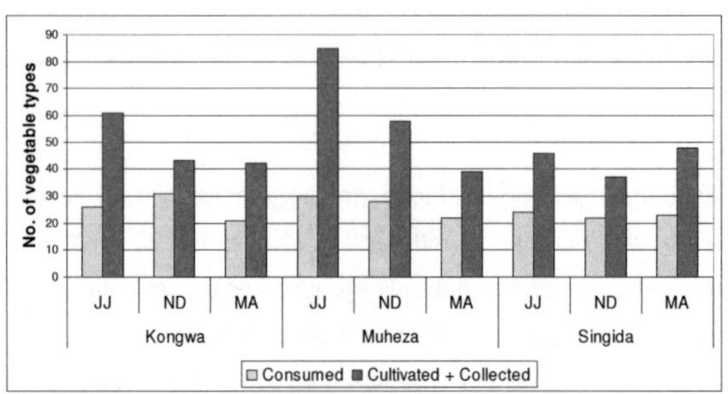

Figure 4.1.2
Total number of vegetable types consumed and produced in three districts during three different seasons in Tanzania

The total number of different vegetable types consumed by study participants was substantially lower than that of vegetable types produced. In some cases, even more than double the number of vegetables was named to be cultivated/collected than consumed. One reason for this difference could be the different data collection methods: while vegetables cultivated or collected "right now" in the garden/field were asked for, the consumption of vegetables only within the last week was requested from participants. Vegetables in the garden could also have been those not yet mature but already growing and nursed in the garden. However, it was striking that the number of vegetable types collected from the wild was much greater than those consumed (Figure 4.1.3), while the number of vegetable types cultivated and consumed was rather similar, especially in Kongwa district (Figure 4.1.4).

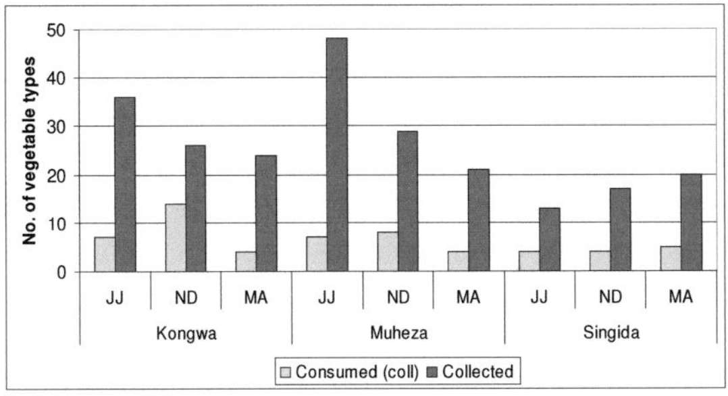

Figure 4.1.3
Total number of wild vegetable types consumed and collected in
three districts during three different seasons in Tanzania

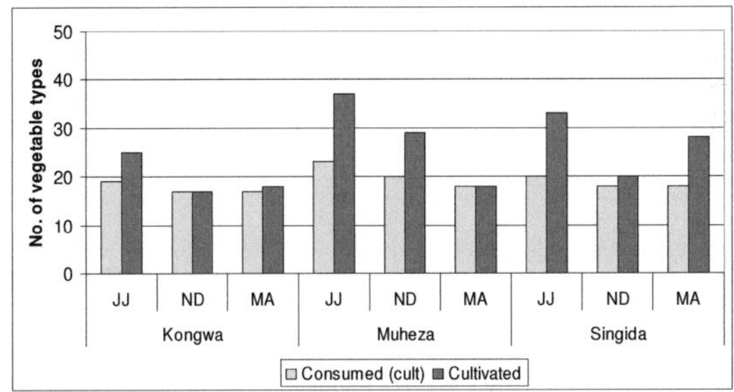

Figure 4.1.4
Total number of cultivated vegetables consumed and cultivated in
three districts during three different seasons in Tanzania

A reason for this imbalance again could have been the data collection method: vegetables collected from the wild "at the moment" were probably collected during this season, yet, not within the last week and, therefore, also not consumed during the last recorded week. Anyway, the data suggests that vegetables collected from the wild were of much less importance for consumption and they were not consumed regularly (here weekly).

Besides the differentiation between cultivated and collected vegetables, traditional (TV) and exotic (EV) vegetable types were also distinguished. It is generally believed that the introduction of exotic vegetables to Africa contributed to the decline in the production and consumption of African indigenous vegetables (Smith and Eyzaguirre 2007). Traditional vegetables, sometimes even called "wayside" or "roadside" species, are often highly important locally (Chweya and Eyzaguirre 1999) but are considered as neglected and underutilised on a global scale, although this observation may be due to scarcity of data (Weinberger 2007). For high agrobiodiversity, traditional crops and foods are especially crucial. Yet, these are often neglected in research and development (Chweya and Eyzaguirre 1999). This is even more surprising as it is known that, for example, traditional vegetables are usually more nutritious compared to exotic, special-bred and high-yielding varieties (IPGRI 2003).

The collected data in Tanzania suggests that there was a noticeable greater diversity of TVs than of EVs both in terms of production (Figures 3.2.5 and 3.2.6) and consumption (Figures 3.3.34, 4.1.5; note that y-axes have different scales). When considering only TVs, clearly more types were produced than consumed (Figure 4.1.5a), while this is not the case for EVs everywhere and in all seasons (Figure 4.1.5b).

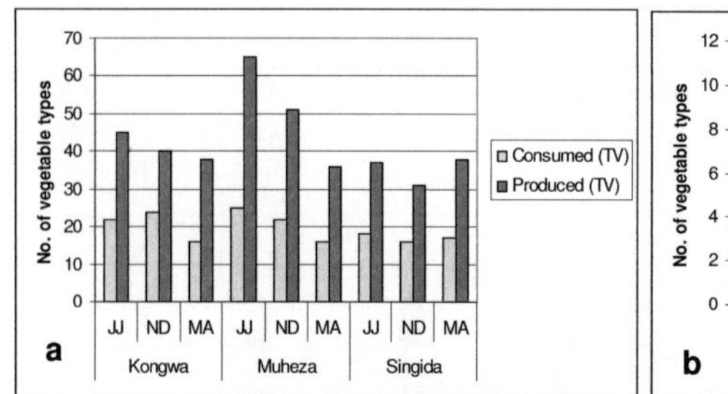

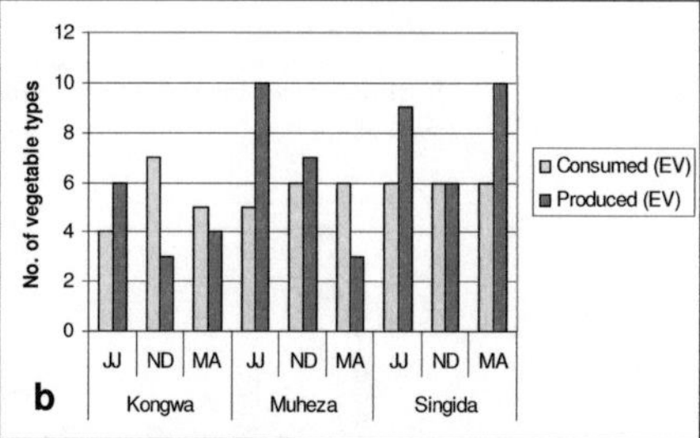

Figure 4.1.5   Total number of vegetable types consumed and produced (cultivated and collected) in three districts of Tanzania during three different seasons; a) traditional vegetables (TV); b) exotic vegetables (EV)

The mean number of TVs and EVs consumed and produced per woman was also of interest. The mean total number of TVs and EVs together did not vary much between seasons and districts (Table 4.1.1). Yet, the ratio between TVs and EVs was highly different in terms of production and also rather high especially for Nov/Dec (SR) (28 times more TVs than EVs cultivated) and for Kongwa district (33 times more TVs than EVs). This ratio in terms of consumption was much lower and also with not much differences between districts and seasons. Consequently, while EVs were obviously seldom cropped by study participants, they were relatively much more consumed in comparison to TVs.

Data on the sources of vegetables confirmed that most exotic vegetables consumed by participants were bought, namely, on average, 95% of onions, 94% of white cabbage, 86% of tomatoes and 57% of Chinese cabbage which explains the discrepancy between production and consumption.

Table 4.1.1   Mean number and ratio of traditional (TV) and exotic (EV) vegetable types produced and consumed <u>per woman</u> during three different seasons and in three districts of Tanzania

	Jun/Jul (DS)	Nov/Dec (SR)	Mar/Apr (LR)	Kongwa	Muheza	Singida
n		252		76	72	104
Production (seasonal or yearly mean)						
TV	8.89	7.17	9.25	6.71	9.43	9.17
EV	0.80	0.25	0.52	0.15	0.40	1.02
Total vegetables	9.70	7.50	9.70	6.90	9.80	10.20
Ratio TV:EV	11.09	28.23	17.92	33.40	23.52	9.02
Consumption (weekly mean)						
TV	4.9	4.4	4.2	3.9	4.1	5.5
EV	1.7	2.2	2.3	2.0	2.2	2.1
Total vegetables	6.6	6.6	6.5	5.9	6.3	7.6
Ratio TV:EV	2.9	2.0	1.8	2.0	1.9	2.6

Note: all species of a vegetable, e.g. *Amaranth* spp., counted

The VDS, counting different vegetable types consumed per week, varied greatly among participants, ranging from one to 15 (Table 3.3.17). This shows that it was not necessarily usual to consume a high vegetable diversity which would include six or more different vegetable types consumed per week. In fact, most women had a medium intake with a VDS of four or five (Figure 3.3.40). The VDS level is, however, difficult to interpret or compare as similar studies on VDS are not available according to the author's knowledge.

Further, the VDS varied only slightly between the different districts, yet, it differed significantly between seasons (Table 3.3.17). In Kongwa and Singida with similar agro-ecological pattern, the highest VDS was found in Mar/Apr (LR), while it was highest during Nov/Dec (SR) in Muheza. In contrary, the number of vegetables produced was both high during Jun/Jul (DS) Mar/Apr (LR) in Singida district and highest during Jun/Jul (DS) in Kongwa. Thus, from the VDS alone it cannot be concluded that vegetable diversity produced is directly influencing that consumed. Yet, this relationship will be investigated further with bivariate and multiple relations, as the focus on vegetable diversity – and not on single vegetable types with, for example, a high content of one specific vitamin – is especially important. Nutrition research and interventions have been and are still focusing strongly on single nutrients, such as protein, vitamin A, iron, iodine or zinc, which are deficient in certain population groups. (DeMaeyer 1989; Waterlow 1992; UNICEF and WHO 1999). Much less is known about the nutrition and health outcomes from a combination of foods and dietary patterns. However, it is becoming increasingly acknowledged that single nutrients alone are not the key to solve nutrition problems or to prevent chronic diseases. Whole foods and nutritionally-wise diets, e.g. a variety of vegetables and fruits as part of a mainly plant-food-based diet, are desirable. In fact, vegetables were found to be especially rich sources of a variety of nutrients and biologically-active compounds (Lampe 1999). In addition to providing several nutrients for a healthy and balanced diet, they exert a health-protective effect attributed mainly to antioxidants and dietary fibre (Eastwood 1999). Dietary patterns are regarded as one approach to consider overall eating habits instead of single nutrients (Hu et al. 1999); this will be discussed in detail in chapter 4.6.

> **LINK between production and consumption**
> (regarding vegetable diversity):
> ✓   yes for TV production and consumption
> ✗✓ partly for EV production and consumption

Amount of vegetables consumed

The daily amount of vegetables consumed varied significantly between districts, however, not between seasons. The median amount consumed by women was relatively high (Table 3.3.15) and even, in some cases, exceeded the suggested optimum of 260g/d (WHO 2003). Interestingly, women in Muheza district with the highest vegetable diversity available, consumed the lowest amount of vegetables. In contrast, women in semi-arid Kongwa and Singida districts, with lower vegetable diversity available compared to Muheza, consumed greater quantities. Apparently, high agrobiodiversity alone does not account for high consumption of this diversity. While some authors state that the decreasing use of traditional vegetables is due to decreasing availability (Adedoyin and Taylor 2000; Okeno et al. 2003), others argue that traditional vegetables are quite available especially during the rainy season, e.g., in Nigeria, but are among the least consumed foods (Maziya-Dixon et al. 2004).

Therefore, it is important not only to look at the diversity of foods produced and consumed but also at further influencing factors such as knowledge, acceptance and possible prejudice of the foods. These factors will be analysed at the end of this chapter.

Traditional vegetables were consumed more often and also in a larger quantity than exotic vegetables. This was probably due to the long distance of consumers to urban markets and little exposure to exotic types. However, it remained open if exotic vegetables were eaten instead of or additionally to traditional ones. Through cluster analysis (see below) it was at least found that wild and exotic vegetables were not consumed together but rather substituted each other.

> **LINK between production and consumption**
> (regarding vegetable quantity):
> ✗   no for general vegetable diversity available and amount consumed
> ✗   "Muheza effect": diversity in the field does not necessarily appear on the plate

4.1.2   Attitudes and knowledge of participants regarding vegetables

One important finding was that there was a significant difference between seasons in terms of number one criteria for choosing vegetables. Again it can be shown that seasonality is of high importance influencing vegetable consumption in the researched districts. In general, taste was more important for participants than health when choosing a vegetable (Figures 3.3.38 and 3.3.39). This highlights that taste is particularly important when e.g. advertising a healthy diet, while health issues regarding vegetables need more promotion.

As it was argued before, also the attitude towards vegetable consumption and the knowledge about vegetables might influence the linkage between vegetable production and consumption, which became, however, not apparent from the present data. In other studies more and even strong evidence was found for knowledge, self-efficacy and social support as predictors of adult fruit and vegetable intake, while variables such as attitudes/beliefs, intentions and barriers showed weaker evidence (Shaikh et al. 2008). Whether the knowledge about vitamin A and iron-rich foods including vegetables was related to vegetable production will be discussed in chapters 4.3 and 4.4, respectively.

The integration of fruits and vegetables as foods rich in micronutrients providing a sustainable strategy to improve human micronutrient status is widely known and acknowledged. However, this information has not yet fully reached consumers, especially in developing countries. While several promotion activities, such as the "5-a-day" programme to enhance vegetable and fruit consumption, are already established in developed countries, fruit and vegetable promotion initiatives are only about to be established in some developing countries (WHO 2003). Yet, these initiatives are badly needed as could be shown in Senegal: when knowledge of dietary and behaviour-related determinants of non-communicable chronic diseases were assessed among urban Senegalese women, the health promoting and protective effects of fruits and vegetables were least understood by study participants (Holdsworth et al. 2006).

4.1.3   Vegetable production and consumption in associations and correlations

Vegetable production and consumption was analysed with regard to different socio-economic factors (Tables 3.2.7 and 3.3.3). While vegetable consumption was never significantly related to wealth in the present study, vegetable production was, however, only during certain seasons. As it was negatively correlated to vegetable collection and positively to the production of exotic vegetables, it is suggested that with increasing wealth there is a shift from collecting traditional to cultivating exotic vegetables. In the researched districts it was already found that traditional vegetables played an important role in resource poor communities (Weinberger and Swai 2006). Education was also not significantly correlated to vegetable consumption, however, positively to general vegetable production during several seasons except exotic vegetable production (Table 3.2.7). During Jun/Jul (DS), for example, a higher education was linked to a more diverse traditional vegetable production. As the latter can be beneficial in many ways the relevance of education was confirmed.

Vegetable consumption was positively correlated to the distance of the village to the next town centre during two seasons (Table 3.3.3) meaning that the further participants lived away from town the greater was the vegetable diversity they consumed. This gives a hint on the 'nutrition transition' (see also chapters 4.2, 4.5 and 4.6), which usually increases dietary diversity in terms of highly processed and fast foods, and decreases diversity in terms of traditional foods (Popkin 2007). Also vegetable cultivation/collection was related to distance during several seasons, however, either positively or negatively. While the production and consumption variables coincide during one season they differ during another and, thus, the linkage between the two needs again to be considered. The correlations with production on the one hand and consumption variables on the other were calculated separately and clearer results might be achieved if both parameters are looked at together in a multiple regression model (see below).

Significant correlations were found during Nov/Dec (SR) and, especially, during Mar/Apr (LR) between vegetable production and consumption quantity (Table A19). During Nov/Dec (SR), vegetable consumption was slightly negatively associated with the production of traditional vegetables, i.e. the more traditional vegetables were produced the less vegetables, in general, were consumed. At this time (Nov/Dec (SR)), quite a number of vegetables were cultivated in the gardens, yet, they were not ready for consumption. If vegetables were consumed at that time, it was rather wild types but not cultivated.

During Mar/Apr (LR), vegetable consumption quantity was significantly and positively related to all production parameters. Consequently, only during this season the present data indicated that an increased vegetable production was accompanied by an increased vegetable consumption. While correlation does not explain one factor influencing the other, it still seems that during this particular season, vegetable consumption was linked to vegetable production, while it was rather influenced by other factors during the other seasons. Therefore, it can be stated that vegetable diversity in the field is related to vegetable diversity on the plate, yet depending on the season.

This seasonal differences were especially true for Kongwa and Singida districts. There, seasons were more distinct than in Muheza and the amount of vegetables consumed was highest during Mar/Apr (LR) (Table 3.3.15). Here, a higher vegetable consumption was probably positively influenced by higher vegetable production. In Nigeria, for example, a high prevalence of micronutrient deficiency was found due to the seasonal variation in vegetable availability as well as inadequate knowledge on processing and preserving vegetables (Hart et al. 2005). While seasonality is obviously an important factor, also knowledge is limiting the adequate consumption of leafy vegetables even if they are available in abundance.

The VDS was significantly correlated to vegetable cultivation and to cultivation/collection of traditional vegetables during all seasons (Table 3.3.18). During Mar/Apr (LR) the relationship was strongest suggesting that, when the number of vegetables cultivated/collected was high, the number of different vegetables consumed was also high. Because of no district differences, the three districts were taken together; still Muheza should also be looked at alone if clearer answers ought to be achieved.

While the consumption quantity was associated with vegetable production only during Mar/Apr (LR), the VDS and, thus, the consumed vegetable diversity was connected to the diversity of vegetable cultivation during all seasons. From this it can be concluded that vegetable diversity on the plate is, in fact, linked to that in the field. Still, besides diversity the amount of vegetables consumed and, of course, also further consumption and processing parameters are of importance and need to be taken into account for a holistic view on nutritional health aspects.

> **LINK between production and consumption:**
> ✓   yes for production and VDS
> ✗✓   partly for production and consumption quantity depending on the season

Multiple relations
When vegetable consumption quantity was chosen as the dependent variable in a multiple regression model, it was controlled for district and age of participants, however, not for education and wealth as these variables were not significantly correlated to the amount of vegetables consumed. While age usually never showed any correlation with the dependent variable, the districts showed consistently significant correlations with the dependent variable during one season or the other.

Besides a "general model" with both consumption and production predictors, also a "production model" and a "consumption model" were calculated. For the "general" regression model only the food groups cereals, oil/fat and pulses were used as they showed the highest correlations with vegetable consumption (Table A37) or, in terms of oil/fat, for logical reasons. The "production model" included only production variables as predictor variables, the "consumption model" only consumption variables. This separation was done because production variables showed poor associations with vegetable consumption, in some cases they had high variance inflation factors (VIF) and were obviously responsible for residuals being not normally distributed.

Both during Jun/Jul (DS) and Mar/Apr (LR) the DDS and VDS were more or less highly correlated to vegetable consumption within the "general model" (Figure 4.1.6). When the mean DDS increased by one (i.e. consumption of one more food group per day) vegetable consumption decreased by about 23 g/d during the Jun/Jul (DS) season, while it decreased by about 35 g/d during Mar/Apr (LR). This negative relationship suggests that the greater the dietary diversity (here in terms of food groups), the lower was the amount of each single food group, or here at least of the food group vegetables. A greater food group variety does, therefore, not mean a better or healthier diet in itself. However, it strongly depends on the food group types and also on the ratio between the food groups in terms of amount consumed. In the present case it seemed that the higher the DDS, the more was the food group vegetable replaced or substituted by another food group (see also chapter 4.5).

When participants were asked for which reason they would not consume any vegetables for a whole day, 30% of participants said that they would break the everyday meal containing vegetables with some more special food if they could afford it such as meat, fish, beans or *makande* (mixture of maize and beans). Thereby, only 4% of participants suggested to break with animal products, while 15% preferred beans or *makande*, thus, other plant products, instead of vegetables. The others (31%) remained undetermined with which of the four foods/food groups to substitute vegetables. Consequently, the most important food groups, which were eaten instead of vegetables, were legumes and starchy staples. Interestingly, it was found in Tanzania that vegetables among other foods have a higher expenditure elasticity relative to pulses and cereals, meaning that the demand for vegetables will increase with rising income (Abdulai and Aubert 2005), which would suggest that vegetables are more valued than pulses, unlike the present findings.

This replacement also does not make sense in terms of nutrients as the content of nutrients is highly different of these food groups compared to vegetables. While it is probably a positive addition if beans or also meat and fish is consumed, this should be, however, done additionally and vegetables should not be replaced by those foods (Burgess and Glasauer 2004). Further popular arguments for eating no vegetables as stated by study participants were being not at home (52%), seasonal availability (35%) and preparation time (34%) (more than one answer possible). The latter argument suggests that there are foods, which are more easy and faster to prepare and, thus, less time and energy consuming. Therefore, when promoting vegetables and, for instance, creating new recipes, the convenience of preparation should be taken into account.

When the FVS instead of DDS was used as a predictor variable in the "general model" it also showed significant and negative correlations with vegetable consumption, however, less strong than the DDS. Still, the same conclusions can be drawn for FVS like for DDS, namely that the higher the food variety (here in terms of food items), the smaller is the actual amount of vegetables consumed.

When the mean VDS increased by one (i.e. consumption of one more vegetable type per week) vegetable consumption during Jun/Jul (DS) increased by about 18 g/d, while it only increased by about 10 g/d during Mar/Apr (LR). Thus, the more different vegetables were consumed, the greater was the amount of vegetables consumed. Thereby, it must be considered that regarding a low VDS, e.g. a change from two to three vegetables, the increase in vegetable consumption quantity might be different than in terms of a high VDS, for instance a change from nine to ten vegetables.

It can be speculated that the amount is even somehow influenced or stimulated by a higher vegetable variety: if a greater diversity of a food group is available for consumption, more of this food group will be eaten – at least up to a certain extent. As for the data from Nov/Dec (SR) residuals were not normal-distributed the regression results are not further discussed.

The "consumption model" additionally showed that significant and positive relationships were found between vegetable consumption and cereal as well as oil/fat consumption, while there was a negative correlation between vegetable and pulses consumption (Table A38). Consequently, while cereals and oil/fat seem to accompany vegetables in consumption, pulses rather replace vegetables as was already stated above. Nut consumption was, in turn, positively correlated with vegetable consumption. This was most likely due to the typical preparation of vegetables, which were often cooked either in coconut milk (Muheza) or together with groundnuts (Kongwa and Singida). Furthermore, during Mar/Apr (LR) fruit consumption was slightly positively correlated with vegetable consumption as well. The latter fact would suggest that fruit and vegetables do not substitute but complement each other. Therefore, when investigating, for example, vitamin A consumption, vegetables and fruits should always be looked at together and not separately as it was also suggested by WHO (2003b).

No significant correlations were identified between vegetable production parameters and vegetable consumption quantity during all seasons. This was true for the "general model" as well as for the "production model" where no significances but one for the district Muheza was assessed (Table A39). While from bivariate correlations it was learned that the number of vegetables cropped/collected and, therefore, the diversity of cropped/collected vegetables was associated with the amount of vegetables consumed during certain seasons, this could be not shown in a multiple regression model. Consequently, other factors than vegetable diversity in the field were responsible, for example, the actual quantity of vegetables cropped and collected (not assessed) or the amount and diversity of vegetables bought additionally to the self-produced ones as well as other factors mentioned before. While within the present study the surveyed socio-economic and knowledge parameters did not show any significant relationships to vegetable consumption, a review on social class and diet quality in developed countries, however, showed that there exist a multitude of epidemiologic data showing that diet quality follows a socio-economic gradient. Thereby, in Europe and Australia, fresh fruits and vegetables were more likely to be consumed by people of higher socio-economic status (Darmon and Drewnowski 2008). This could not be confirmed for study participants in Tanzania and it is suggested that for a developing country with a high consumption rate of domestically produced foods, diet also follows a socio-economic gradient, however, the other way round with fruits and vegetables being more likely consumed by the lower socio-economic class. The latter needs to be considered more strongly next to food diversity in production.

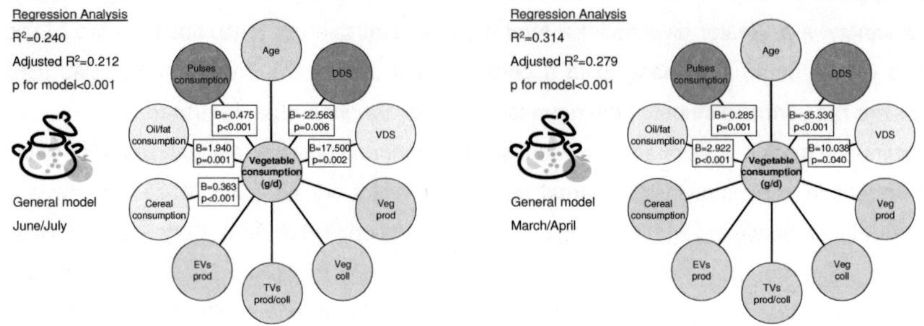

Figure 4.1.6   Results of the multiple regression analysis "general model" during June/July and March/April with vegetable consumption (g/d) as dependent variable; n=252 women from three districts in Tanzania

<u>LINK between production and consumption:</u>

✘   link between different consumption parameters only such as among
  - DDS and vegetable consumption quantity (negative)
  - VDS and vegetable consumption quantity (positive)

<u>Cluster analysis</u>

Besides examining correlations between vegetable production and consumption for the whole group of study participants it was also of interest to identify typical characteristics of women who consumed or produced rather the one or the other vegetable or vegetable group. Thus, not the variables but the cases were focused on and patterns among similar responses were aimed to identify to form groups. The method of choice was, therefore, cluster analysis to group participants based on the similarity of their responses to several variables, here production and consumption of vegetables.

Regarding consumption (Figure 3.3.37, Table 3.3.13), clusters 1 and 2 were not only characterised by a higher vegetable diversity, but also by both cultivated and wild vegetables (Table 3.3.13). Cluster 3 was dominated by wild vegetables, whereby amaranth can be both cultivated or collected from the wild. Cluster 4 and 5 comprised only cultivated vegetables with cluster 5 having especially low diversity.

Amaranth as a very popular vegetable showed some affiliation to each cluster, while other vegetables such as wild simsim was only found in clusters 2 and 3. It is assumed that no direct relationships can be drawn from consuming mainly one or another vegetable to other characteristics of participants. However, the vegetables or vegetable patterns consumed can be used as an indicator standing, for instance, for typical characteristics of a participant or a group of participants.

From those variables which showed significant differences between vegetable consumption clusters (Table 4.1.2) only some will be discussed. Regarding the three different districts, some clusters were only found in one district such as cluster 1 in Kongwa and cluster 2 in Singida. Cluster 3 was also mainly located with Singida, cluster 5 mainly consumed by participants from Muheza, and cluster 4 by women from all three districts. These clear differences between districts indicate that vegetable consumption depended on regional preferences and also availability. For example, in a study on fruit and vegetable consumption of male adolescents in the UK, an association with the distance to food stores was found which was partially mediated through vegetable preferences (Jago et al. 2007). While the comparison between Tanzanian women and UK boys should, of course, be considered with caution, still the fact that vegetable consumption is influenced by the place of residence as well as either regional or individual preferences, seems to be comprehensible regardless of nationality, sex and age. Just as districts, also ethnic groups showed significant differences between clusters.

An interesting fact concerning the different occupation of women was that farmers additionally running a business or being employed in service were mainly consuming vegetables from cluster 5 and also cluster 4, thus, mainly cultivated and no wild vegetables. Women being involved in business or service definitely earn some cash and instead of collecting traditional vegetables from the wild obviously preferred to either invest in cropping their own traditional or exotic vegetables or buy them from the market. Thus, along with a usually desired development of communities and increasing wealth, there exists a trend for wild traditional vegetables and indigenous foods in general to get lost; this trend could, of course, be prevented by the collection and documentation of indigenous knowledge and indigenous genetic resources (Kuhnlein et al. 2009).

From the health parameters, only Hb showed significant differences between vegetable consumption clusters, with a mean Hb being highest for clusters 2 and 3 and lowest for cluster 5. Further, the share of anaemic participants was highest within cluster 5. It can be argued that wild vegetables consumed within clusters 2 and 3 are likely to have a higher iron content than cultivated vegetables consumed within cluster 5. While the iron content in plants can be highly variable and is influenced by many factors, data from the literature approve that the wild vegetables false sesame with an average amount of 3.2mg/100g (Bedigian and Adetula 2004) and jute mallow with 7.2 mg/100g (Fondio and Grubben 2004) have substantially higher iron contents than the cultivated vegetables okra with 1.1 to 1.5mg/100g (Siemonsma and Kouamé 2004) and sweet potato leaves with about 1.8mg/100g (Islam 2007). Consequently, while workload is decreasing through purchasing or cultivating vegetables instead of collecting from the wild, the positive contribution wild vegetables can make to nutritional health gets lost. If participants within clusters 2 and 3, in fact, have a higher iron intake depends, however, on the actual amount of vegetables consumed, processing techniques, and further meal ingredients enhancing or inhibiting iron intake and bioavailability. As clusters 2 and 3 mainly include women from Singida district and cluster 5 from Muheza district, malaria, being more prevalent in Muheza, will also play a role in Hb values of study participants, which can not be solely traced back to vegetable consumption.

Food consumption in general was also related to vegetable consumption, e.g., dietary diversity and food variety showed significant differences between clusters. The mean DDS and FVS were lowest for participants within clusters 1 and 2 and increasing in hight for clusters 3, 4 and 5. Consequently, most women with a high DDS/FVS were consuming vegetables according to clusters 5 and also 4. Besides other influencing factors on dietary diversity and food variety, there seemed to be a link between the consumption of wild vegetables and a lower variety of foods as well as between the consumption of cultivated vegetables and a higher variety of foods (see also chapter 4.5). As stated before, wild vegetable consumption is an indicator for low wealth and low or no purchase of additional food and, consequently, low food diversity. As soon as food is cultivated or even bought and one does not need to rely on food gathered from the wild, food diversity is increasing. Still, it is also argued that a shift from using wild and traditional foods to purchasing foods in the market, e.g., through urbanisation, may negatively impact on nutrition by narrowing food choices (Delisle 1990).

While the collection of wild vegetables and the sales of vegetables was not related to vegetable consumption clusters, the cultivation of vegetables was. The number of vegetables cultivated per person increased from cluster 1 to cluster 5. The diversity of vegetables consumed was highest in clusters 1 and 2 and also 4, according to the VDS it was highest in clusters 4 and 5. From this data for the whole year a high level of vegetable diversity produced coincided with a high level of vegetable diversity consumed. The figures do especially make sense as only vegetable cultivation is connected to the clusters and not vegetable collection, because in clusters 1 and 2 the high diversity is mainly due to wild vegetables.

Table 4.1.2   Characteristics of study participants from three districts in Tanzania belonging to a certain vegetable consumption cluster (n=252 except Hb data: n=185; cluster generated from number of days each vegetable was consumed per week)

	$P (\chi^2)^*$	Cluster 1 n (%)**	Cluster 2 n (%)	Cluster 3 n (%)	Cluster 4 n (%)	Cluster 5 n (%)
Socio-economic parameters		n=28	n=16	n=46	n=67	n=95
District	<0.001					
Kongwa		28 (100)	0 (0)	8 (17.4)	27 (40.3)	9 (9.5)
Muheza		0 (0)	0 (0)	0 (0)	9 (13.4)	67 (70.5)
Singida		0 (0)	16 (100)	38 (82.6)	31 (46.3)	19 (20.0)
Religion	<0.001					
Christian		28 (100)	3 (18.8)	28 (60.9)	41 (61.2)	61 (64.2)
Muslim		0 (0)	13 (81.3)	18 (39.1)	26 (38.8)	34 (35.8)

	P (χ^2)*	Cluster 1 n (%)**	Cluster 2 n (%)	Cluster 3 n (%)	Cluster 4 n (%)	Cluster 5 n (%)
Ethnic group	<0.001					
Bondei		0 (0)	0 (0)	0 (0)	4 (6.0)	15 (15.8)
Gogo		23 (82.1)	0 (0)	4 (8.7)	5 (7.5)	4 (4.2)
Kaguru		3 (10.7)	0 (0)	4 (8.7)	13 (19.4)	4 (4.2)
Nyaturu		0 (0)	16 (100)	38 (82.6)	31 (46.3)	19 (20.0)
Shambaa		0 (0)	0 (0)	0 (0)	4 (6.0)	20 (21.1)
Other		2 (7.1)	0 (0)	0 (0)	10 (14.9)	33 (34.7)
Occupation	0.011					
Crop farmer		7 (25.0)	1 (6.3)	5 (10.9)	6 (9.0)	8 (8.4)
Crop and livestock farmer		20 (71.4)	14 (87.5)	40 (87.0)	53 (79.1)	67 (70.5)
Farmer + business/service		1 (3.6)	1 (6.3)	1 (2.2)	8 (11.9)	20 (21.1)
Distance village-town (km)	<0.001					
Short (5-20 km)		21 (75.0)	0 (0)	5 (10.9)	10 (14.9)	49 (51.6)
Medium (21-34 km)		7 (25.0)	4 (25.0)	17 (37.0)	26 (38.8)	22 (23.2)
Long (35-67 km)		0 (0)	12 (75.0)	24 (52.2)	31 (46.3)	24 (25.3)
Distance village-town (min)	<0.001					
Short (20-45 min)		21 (75.0)	0 (0)	5 (10.9)	8 (11.9)	27 (28.4)
Medium (60-90 min)		0 (0)	0 (0)	4 (8.7)	13 (19.4)	38 (40.9)
Long (120-240 min)		7 (25.0)	16 (100)	37 (80.4)	46 (68.7)	30 (31.6)
Marital status	0.009					
Single		3 (10.7)	1 (6.3)	2 (4.3)	11 (16.4)	23 (24.2)
Married		20 (71.4)	14 (87.5)	42 (91.3)	48 (71.6)	59 (62.1)
Widowed		1 (3.6)	0 (0)	1 (2.2)	4 (6.0)	3 (3.2)
Divorced		4 (14.3)	0 (0)	0 (0)	1 (1.5)	3 (3.2)
Separated		0 (0)	1 (6.3)	1 (2.2)	3 (4.5)	7 (7.4)
Health status						
Hb	0.003	n=20	n=14	n=34	n=46	n=71
Severe anaemia		1 (5.0)	0 (0)	0 (0)	0 (0)	0 (0)
Moderate anaemia		0 (0)	0 (0)	0 (0)	0 (0)	4 (5.6)
Mild anaemia		4 (20.0)	0 (0)	2 (5.9)	16 (34.8)	19 (26.8)
Normal iron status		15 (75.0)	14 (100)	32 (94.1)	30 (65.2)	48 (67.6)
Mean (mg/l)	0.002	12.7 (±1.6)	13.7 (±0.8)	13.3 (±1.2)	12.6 (±1.3)	12.4 (±1.6)
Food consumption		n=28	n=16	n=46	n=67	n=95
DDS	<0.001					
Low DDS (2-4 food groups)		17 (60.7)	15 (93.8)	22 (47.8)	18 (26.9)	12 (12.6)
Medium DDS (5-6 food groups)		8 (28.6)	1 (6.3)	21 (45.7)	29 (43.3)	16 (16.8)
High DDS (7-11 food groups)		3 (10.7)	0 (0)	3 (6.5)	20 (29.9)	67 (70.5)
Mean	<0.001	4.4 (±1.3)	3.5 (±0.7)	4.7 (±1.2)	5.8 (±1.6)	7.1 (±1.6)
FVS	<0.001					
Low FVS (2-6 foods)		19 (67.9)	13 (81.3)	17 (37.0)	12 (17.9)	10 (10.5)
Medium FVS (7-9 foods)		8 (28.6)	3 (18.8)	28 (60.9)	35 (52.2)	24 (25.3)
High FVS (10-14 foods)		1 (3.6)	0 (0)	1 (2.2)	20 (29.9)	61 (64.2)
Mean	<0.001	5.9 (±1.8)	5.5 (±1.2)	6.9 (±1.5)	8.4 (±1.9)	9.8 (±2.0)
VDS	<0.001					
Low VDS (1-3 vegetables)		6 (21.4)	6 (37.5)	16 (34.8)	8 (11.9)	29 (30.5)
Medium VDS (4-5 vegetables)		16 (57.1)	9 (56.3)	28 (60.9)	32 (47.8)	48 (50.5)
High VDS (>6 vegetables)		6 (21.4)	1 (6.3)	2 (4.3)	27 (40.3)	18 (18.9)
Mean	<0.001	4.7 (±1.1)	4.0 (±1.0)	4.3 (±0.9)	5.5 (±1.3)	4.6 (±1.6)
Starchy foods	<0.001					
Below 300 g/d		10 (35.7)	0 (0)	6 (13.0)	10 (14.9)	6 (6.3)
300-500 g/d		14 (50.0)	10 (62.5)	33 (71.7)	35 (52.2)	44 (46.3)
Above 500 g/d		4 (14.3)	6 (37.5)	7 (15.2)	22 (32.8)	45 (47.4)
Fruits and vegetables	0.010					
Below 300 g/d		16 (57.1)	2 (12.5)	17 (37.0)	19 (28.4)	25 (26.3)
300-500 g/d		10 (35.7)	7 (43.8)	18 (39.1)	36 (53.7)	53 (55.8)
Above 500 g/d		2 (7.1)	7 (43.8)	11 (23.9)	12 (17.9)	17 (17.9)
Oil/fat	<0.001					
No consumption		7 (25.0)	2 (12.5)	1 (2.2)	0 (0)	0 (0)
Below 15 g/d		13 (46.4)	6 (37.5)	17 (37.0)	21 (31.3)	27 (28.4)

	P (χ^2)*	Cluster 1 n (%)**	Cluster 2 n (%)	Cluster 3 n (%)	Cluster 4 n (%)	Cluster 5 n (%)
15-30 g/d		7 (25.0)	7 (43.8)	22 (47.8)	29 (43.3)	38 (40.0)
Above 30 g/d		1 (3.6)	1 (6.3)	6 (13.0)	17 (25.4)	30 (31.6)
Animal products	0.014					
No consumption		15 (53.6)	6 (37.5)	16 (34.8)	24 (35.8)	14 (14.7)
Below 50 g/d		5 (17.9)	3 (18.8)	13 (28.3)	18 (26.9)	28 (29.5)
50-150 g/d		7 (25.0)	4 (25.0)	12 (26.1)	22 (32.8)	41 (43.2)
Above 150 g/d		1 (3.6)	3 (18.8)	5 (10.9)	3 (4.5)	12 (12.6)
Nuts and pulses	<0.001					
No consumption		1 (3.6)	6 (37.5)	10 (21.7)	3 (4.5)	2 (2.1)
Below 50 g/d		12 (42.9)	4 (25.0)	12 (26.1)	15 (22.4)	12 (12.6)
50-200 g/d		14 (50.0)	5 (31.3)	21 (45.7)	37 (55.2)	31 (32.6)
Above 200 g/d		1 (3.6)	1 (6.3)	3 (6.5)	12 (17.9)	50 (52.6)
Cereals – mean intake (g/d)	0.025	347 (±102)	457 (±123)	363 (±93)	362 (±99)	352 (98)
Bread/cakes – mean intake (g/d)	<0.001	4 (±19)	0 (±0)	4 (±18)	28 (±63)	59 (±62)
Fruits – mean intake (g/d)	<0.001	17 (±34)	0 (±0)	36 (±71)	51 (±101)	117 (±118)
Vegetables – mean intake (g/d)	<0.001	293 (±125)	510 (±183)	358 (±133)	325 (±137)	267 (±104)
Nuts – mean intake (g/d)	<0.001	28 (±29)	1 (±3)	10 (±41)	36 (±63)	92 (±94)
Pulses – mean intake (g/d)	<0.001	42 (±52)	55 (±69)	62 (±77)	77 (±70)	119 (±96)
Starchy plants – mean intake (g/d)	<0.001	1 (±7)	12 (±32)	47 (±80)	53 (±75)	92 (±122)
Tea – mean intake (ml/d)	<0.001	67 (±149)	15 (±58)	93 (±174)	194 (±199)	316 (±197)
Oil/fat – mean intake (g/d)	<0.001	10 (±9)	14 (±13)	18 (±9)	25 (±19)	25 (±15)
Sugar – mean intake (g/d)	<0.001	5 (±5)	1 (±4)	6 (±11)	10 (±12)	13 (±15)
Fish – mean intake (g/d)	<0.001	6 (±13)	18 (±30)	14 (±21)	21 (±33)	57 (±55)
Nutrient intake						
Energy	0.005					
600-1799 kcal/d		22 (78.6)	8 (50.0)	28 (60.9)	33 (49.3)	30 (31.6)
1800-2599 kcal/d		6 (21.4)	7 (43.8)	17 (37.0)	31 (46.3)	61 (64.2)
2600-3500 kcal/d		0 (0)	1 (6.3)	1 (2.2)	3 (4.5)	2 (2.1)
Above 3500 kcal/d		0 (0)	0 (0)	0 (0)	0 (0)	2 (2.1)
Mean intake (kcal/d)	<0.001	1645 (±369)	1803 (±457)	1719 (±437)	1850 (±396)	2021 (±457)
Protein	0.002					
Below 45 g/d		4 (14.3)	5 (31.3)	13 (28.3)	13 (19.4)	10 (10.5)
45-65 g/d		17 (60.7)	7 (43.8)	26 (56.5)	40 (59.7)	41 (43.2)
Above 65 g/d		7 (25.0)	4 (25.0)	7 (15.2)	14 (20.9)	44 (46.3)
Mean intake (g/d)	<0.001	58.2 (±13.3)	59.6 (±24.4)	53.2 (±18.3)	55.0 (±13.6)	65.1 (±19.0)
Vitamin A	<0.001					
1-499 µg/d		4 (14.3)	13 (81.3)	29 (63.0)	23 (34.3)	20 (21.1)
500-850 µg/d		11 (39.3)	3 (18.8)	5 (10.9)	9 (13.4)	12 (12.6)
850-7500 µg/d		13 (46.4)	0 (0)	12 (26.1)	35 (52.2)	63 (66.3)
Mean intake (µg/d)	<0.001	889 (±417)	415 (±173)	802 (±1011)	1096 (±866)	1313 (±915)
Iron	<0.001					
1-14.99 mg/d		4 (14.3)	5 (31.3)	16 (34.8)	31 (46.3)	56 (58.9)
15-30 mg/d		4 (14.3)	10 (62.5)	19 (41.3)	26 (38.8)	38 (40.0)
30-90 mg/d		15 (53.6)	1 (6.3)	10 (21.7)	8 (11.9)	1 (1.1)
Above 90 mg/d		5 (17.9)	0 (0)	1 (2.2)	2 (3.0)	0 (0)
Mean intake (mg/d)	<0.001	56.1 (±37.1)	20.0 (±7.1)	24.1 (±21.5)	22.7 (±20.0)	14.4 (±4.6)
Vegetable production		n=28	n=16	n=46	n=67	n=95
Production	0.015					
No vegetable production		0 (0)	0 (0)	2 (4.3)	0 (0)	0 (0)
Veg prod during 1 season		2 (7.1)	3 (18.8)	1 (2.2)	3 (4.5)	2 (2.1)
Veg prod during 2 seasons		8 (28.6)	4 (25.0)	5 (10.9)	12 (17.9)	14 (14.7)
Veg prod during 3 seasons		18 (64.3)	9 (56.3)	38 (82.6)	52 (77.6)	79 (83.2)
Mean (number of vegetables)	<0.001	2.7 (±1.1)	3.0 (±1.8)	3.9 (±1.7)	4.0 (±1.7)	4.3 (±1.6)

* p for differences between groups (rows) and for differences between mean/median values of cluster groups according to K-W test if applicable

** absolute number of participants and percent within cluster, if not mean/median

In the cluster obtained from vegetable production (Table 3.2.2, Figure 3.2.2), tomatoes and onions were of no consequence, despite their importance in consumption. On the other hand, amaranth like for consumption, occurs in each cluster. Further, wild simsim and, to a certain extent, also wild cucumber dominate clusters 1 and 2. Cluster 3 is rather dominated by false sesame, cowpea leaves and also Mhilile and, again, in all first three clusters wild vegetables play an important role. These wild vegetables can usually cope with drought while the vegetables within clusters 4 and 5 rather need regular water supply, especially those in cluster 5. Further, while cluster 4 is dominated by cultivated vegetables, in cluster 5 both wild and cultivated vegetables occur. If cultivated vegetable diversity was higher in one or the other cluster does not become evident from these figures.

Because vegetable production data was based only on the number of seasons a vegetable was produced per year (0 – 3) and not on more exact data of yield, area under cultivation among others, it was decided to relate these results only to a few characteristics of participants, namely only to districts, VDS, vegetable intake and wealth. Most important differences were found especially between the three districts (Table 4.1.3) with cluster 3 occurring only in Kongwa and cluster 5 only in Muheza. Similarly to vegetable consumption, production of vegetables was locally different in terms of vegetable types. The mean intake of vegetables per day also differed significantly between the production clusters, which suggests that vegetable consumption is very well influenced by vegetable production parameters. The connection of the production clusters and the VDS was less strong but still showed differences; especially participants within cluster 4 ate a greater vegetable diversity per week than those within cluster 2, supporting the fact that diversity in vegetable consumption increases when shifting from wild to cultivated vegetables – even in production. Wealth was not significantly related to the vegetable production clusters.

Table 4.1.3 Characteristics of study participants from three districts in Tanzania belonging to a certain vegetable production cluster (cluster generated from number of seasons a vegetable was produced per year)

	$P(x^2)$	Cluster 1 n=75	Cluster 2 n=30	Cluster 3 n=56	Cluster 4 n=18	Cluster 5 n=73
District	<0.001					
Kongwa n (%)		1 (1.3)	1 (3.3)	56 (100.0)	14 (77.8)	0 (0.0)
Muheza n (%)		0 (0.0)	0 (0.0)	0 (0.0)	3 (16.7)	73 (100.0)
Singida n (%)		74 (98.7)	29 (96.7)	0 (0.0)	1 (5.6)	0 (0.0
Vegetables – mean intake (g/d) (± sd)	<0.001	345 (123)	424 (197)	303 (119)	379 (160)	242 (87)
VDS mean (± sd)	0.027	4.7 (1.2)	4.3 (1.1)	4.8 (1.2)	5.5 (1.2)	4.7 (1.8)

In order to further investigate differences of production and consumption concerning vegetable species, not only participants were clustered but also villages. For this, again, only the 20 most common vegetables were chosen and differences can be viewed in the dendrograms for consumption (Figure 4.1.7) and production (Figure 4.1.8).

In both dendrograms, three clusters were formed out of the 18 villages, which mainly comprise villages from one district. Regarding vegetable consumption (Figure 4.1.7), the second cluster combines five villages from Singida and two from Kongwa, while one Singida village (No. 14) stands quite alone. Obviously in each district, vegetable types were consumed in a certain combination, which was unique and not shared between the districts, except partly between Kongwa and Singida. From the distances, also the differences between villages can be accounted for: long distances mean that cases, here villages, were more different from each other. Therefore, it is suggested that among Muheza villages, vegetable consumption patterns were more similar, while in Singida, villages vary more strongly. This might be influenced by long distances between the villages, however, can not be due to ethnic group because most participants in Singida were Nyaturu, while in Muheza many different ethnic groups were found among participants.

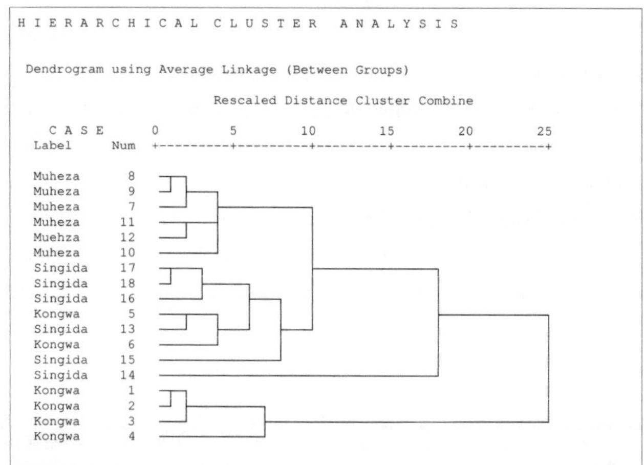

Figure 4.1.7   Dendrogram for 18 villages from three districts in Tanzania clustered according to <u>consumption</u> data of 20 vegetables; n=252 women; mean values across three seasons

Village No.14 in Singida district, Mwakiti, is characterised by having the longest distance to the next vegetable market (15km compared to 0.5-10km for the other villages) and also being located furthest north and east compared to all other villages (according to GPS data). Especially the fact of being far away from vegetable trading could explain Mwakitis' special position regarding vegetable consumption. The two villages of Kongwa district, No.5 Tubugwe A and No.6 Mlali-Iyegu, that mingle with the Singida villages within one cluster, both apply irrigation for vegetable cultivation. All other villages in Kongwa rely on rainfed horticulture. While in Singida district irrigation of vegetables is not the norm, irrigation still makes the difference for the two Kongwa villages, so that vegetable consumption is rather equal to those in another district than in their own. Here, conditions of vegetable cultivation seem to actually influence vegetable consumption. In southeastern Tanzania it was observed that subsistence farmers adapted their food production to various conditions, irrespective of their tribal origins (Zehnder et al. 1987). Consequently, with similar cropping conditions, e.g. in Singida and Kongwa, also consumption, here of vegetables, becomes more similar.

In terms of vegetable production, all three districts form one solid cluster each without any interlinkages between Kongwa and Singida. Yet, Singida and Kongwa were more closely related to each other than Muheza district. Compared to the consumption dendrogram, distances were much shorter between villages of one district within the production dendrogram. Apparently, while production of vegetables was rather similar in different villages of a district, weekly consumption of vegetables differed more between villages, especially in Singida district. In other words, within one district similar vegetables were produced but not consumed, which approves no clearl link between vegetable consumption and production.

Consumption and production regarding vegetable types was also compared with Sørensen Index both between districts and seasons (Table A40). However, an adverse trend then through cluster analysis was found, namely that consumption was more similar in two districts compared than production. The Sørensen Index, comparing only the whole districts and not single villages, might give too simplified results and will be, therefore, not discussed any further.

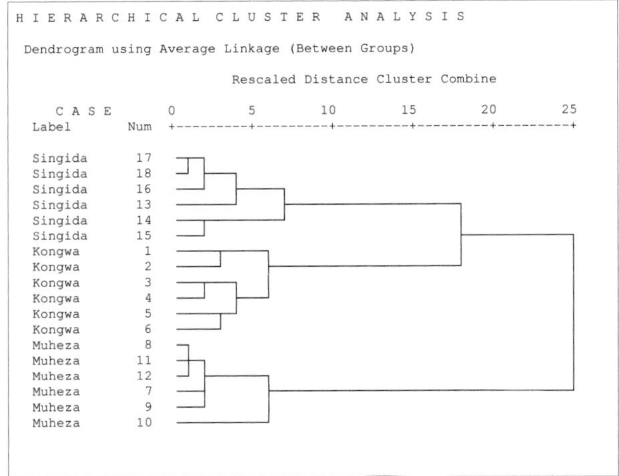

Figure 4.1.8   Dendrogram for 18 villages from three districts in Tanzania clustered according to <u>production</u> data of 20 vegetables; n=252 women; mean values across three seasons

<u>Factor analysis</u>

In order to find even more differences and/or linkages between vegetable production and consumption, a principal component analysis (PCA) was carried out. Results of a PCA with consumption data of 18 vegetables (onions and tomatoes excluded) and four, five and six components was compared and found that five components would provide best interpretable results (Table 3.3.14).

Like in cluster analysis, the consumption of wild and exotic vegetables is obviously not combined but they are consumed separately (either - or) and, thus, substitute rather then complement each other. Further, vegetables from different agro-ecological zones, here dry or humid, were not consumed together. This again shows the differences between districts and, thus, regional differences, and that obviously no or only few interchanges occurred between the districts in terms of consumption. While it is argued that the 'nutrition transition' has started already in the researched areas of Tanzania (chapters 4.2 and 4.6), the consumption of vegetables is at least not standardised totally over the three districts studied, but it is still following "local patterns".

The factor scores were also related to the age of participants, which showed some significant correlations ($p \leq 0.05$) with component 1 (negative), and components 2 and 5 (positive). This would suggest that rather young people consumed vegetables from component 1, while older people prefered the vegetables that characterised component 2 and 5. That older women rather preferred wild and bitter vegetables from component 2 coincides with findings from Kenya, namely that adults tolerate sour and bitter tasting leafy vegetables, while youth prefers vegetables with milder taste (Owuor and Olaimer-Anyara 2007). This is a relevant issue when vegetables are promoted for consumption or when new types are introduced.

Wealth of participants was significantly and negatively correlated to both components 1 and 2 ($p \leq 0.01$) as well as positively to component 4 ($p \leq 0.05$). Those women with lower wealth rather go for wild vegetables, which characterise component 2 and also partly 1 (*Mhilile*). Higher wealth was, at the same time, associated with exotic vegetables from component 4. While so far this relationship was assumed but could not be demonstrated, this shows for the first time within this study that there is a shift from wild and traditional to exotic vegetables with increasing wealth. It was already found in Tanzania that poor households rely more on traditional vegetables than more wealthy households (Weinberger and Msuya 2004), while here it additionally became apparent that with higher wealth, exotic vegetables were preferred and used in consumption.

When participants' factor scores were correlated to production issues, it appeared that the production of vegetables was significantly but negatively correlated to component 1 and 3 and positively to component 4, while the collection of vegetables was significantly and positively related to component 2 ($p \leq 0.01$). The latter is obvious, as component 2 is characterised by the consumption of mainly vegetables gathered from the wild, yet, this is also the case for component 3. Vegetable cultivation and collection combined was significantly correlated to all components but number 5, negatively to components 1 and 3 and positively to components 2 and 4 ($p \leq 0.05$). From this, no clear relationship between vegetable production and consumption can be observed.

> <u>LINK between production and consumption:</u>
> ✓   yes for production diversity and
>   consumption diversity (VDS)
> ✓   consumption diversity increases when production
>   shifts from wild to cultivated

4.1.4   Conclusions

In general, while the relationship between vegetable consumption and health was often the central topic of investigations, the link between production and consumption seems to be less studied. At the same time, it was found with the present data that this link is not easy to establish. While in this study the data on vegetable consumption was collected in much detail and with different tools, the data on vegetable production was surveyed less closely (e.g. number of vegetable types produced during one season but not cropping area, yield etc.). It is, therefore, suggested that for investigating the linkages between vegetable production and consumption data need to be surveyed more equally and concerted.

Still, linkages could be detected in some cases, such as during certain seasons, within certain vegetable groups (TV/EV), or with certain tools such as bivariate correlation and cluster, yet, not with PCA. It can be concluded that there is a clear link between vegetable diversity produced and consumed (VDS) in general, while there is no clear relationship between production and consumption quantity. The following influencing factors have to be considered:

- Purchase of additional vegetables next to homegrown ones (especially for exotic vegetables);
- Preservation and storage (especially for wild vegetables);
- Seasonality;
- Knowledge about both production and consumption of vegetables;
- Taste is more important than health issues when choosing a vegetable;
- Convenience of preparation is decisive when choosing a food type;
- A shift from low to high wealth can result in a shift from indigenous wild to exotic vegetables in consumption.
- "Muheza effect": even if a great diversity is available naturally, it does not mean that it is used by the local population; the knowledge about this diversity and why it should be used as well as attitudes and believes are similarly important like diversity itself.

The VDS was significantly correlated to the number of vegetables produced and, especially, to the number of traditional vegetables collected. While this relationship could not be found for exotic vegetables, it is suggested that, as long as subsistence farming is dominating and traditional vegetables are favoured and used, the diversity of vegetables in the field can very well influence vegetable diversity on the plate of women in rural Tanzania and, consequently, their nutritional health.

FOCUS A: The success story of tomatoes and onions

Tomatoes

Tomatoes and onions are among the exotic vegetables in Tanzania which are consumed by nearly everyone on a daily basis. Today the tomato (*Lycopersicon esculentum*) is, in fact, one of the most important vegetables worldwide and has its origin in the South American Andes. It was brought to Europe in the 16th century, from there introduced to southern and eastern Asia in the 17th century and subsequently to the United States, Africa and the Middle East (van der Vossen et al. 2004).

In south-eastern Tanzania, it was found among subsistence farmers that they used irrigation for dry season cropping in addition to the prevailing wet season farming. Through this, the cultivation of marketable crops was possible, which were mainly tomatoes (Zehnder et al. 1987). Already in the 1980s, tomatoes were cash crops providing additional income to Tanzanian farmers. From 1990 to 2004 the area under cultivation increased from 14 to 19,000 ha. Tomato yield increased during the same time only from 7.5 to 7.6 t/ha. In fact, the world's average tomato yield in 2001 was 27 t/ha, but in tropical Africa only 8 t/ha of fruits are obtained on average from an open-field tomato crop (van der Vossen et al. 2004).

Only seven tomato varieties are mainly cultivated by farmers in northern Tanzania (Arusha district); on the Arusha market even only three main varieties can be found. Furthermore, despite high costs, an intensive use of fertilisers and pesticides is common (SLE 2008). Thus, tomato production is, compared to that of traditional vegetables, rather intensive and can be a hazard for the environment. Cultivation should, therefore, be sustainable and plant diversity should be enhanced by tomato breeders, e.g. at AVRDC-RCA.

Onions

The onion (*Allium cepa)* is only known from cultivation and originates most likely from Central Asia, where some of its relatives still grow in the wild. Traditional tropical African cultivars may have been introduced either from southern Egypt, or from India via Sudan to Central and West Africa (Messiaen and Rouamba 2004).

The area under onion cultivation is expanding, especially in northern Tanzania (Karatu district, Arusha region), which is known for good quality onions. Yet, also for onions high amounts of chemical pesticides and fertilizers are used. Diversity is low with only one red onion variety being mainly cultivated in Tanzania, while in Karatu district a second red variety is of high importance (SLE 2008).

<u>Nutrition</u>

In Tanzania, tomatoes and onions are used as condiments in nearly every meal, yet, only in very small quantities. Tomato fruits are mainly cooked in sauces and used as a flavouring in soups and meat or fish dishes and are rather seldom consumed fresh in salads. Onions are an essential ingredient in many sauces and relishes consumed usually in small amounts for their pungency (Messiaen and Rouamba 2004).

Though being exotic and introduced vegetables, onions and tomatoes do not actually compete with the traditional vegetables in Tanzania. The latter are mainly leafy vegetables, which are even combined in meals with onions and tomatoes as condiments. While there is no comparable traditional vegetable to onion in Tanzania, African eggplant as a traditional vegetable might be displaced by or has to compete with the exotic tomato. However, while tomato is, above all, used as a condiment or spice in sauces accompanying other foods, African eggplant is usually cooked whole and consumed as a vegetable side dish.

When nutritional contents of the introduced tomato and the traditional African eggplant are compared it shows that African eggplant contains more protein, calcium and iron while the ascorbic acid content of tomato is higher (Table 4.1.4). Further, an important advantage of African eggplant is that the leaves are also edible as dark green leafy vegetables and, thus, it is a multipurpose crop. Consequently, next to the popular tomatoes, traditional vegetables should receive similar attention in breeding and research.

Table 4.1.4   Nutritional contents of different fruits per 100g edible portion

	Tomato (*Lycopersicon esculentum*)*	African eggplant (*Solanum aethiopicum*)**	African eggplant (*Solanum macrocarpon*)***
Protein (g)	0.7	1.5	1.4
Carbohydrates (g)	3.1	7.2	8.0
Calcium (mg)	7.0	28.0	13.0
Carotene (mg)	0.6	0.4[†]	n.a.
Iron (mg)	0.5	1.5	n.a.
Ascorbic acid (mg)	17.0	8.0	n.a.

* van der Vossen, Nono-Womdim and Messiaen, 2004; ** Lester and Seck, 2004; *** Bukenya-Ziraba and Bonsu, 2004; [†] β-carotene; n.a.=not available

4.2   Linking overweight/obesity, food consumption and attitudes: "how the 'nutrition transition' is on the rise in rural Tanzania"

To measure overweight or obesity most often the body mass index (BMI) is used for which the body weight is put into relation with body height (kg/m^2). Obesity defined as a BMI of 30 kg/m^2 or higher is associated with illnesses such as high blood pressure, heart and vascular diseases and diabetes. An estimated 1.3 billion people is overweight worldwide (Popkin 2007) and, from those, at least 300 million are obese (WHO 2003a). While micronutrient deficiencies remain widespread, the named chronic diseases are increasing globally and caused 60% of all deaths in 2005 (WHO 2007). About 80% of the deaths from chronic diseases occur in low- and middle-income countries and, as overweight and obesity are no longer a problem in developed countries alone. Therefore, more attention on this 'double burden' of malnutrition, namely under- and overnutrition occurring in the same country, region and even household, is required (WHO 2003c).

Many developing countries are nowadays simultaneously experiencing a demographic, an epidemiological and a 'nutrition transition'. The latter is said to be mainly caused by rapid socio-economic development, globalisation of food markets, urbanisation and adoption of so-called western diets and lifestyles (Popkin 2007). Thereby, an early stage of the 'nutrition transition' is characterised by an increased consumption of cheap vegetable oils that are rapidly integrated as additional food items into local diets and, which is the typical trend in developing countries. At a later stage, the 'nutrition transition', as it usually appears in wealthy countries, is marked by an increased use of meat, milk, processed food and soft drinks as well as by a rising food consumption away from home (Drewnowski 2000).

The BMI should be used only as an estimation for overweight and obesity as it does not necessarily correlate with body fat (WHO 2008). Furthermore, the optimal BMI associated with the highest life expectancy is both age- and sex-related (WHO 2000; Visscher et al. 2001). The National Research Council of the USA suggested already in 1991 an optimal BMI between 24 and 29 kg/m^2 for people aged 65 years and above. Also, the suggested cut-off points by WHO are not necessarily applicable for all population groups. In several Asian countries, for example, a BMI of 23 kg/m^2 is the threshold value for overweight (WHO 2004; Caballero 2007). On the other hand, it is suggested that a BMI of 23 kg/m^2 is the threshold to provide benefits to adults in developing countries (WHO 1998).

4.2.1   Body Mass Index (BMI) values of the study population

In the present study, the mean BMI was with 22.7 kg/m^2, median 21.7 kg/m^2, well within the range of normal weight. Still, 7% of all participants were underweight, 16% overweight and 6% obese, meaning that nearly as many women were obese as underweight (Figure 3.4.6). Taking overweight and obesity together, the share of participants with a BMI above 25 kg/m^2 was even three times higher than that for participants with a BMI below 18.5 kg/m^2. This means that, in the rural areas of Tanzania researched, more women were overweight or obese than underweight, which will be discussed in more detail below.

Absolute BMI values were marginal significantly different between the three seasons, yet, not between the three districts (Table 3.4.7). On the one hand, it was expected that the BMI as a medium term measure does not vary significantly within one year. However, in Senegal it was found that season was a major determinant of the anthropometric status of rural women, whereby body weight was reduced from the onset of agricultural labour due to a negative energy balance (Simondon et al. 2007). A moderate but significant trend was also found for body weight changes of women in rural Ethiopia (Ferro-Luzzi et al. 1990). Still, these changes were all in terms of body weight, which, of course, varies to a greater extent than the BMI itself. As in the present study the median BMI was in fact the same for Jun/Jul (DS) and Mar/Apr (LR) (21.7 kg/m^2), while it was only slightly higher during Nov/Dec (SR) (21.9 kg/m^2), seasonal differences can be described as minor.

In general, data on overweight and obesity in Sub-Sahara Africa is scarce. In Tanzania, most studies that deal with this topic focus on obesity in urban areas or rural and urban areas in comparison. For example, Hoffmeister et al. (2005) found in Moshi, a town in the Kilimanjaro region, that 70% of 50 surveyed diabetes patients were overweight. In Morogoro town in central Tanzania, 100 adults (19 to 50 years) and 40 pupils (14-18 years) from four educational institutions were examined and a prevalence of overweight and obesity of 25% was found. Thereby, prevalence of obesity increased with increased age and employed subjects had higher rates than pupils or students (Nyaruhucha 2003). A cross-sectional epidemiological study with 545 men and women aged between 46 and 58 found that the prevalence of overweight and obesity were significantly higher in urban (Dar es Salaam) than in rural (Handeni and Monduli) areas in both men and women (Njelkela et al. 2002). In a cross-country study with the focus on urban Africa, recent analysis of national data on BMI from women showed that the prevalence of BMI ≥25 kg/m^2 exceeded that of BMI <18.5 kg/m^2 in 17 of 19 countries (Mendez et al. 2005). In the present study exactly this situation of the number of overweight women exceeding those of underweight was found. Whether a similar trend exists even for obesity for Africa is not known, yet, in Tanzania a large and rapid increase in the prevalence of obesity, from 3.6% to 9.1% during a 10-year period, in women attending antenatal clinics in Dar es Salaam was observed (Villamor et al. 2006).

The ratio of overweight/obese to underweight study participants of 3.1 (Table A41) is similar to the ratio of 3.3, which was surveyed in 1996 in Tanzania for a group of urban women aged 20-49 years, while the ratio for rural women was only 1.2 (Mendez et al. 2005). The small sample size within the present study may be prone to give a biased picture, yet, the overall trend of higher prevalences of overweight than underweight in women in most developing countries (Mendez et al. 2005) can be confirmed.

Reasons for this trend are expected to be not only nutritional habits but also the level of wealth and education, physical activity as well as attitudes and culture/ethnicity. However, within the present study the BMI was not significantly correlated to any of the socio-economic parameters. Yet, in another study examining women in Dar es Salaam, the largest city in Tanzania, obesity appeared to be a particular serious problem in women of the higher socio-economic strata as it was positively related to the level of education, the amount of money spent on food, and the possibility of earning extra income outside the household (Villamor et al. 2006). These findings were consistent with an across-country analysis that documented how a low socio-economic status is "protective" against obesity in low-income countries, while it is, at the same time, a risk factor in upper-middle income developing countries (Monteiro et al. 2004). While in developed countries, a high BMI is also associated with low education, income and social status (Mensink 2005; Popkin 2007; MRI 2008), it is questionable whether education and nutritional knowledge alone can have an influence on BMI. Culture, attitudes and the stage of development of a country also seem to play important roles.

4.2.2 Attitudes towards overweight

It was not clear whether, among study participants, a corpulent body was associated with wealth, health and also beauty as it is suggested by different studies from Africa. In Mauritania, for example, it was a custom for a long time to fatten girls with rich camel or cow milk so that they became obese very fast and could represent their family's wealth and the ideal of the Mauritanian woman and feminine beauty (Resnikoff 1980). In Morocco, to gain weight, women even used appetite enhancers and steroid hormones during a special fattening period accompanied by overeating and reduction of physical activity (Rguibi and Belahsen 2006). Also in Tanzania, it is suggested that especially overweight and obese women are perceived as beautiful, they are admired and respected, while skinniness is still associated with illnesses, especially HIV/Aids and, therefore, not desirable (Maletnlema 2002).

According to this literature it was expected that more positive than negative characteristics would be named by study participants. However, 71% of participants named more negative than positive characteristics for a corpulent person, only 5% named more positive than negative, while 24% were undetermined (Figure 3.4.12). Consequently, only one third of study participants mentioned the common link of overweight and beauty and even less the also often mentioned association of overweight and good health. Circumstances under which interviews were performed must be regarded carefully as participants may have given answers, that they expected the interviewer wanted to hear as may happen in any interview. Nevertheless, as participants mentioned a number of "technical terms", they must previously have heard about these problems and were, apparently, already sensitised for this topic.

In fact, the most often named negative characteristics were „person has/can get high blood pressure, heart disease, stroke" (24%) and „person can not walk, run, climb, sit and is not fit" (22%) (Figure 3.4.10). The most often named positive characteristics were „person is attractive, beautiful, looks good" (34%) and „person has good health, no disease, much blood" (20%) (Figure 3.4.10). Assuming that all women told what they thought, the question arises why the majority of these study participants from rural Tanzania did not think any longer in the traditional way, yet, were already sensitised to obesity as a health problem.

Though the BMI itself was not significantly correlated to women's attitudes towards overweight and obesity, the number of positive characteristics for a corpulent person itself was highly correlated with some other variables (Table 3.4.10). Interpretation, however, is difficult: it can be argued that vegetable collection, which was positively correlated to the number of positive characteristics for a corpulent person, is a traditional activity and to think positively about obesity is also rather traditional. At the same time a positive attitude towards overweight was, according to the data, associated with living relatively close to town where women will be somehow exposed to new lifestyles, new food habits (WHO 2003a) and, probably, also new ideas of a different body image.

The number of negative characteristics was positively correlated to the consumption of pulses. The latter is a traditional food and it is not clear why its consumption should go along with the awareness of negative effects of overweight and obesity. Furthermore, the number of negative characteristics was positively associated to the food pattern "animal products". However, especially meat consumption is still determined by wealth (Speedy 2003) and obesity appeared to be particularly a problem in women of the higher economic strata in Tanzania (Villamor et al. 2006) and is still rather appreciated as a sign of wealth (Maletnlema 2002). Thus, the association between a relatively high consumption of animal products and a negative attitude towards a corpulent person does not appear logical.

Regarding the correlations between women's attitude and ethnic group, it is suggested that women with different ethnic background have also different attitudes towards body image and overweight. Different body image perceptions were also identified among Australian school children from varying ethnic groups (O'Dea 2008). Similarly, a cross-cultural study with participants from Japan, India, Oman, Europe, USA and the Phillipines also found differences in the drive for thinness as well as eating attitudes among the different cultural groups (Kayano et al. 2008). These different perceptions by ethnic groups coincided with the fact that the attitudes also differed between the districts studied in Tanzania, which are inhabited by different ethnic groups. Consequently, even within one country and when investigating rural areas only, ethnical differences need to be considered in particular.

4.2.3   BMI in associations and correlations

<u>Associations of BMI with vegetable production variables</u>

Only the cultivation of exotic vegetables (number of types cultivated per women) was negatively associated with BMI, indicating that the cultivation of exotic vegetables was accompanied by a tendency towards lower BMI values. This association was not very strong and, in fact, it was rather puzzling as the cultivation of exotic vegetables is rather associated with knowledge and a certain degree of wealth, because seeds and sometimes further inputs have to be purchased (see chapter 4.1). Thus, it would be rather expected that a higher degree of exotic vegetable production goes along with a higher BMI. A relevant question is, in fact, whether exotic vegetables are mainly or even exclusively sold and do not directly impact on household food and nutrition security, but only contribute to the general wealth status of a family. While many studies suggest an association between wealth and BMI in developing countries (Mendez et al. 2005; Villamor et al. 2006), this could not be confirmed within the present study.

<u>Associations of BMI with nutrition variables</u>

BMI values were related to some of the food groups (intake per day) and to both DDS and FVS. This relationship showed that great dietary diversity and, especially, food variety were associated with higher BMI values (Figures 3.4.7 and 3.4.8). Of course, it must be distinguished if an increased BMI means a change from underweight to normal weight – which would be desirable – or if the change is from normal weight to overweight or even obesity. Thus, it is suggested that increasing food diversity not necessarily goes along with better health, here in terms of BMI.

In the present study, a low diversity was rather characterised by a simple but not necessarily unhealthy diet consisting mainly of cereals, vegetables and pulses (Figure 3.3.26). Not only in Tanzania, but in many Sub-Sahara African countries vegetables, legumes and fruits are culturally less desired and especially indigenous vegetables are seen as survival food for poor people (Renzaho 2004). When food diversity increased in this study, it was, e.g., sugar, beverages (tea), or animal products that were additionally consumed (see also chapter 4.5). Therefore, a high dietary diversity does not *per se* secure a healthy diet but the types of food must be considered as well as the amount consumed (see also chapter 4.5). The shift towards higher consumption of animal products such as meat and milk reflects, to a certain level, a desirable nutritional development. Yet, the benefits of higher dietary availability of protein and micronutrients, especially iron, decline rapidly as intake levels rise. For example, high intakes of red meat are associated with an increased risk of colon cancer, while increased intakes of saturated fat and cholesterol from meat, dairy products and eggs increase the risk of coronary heart diseases (Schmidhuber and Shetty 2004).

Still, in terms of BMI it can be questioned what "good health" really means and if we can talk about "bad health" considering a BMI 25 kg/m² and upwards. In fact, in the USA the link between weight and mortality has recently been assessed, and it was observed that underweight, obesity and especially extreme obesity (BMI of 35 kg/m² and above) was associated with an increased mortality – while overweight (BMI 25 to <30 kg/m²) was not (Flegal et al. 2005). Thus, while obesity is a clear health risk this is not necessarily true for overweight depending on the age, sex and origin of a person. Especially in countries or regions with recurrent food shortages and high disease pressure, it should be considered that people with slight overweight may be healthy and have higher chances of longevity than lean people.

When considering the consumption of different food groups (g/d) (Table 3.3.1) it was found that only some groups were positively correlated to BMI. The food group bread/cakes comprising either bought or home-made different types of cakes and breads, usually fried or baked in oil, represents products, which contain a high amount of fat. Therefore, it was not surprising that participants who consumed a high amount of this food group also had a higher BMI. Black tea was mostly drunk with a high amount of sugar in it, and both food groups were consumed to a great extent by participant's with a high BMI. Regarding sugar, this again is not surprising as the amount of calories consumed was, most likely, often exceeding the needs, which leads to an imbalance of energy input and output and, in the long run, to the accumulation of weight. In a study among women in the USA, for example, it was found that higher consumption of sugar-sweetened beverages was associated with a higher degree of weight gain and, at the same time, with an increased risk for development of type 2 diabetes mellitus (Schulze et al. 2004). Thereby, not only the amount of sugar is decisive but also that the sugar in soft drinks, similarly to the sugar consumed by the Tanzanian participants with tea, is rapidly absorbable. In general, an increasing intake of soft drinks can be also observed in developing countries (Popkin 2007), which was, however, not confirmed within this study.

Furthermore, food patterns generated through PCA and cluster analysis (chapter 4.6) were related to BMI. Only one food pattern was significantly related to BMI values (ρ=0.192; p=0.005), namely the so called "purchase" pattern No. 2 (see chapter 4.6). Overweight women had overall higher factor scores for pattern 2 and, especially, the mean value for obese women was highest, meaning that they followed the "purchase" pattern to a great extent (Figure 4.2.1; for scatterplot with continuous BMI values and pattern 2 see chapter 4.6). As this pattern was characterised by the food groups bread/cakes (usually baked or fried in oil), sugar and tea, the same associations occurred as when single food groups were analysed. The association of BMI values to dietary patterns is more meaningful as the latter are medium-term values compared to the short-term intake data of different food groups.

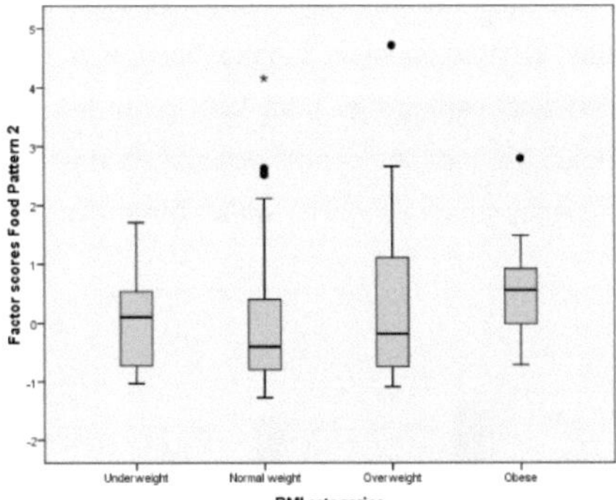

Figure 4.2.1
Association between four body mass index (BMI)
categories and factor scores of food pattern 2
"purchase"; p=0.005; n=210 women from three
districts in Tanzania, interviewed during three different
seasons

<u>Associations of BMI with further health variables</u>

With rising or in general high BMI values, as detected within this study, a premature conclusion could be drawn on diets moving away from problems of insufficiency towards problems of excess (Eckhardt et al. 2008). Nevertheless, the energy that increasingly is available usually comes from energy-dense sources, which are, in addition, micronutrient-poor such as sugars and edible oils; simultaneously, intakes and availability of micronutrient-rich foods such as vegetables, fruits and high-quality animal products often remain low (Drewnowski and Popkin 1997; McIntyre et al. 2001; Murphy and Allen 2003). This situation gives rise to the so-called double burden of nutrition in developing countries, where micronutrient malnutrition remains highly prevalent while, at the same time, overweight, obesity and related chronic diseases appear as new serious health risks. In fact, it was found that iron needs of overweight women in developing countries are not necessarily being met, thus, diet quality remains an important issue even among women with sufficient energy intakes (Eckhardt et al. 2008).

Therefore, BMI values of study participants were related to haemoglobin (Hb) values (n=175) as well as to retinol-binding protein (RBP) values (n=119; data only from Mar/Apr (LR)) in order to check for existing associations. However, neither bivariate correlations between BMI and the other two health variables, nor differences or a trend between BMI categories in terms of Hb or RBP values could be detected. Still, when iron status of women was divided into two categories only, namely normal iron status and iron deficiency anaemia (IDA), the share of obese and overweight women having IDA was larger than that of underweight women (Figure 4.2.2).

In other studies, overweight women were found to have either lower odds of anaemia (determined also by Hb values) than non-overweight women but, at the same time, suffer from anaemia at high rates (Egypt and Peru), or overweight women were equally as likely to suffer from anaemia as non-overweight women (Mexico). In these studies, sample sizes were 6,841 in Egypt, 5,078 in Peru and even 11,965 in Mexico with participants from both rural and urban areas (Eckhardt et al. 2008). Due to the low sample size and the unequal distribution of participants between normal iron status (75%) and iron deficiency anaemia (25%) within the present study, no significant results could be obtained.

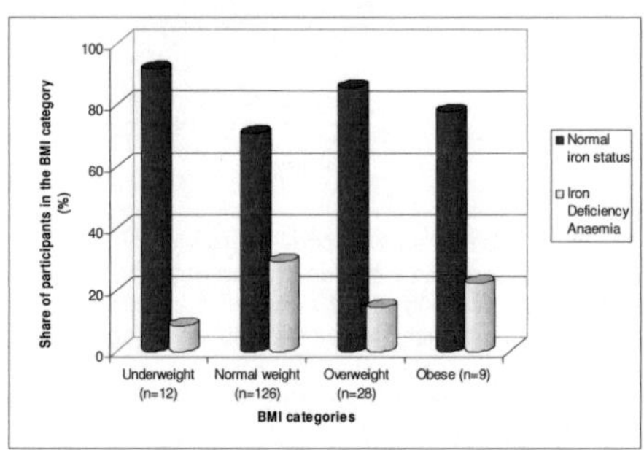

Figure 4.2.2
Share of participants within four body mass index (BMI) categories and their iron status according to Hb values; n=175 women from three districts in Tanzania, interviewed during three different seasons

<u>Multiple relations</u>

For a multiple regression model the BMI was chosen as the dependent variable and age, education, dietary pattern No.2 "purchase" (see chapter 4.6), and FVS were considered as predictors. The food groups (amount consumed per day), though significantly correlated to BMI were not considered as they were regarded as short-term measures and, therefore, less adequate to explain the BMI as a medium-term measure. In contrary, food scores and dietary patterns were considered better suitable as food intake variables to possibly influence the BMI.

The residuals of the present model were not normally distributed (K-S-test 0.119; S-W-test 0.941) and, therefore, the BMI values were converted with a log transformation. As a result, the residuals of the regression model with ln (BMI) were normally distributed (according to K-S-test 0.095; S-W-test 0.971; and Q-Q-plots, not shown). Results of the multiple regression model with lnBMI are shown in Table 4.2. As the BMI is ln-transformed, the B values need to be re-transformed with the e-function (last column). R-Square for the multiple regression model was 0.086, the adjusted R-Square was 0.068 and the significance of the whole model 0.001, meaning that the chosen predictor variables explain 8.6% or 6.8%, respectively, of the variation in the response variable BMI.

Table 4.2 Results of multiple regression analysis with ln(BMI) as dependent variable and four predictor variables; n=210 women from three districts of Tanzania, interviewed during three different seasons

	Unstandardized Coefficients		Standardized Coefficients			
	B	Std. Error	Beta	t	Sig.	e^B
(Constant)	3.090	0.098		31.651	<0.001	
Age	0.002	0.002	0.106	1.581	0.116	1.002
Educ	-0.070	0.035	-0.136	-2.013	0.045	0.932
Pattern2	0.023	0.013	0.143	1.776	0.077	1.023
FVS	0.009	0.006	0.126	1.556	0.121	1.009

When education – the only variable, which was significantly related to BMI – will increase by one unit, e.g. from no education to completed primary school, the mean BMI will decrease by 6.8%. As discussed before, education is usually positively associated with the BMI in developing countries. However, in the present study, women with higher education had a lower BMI than those with lower or even no education. In general, it is assumed that education will lead to normal weight independent from the initial situation of being either over- or underweight, and, in fact, all women with a high education within the present study had a normal weight.

A problem with the present data was, however, the poor distribution of the 252 participants between the categories of low (8%), medium (89%) and high (3%) education. Most likely, other factors than education influenced the BMI more strongly in this setting. In general, the results of this multiple regression analysis were not very convincing with only education being significant. While all predictor variables showed significant associations with BMI on a bivariate level, in combination with other variables their effect is obviously less strong and, e.g., consumption according to dietary pattern No. 2 does not influence the BMI when it was controlled for age, education and FVS.

For a second regression model, the residence of participants (districts) as a nominal variable were transformed into dummy variables to include them in the model and control for them. As education is an ordinary variable but not metric, it was decided to also develop dummy variables for it and use those for the multiple regression model (Table A42). The significance of this regression model was p=0.005 whereas R-Square was 0.096 and the adjusted R-Square 0.064, meaning that the chosen predictor variables explained 9.6% or 6.4%, respectively, of the variation in the response variable BMI.

Within this model, only the FVS was significantly correlated to BMI (Figure 4.2.3). If the FVS, e.g., increased by five, the mean BMI increased by 7%. In fact, it was found that the 'nutrition transition' is not only associated with an increase in obesity but also with greater dietary variety and, in addition, with an improved access to cheap energy-dense foods (Drewnowski 2000) (see also chapter 4.6). Consequently, a diverse diet characterised by a high FVS does not simultaneously translate into a healthy and balanced diet. In general, it must be stressed that while obesity and related chronic diseases become more and more serious health problems in developing countries, at the same time, the prevalence of micronutrient malnutrition is likely to remain high (Eckardt et al. 2008).

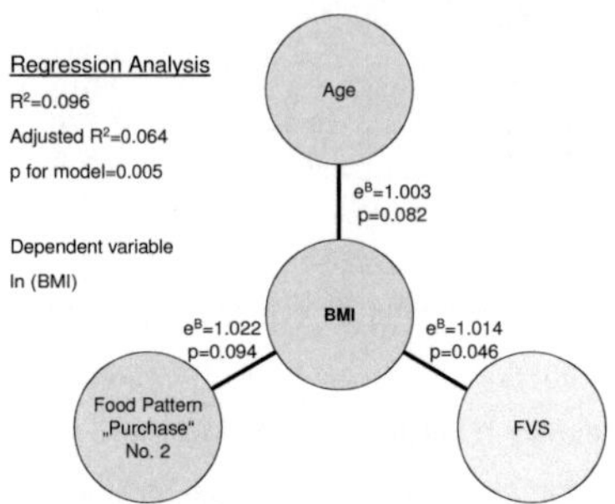

Figure 4.2.3 Association of different variables with the BMI of women (n=210) when controlled for education and district; data from three districts in Tanzania surveyed during three different seasons

From these two multiple regression models it appears that depending on the choice and combination of predictor variables, associations and dependencies can be different. In fact, the purpose of multiple regression analysis is to identify, which combination of variables best predicts the response variable (Lang and Secic 2006). For the present case, R^2, the coefficient of multiple determination, is slightly larger for the second model; further, the second model controls for more variables, results are better interpretable and, therefore, this model and its results are preferred over the first one.

4.2.4   Conclusions

With the present study, it has been shown that a certain dietary pattern and food variety had an influence on the BMI of women (bivariate correlations). At the same time, overweight and obesity were more prevalent than underweight within the study population in rural Tanzania. From the three main reasons for the obesity epidemic in Tanzania or, in general, in developing countries, namely changing food habits, cultural attitude towards overweight/obesity and decreasing physical activity, only the first could be confirmed in the present study. The fact that participants had more negative than positive associations with overweight and obesity indicated some awareness of the problem. Still, all three areas may be considered as starting points for public health interventions aiming at prevention.

While it can be argued that high BMI values of people in urban areas of Tanzania are, most likely, due to changing lifestyles, food habits and physical activity patterns, it still needs to be determined how a similar trend is possible in rural Tanzania. Without the influence of Western fast food chains, however, with an increased consumption of cheap vegetable oil, e.g. through the dietary pattern No. 2 "purchase", an early stage of the 'nutrition transition' has obviously started even in rural areas (see also chapter 4.6). At the same time, wheat and rice replace the traditional starch sources millet and cassava. While, of course, the consumption of wheat and rice *per se* does not lead to weight increase, this change implicates new processing techniques. Thereby, cooking in water is replaced by frying in oil, and products are often processed further with any kind of fat (Raschke and Cheema 2007). Similarly, in China a vegetable-rich diet was unexpectedly found to be associated with obesity. The explanation for this was that the vegetables were stir-fried in oil (Shi et al. 2008).

While the consumption of large amount of sugar was widespread, usually in tea, which sometimes even substituted a whole breakfast, the consumption of carbonated soft drinks was not pronounced among study participants. However, highly sugary carbonated soft drinks are becoming available even in the smallest corner shop in remote villages nowadays. Soon, sugar-sweetened beverages and, particularly, carbonated soft drinks may become a key contributor to the epidemic of overweight and obesity as elsewhere (Malik et al. 2006; Popkin 2007). While most studies on sugary drinks were carried out in developed countries sufficient evidence exists already for this fact, which could be translated to developing countries and consequently induce public health strategies to discourage consumption of sugary drinks as part of a healthy lifestyle (Malik et al. 2006). Furthermore, it is of concern that only few multinational corporations govern the globalised food system and are responsible for an eradication of quality whole foods and for the widespread dissemination of low-quality processed foods (Raschke and Cheema 2007). If the latter are easier available, have a certain image, and may be even cheaper than high-quality foods, it is questionable if nutritional knowledge alone can achieve that consumers follow a balanced and healthy diet.

Is now obesity only a temporary problem of a few rich people as it is argued by some studies, or will this health risk expand to less wealthy members of the society and to rural areas like it was found already in the present study? In all countries with an annual GDP of more than 2500 US$ per capita, the proportion of poor women with obesity was found to be higher than that of women in higher socio-economic levels (Popkin 2007). Countries in Sub-Sahara Africa with an explicitly lower GDP, consequently, should be excluded from this trend. However, there is no doubt that the obesity epidemic is on the rise in developing countries and even in rural, poor and underdeveloped regions, fuelled primarily by increased consumption of cheap oil, sugary drinks, animal products and a decrease in physical activity (Popkin 2007). Furthermore, based on the outcomes of the present study, it needs to be understood that increased production and consumption of vegetables, TV or EV, does not *per se* reduce the risk of overweight and obesity in Tanzania. It obviously needs to be accompanied by nutrition and health information and education already at school level.

This is even more needed as many people in Tanzania are not aware of overweight and obesity as a health hazard and can not assess their own weight (Maletnlema 2002). In fact, even health staff in the researched districts stated, for instance, in Kongwa that they did not look at obesity as a health problem, that fat people were regarded as healthy, that obesity appeared only in those who were employed ("paper-pusher"), and that it was in general not a problem in villages because people worked hard and food was 'simple' (Fatuma Mganga, District Medical Officer, personal communication, 2007). Also in Singida, the health personnel argued that overnutrition was negligible in villages, that it was not a problem in Singida but only in big towns (Steven Msambu, Malaria Coordinator, personal communication, 2007).

Therefore, next to nutrition education and counselling about the influence of food intake and physical activity on body weight, creating awareness for the problems associated with obesity is essential. While the attitude that obesity stands for beauty and health should be further changed, it is also important that this trend is not just reversed favouring a BMI below 18.5 kg/m^2 and, thereby, creating new public health problems in Tanzania through behavioural eating disturbance.

An integrated approach is especially important where so-called dual burden households exist in which both underweight and overweight persons live next to each other. Here, the health risks of all persons need to be considered before any dietary recommendations are made that involve drastic changes to the household's diet. For example, energy restrictions need to be individualised and dieting parents must be aware of the energy density needs of their children. Therefore, general lifestyle recommendations must contribute to the optimal weight and good health of all persons in the household and should focus on increasing fruit and vegetable intakes – processed in a healthy way – as well as physical activity (Doak et al. 2005).

FOCUS B: Obesity in developing countries

In developing countries, especially in urban areas, profound changes in the society and in behavioural patterns of communities took place during the last decades influencing, besides others, nutritional patterns. The genes of each human being can be decisive for his or her susceptibility to weight gain, yet, the actual energy balance of a body is mainly determined by calorie intake and physical activity. Thus, the so-called obesity epidemic is mainly influenced by socio-economic and cultural changes accompanied by the 'nutrition transition'. These changes, in turn, are stimulated by globalisation of food markets, urbanisation and economic growth (WHO 2003a).

In developing as well as developed countries, the population urbanises with increased income and wealth. Simultaneously, traditional foods usually rich in carbohydrates are replaced by a more diverse diet characterised by a high proportion of fat, especially saturated fats, and sugar. In addition, lifestyles and occupations are more and more sedentary, which is supported by automatic transport and (processing) technologies within households, leading to a further decrease in physical activity.

A special challenge for developing countries is the association between foetal/infantile malnutrition and adult obesity: persons who face prenatal or infantile malnourishment and/or a low birth weight and who become obese as adults, especially tend to develop chronic diseases associated with obesity such as high blood pressure, heart and vascular diseases or diabetes mellitus; these are usually developed earlier and more heavy than in persons who never faced malnutrition (Ravelli et al. 1976; Barker 1992).

Next to this association, the coexistence of undernutrition in terms of micronutrients and overnutrition in terms of calories leads to a double burden of malnutrition not only on a population level (Popkin 1994) but also on a household level (Popkin 2000; Doak et al. 2005). For many developing countries, obesity and sequelae demonstrates a similar challenge as hunger and undernutrition (SCN 2006; FAO 2006). A further alarming factor is that obese people in Africa, South America and South Asia tend to develop high blood pressure or diabetes sooner than obese people from Europe. A reason for this is the genetically determined feature to store fat more efficiently, usually to better survive hunger periods (Popkin 2007).

4.3   Linking iron status, food consumption and nutritional knowledge: "dietary diversity versus health issues"

Iron deficiency anaemia (IDA) in developing countries is clearly more prevalent in pregnant (56%) and non-pregnant women (43%) than in men (35%) (ACC/SCN 2000). In Sub-Sahara Africa, the prevalence of IDA in pregnant women (Hb < 11 g/dL) even rose from 47.9% in 1990 to 48.2% in 2000 (Mason et al. 2005). In Tanzania, the prevalence of IDA in pregnant women in 2000 was estimated to be 49.5% and 44.7% for non-pregnant women (Mason et al. 2005). IDA is defined as iron deficiency and, concurrently, a low haemoglobin (Hb) level (Zimmermann 2008). Further, it is not only determined by dietary but also health issues, such as malaria or parasite infections. Consequently, to link the iron status of women within this study to their food consumption and nutritional knowledge characteristics was only one side of the story. It became apparent that health issues need to be considered as well when, e.g., looking at dietary diversity. Still, food-based strategies for improving micronutrient malnutrition are acknowledged to be the best long-term and sustainable strategy (Tontisirin et al. 2002) and were, therefore, of special interest within the context of this study.

4.3.1   Iron status of the study population

In contrary to the BMI of participants, the Hb was clearly and significantly different both between the three districts and the three seasons (Table 3.4.2). This was more or less expected as BMI is a medium or long-term measure, while Hb is rather a short-term measure and is subject to fluctuations in short time. This was already demonstrated in earlier studies where Hb was used to reveal seasonal variations of iron status (Futatsuka et al. 1985; Lee et al. 1987). The mean Hb level for the overall year and all districts was, with 12.7 g/dl (±1.5 g/dl), above the cut-off of 12 g/dl. Yet, it was below this cut-off in Muhezau district (11.9 g/dl) with a share of 48% of women having an Hb below 12 g/dl. This share was still lower in Kongwa and especially in Singida, suggesting that the iron status was differently influenced depending on the district (Figure 3.4.3).

In terms of seasons, Hb values were clearly higher during Jun/Jul (DS) than during Nov/Dec (SR) and Mar/Apr (LR). Besides food intake and vegetable production during the respective season, also malaria might have influences on the seasonal changes in Hb depending especially on the rainfall pattern. In fact, next to iron deficiency in general, malaria, hookworms, and other infections were identified as the major causes of anaemia in Tanzania (Massawe et al. 1999). Also in Africa in general, malaria was frequently reported as a major cause of anaemia (Garner and Gulmezoglu 2000), and it was found to be a major contributor to anaemia in Tanzanian women with Hb below 9 g/dl (Hinderaker et al. 2002).

In all three districts in Tanzania researched for this study, malaria was named to be the number one disease among the top 10 diagnoses. In fact, in whole Tanzania the rate of deaths among children under five years of age due to malaria (23%) was much higher than due to diarrhoea (17%) in 2000-2003 (WHO 2006). In Kongwa, it was stated that the rate of malaria cases even doubles during the rainy season. However, in Muheza district malaria as well as anaemia cases were highest (Figures A24-A26). Next to campaigns such as the Tanzanian National Voucher System (TNVS) (Table A 53), it is suggested that education of both school children, as in Kongwa and Singida districts, and parents/caregivers can contribute considerably and sustainably to reduce malaria pressure.

Transferrin receptor (TfR) values, measured only during Mar/Apr (LR), were significantly associated to Hb values during this season and also showed significant differences between the three districts (Table 3.4.3). The share of participants classified as having iron deficiency by this parameter was again highest with 48% in Muheza district, followed by Kongwa with 24% and Singida with only 20% (Figure 3.4.4). These values were similar to the overall country data (IDA prevalence of 44.7% for non-pregnant women) only for Muheza, while in the other two districts IDA prevalence was much below this rate. Yet, as the study population was not representative for the districts, no conclusion can be drawn from this comparison.

4.3.2   Knowledge about iron in nutrition

The largest share of participants giving the correct answer regarding the function of iron in the human body were those from Muheza district (Figure 3.4.18). This was striking as in Muheza district the mean Hb was lowest and the share of women with iron deficiency was highest. Regarding foods that contain iron, more women from Singida district gave correct answers than from Muheza or Kongwa (Figure 3.4.19). Yet, less than one third of women had some knowledge about iron-rich foods, which most likely affected the iron status of the study population in general.

Significant correlations between the knowledge on iron and further variables were only found between the own statement of participants if they knew the function of iron in nutrition and the consumption of nuts/pulses (Figure 3.4.22). Thereby, nuts were consumed by nearly all participants who stated to know the function, while from those who did not know the function 16% did not consume any nuts. This can lead to the assumption that those women knowledgeable about the function of iron were, simultaneously, eating more iron-rich foods, though nuts are not the best source of iron. Yet, the knowledge about iron-rich foods was not correlated to other variables. While an improvement in the educational status is one prerequisite for assuring nutrition security, it is also known that knowledge alone does not necessarily lead to a healthy lifestyle but that other factors are important to ensure a balanced nutrition. For example, it could be shown in Tanzania that the highest prevalence for nutrition-related diseases, such as diabetes and hypertension, was found among high-ranking executives (Maletnlema 2002).

4.3.3   Iron in associations and correlations

<u>Associations of iron status with socio-economic variables</u>
While Hb was not associated to any socio-economic variable, TfR and wealth were negatively correlated, meaning that the higher the wealth status the more likely was it for a participant not to suffer from iron deficiency. In Timor-Leste, the opposite was found in children under five years of age, namely that those from the poorest households had higher Hb levels than those from the richest and middle-class households (Aqho et al. 2008). However, it is conceivable that women in the present study with a higher wealth status were more able to protect themselves against malaria infections (e.g. with insecticide-treated nets) and to consume more animal products, which are usually more expensive but, simultaneously, provide the better absorbable heme iron.

<u>Associations of iron status with vegetable production variables</u>
Associations between Hb and vegetable production (Table 3.4.4) were significant but negative during Jun/Jul (DS) and Nov/Dec (SR) meaning that a high Hb was related to low vegetable production in terms of number of vegetable types cultivated/collected. In Mar/Apr (LR), however, the association between Hb and vegetable production was positive, i.e., women with a high Hb were, at the same time, cultivating/collecting more different vegetable types. While certainly a great number of additional factors stands between vegetable production and iron status of a person, still in a context with a high degree of subsistence horticulture, a direct link between these two areas is possible. Additionally, dark green leafy vegetables are a major source of dietary iron in a setting where foods from animal sources are expensive and not accessible for everybody. However, only diversity of vegetables produced and not, for instance, cropping area or yield were surveyed; moreover, the diversity was not clearly associated to Hb but varied between seasons. Thus, the causes for seasonal variability of Hb levels must be found elsewhere than with the data on produced vegetable diversity. While it is acknowledged that micronutrient malnutrition can be combated with dietary diversification and this, in turn, requires an adequate supply and access (Tontisirin et al. 2002), the focus on vegetable diversity alone seems to be not broad enough but overall food diversity in production needs to be taken into account.

<u>Associations of iron status with nutrition variables</u>
No significant correlation was found between Hb values and iron intake. This is not surprising as Hb has low specificity in determining IDA as there are other causes of anaemia than iron deficiency alone (Zimmermann 2008). Yet, the Hb level of participants was significantly correlated to vitamin C intake (from all foods) for the whole year, during Jun/Jul (DS) and during Mar/Apr (LR), however, always negatively (chapter 3.4.2). This finding was rather strange as vitamin C is known to increase iron absorption (Bender 2002). Again, an increased iron absorption and in general a high iron intake is not an exclusive factor influencing the Hb level of women (Zimmermann 2008).

In general, iron intake as measured through 24-h recalls must be interpreted with caution. Foods surveyed with the 24-h recall are only a surrogate for the variable of interest, in this case dietary iron, and the analysis of actual iron intake relies on the accuracy and completeness of food composition databases (Patterson and Pietinen 2004), which was rather poor in the present case for Tanzanian foods. In fact, iron content of foods and especially vegetables can vary to a great extent and the iron content of leaves is affected by stage of maturity, conditions of growth, fertilisers used and nature of soil, among others (Schmidt 1971; Singh et al. 2001). Therefore, it is again suggested to refrain from analysing single nutrient intake in more detail but rather concentrate on whole foods, food groups, dietary diversity (chapter 4.5) and dietary patterns (chapter 4.6).

The intake of several food groups was also negatively correlated to Hb with the only exception of animal products (Table 3.4.5). The Hb level was high when also the intake of animal products was high; this might be due to foods of animal origin containing about 40% heme iron (Strain and Cashman 2002), which is more readily absorbed than non-heme iron from plant sources. The consumption of black tea was highly negatively correlated to Hb level, however, also those of bread/cakes and sugar as well as starchy plants, nuts, fish and fruits. This coincides with the finding of correlations between food patterns and Hb, which will be discussed below.

Both DDS and FVS were highly negatively correlated to Hb during all seasons (Table 3.4.5; Figure 3.4.5), meaning that the more diverse food consumption was, the lower was the Hb of the particular person. Similarly, according to TfR values iron deficiency were associated to a high DDS (Table 3.4.6). That a higher dietary diversity does not necessarily result in a better health status will be discussed in detail in chapter 4.5. The VDS was, similar to DDS and FVS, significantly and negatively associated to the Hb level of women during Mar/Apr (LR). Consequently, an increased variety of vegetables in consumption during that time was associated to a low Hb, which is, however, difficult to interpret.

Hb values were also significantly correlated to three of the five dietary patterns, which were determined by PCA (Figures 4.6.8 – 4.6.10 in chapter 4.6). Similarly, also TfR values were positively correlated to food pattern 1 (rho=0.227; p=0.009), supporting the relationship of a poor iron status with a high consumption of this food pattern. The more patterns 1 and 2 were followed, the lower were Hb values of participants. Consequently, these seem to be no favourable dietary patterns in order to maintain a normal iron status, or they might even have a negative impact on the Hb level and, therefore, the iron status. Pattern 1 was characterised by a high consumption of the food groups fruits, nuts, starchy plants and fish, while pattern 2 was dominated by bread/cakes (usually fried in oil), sugar and black tea. The negative association to the latter pattern can be, thus, due to a high consumption of black tea, which can inhibit the bioavailability of iron in the food and, consequently, the intake of iron into the body (King and Burgess 1992). The negative association to pattern 1 is more difficult to explain. As both patterns are highly favoured by participants living in coastal Muheza where, in addition, the prevalence of malaria is high, it is suggested that the iron status of women is probably more influenced by the infection than by nutrition. The positive relationship between Hb values and food consumption according to pattern 5 dominated by animal products, can be traced back to the fact that iron can especially easily be absorbed from foods of animal origin (Strain and Cashman 2002). The Hb value for participants from different dietary pattern quintiles is discussed in detail within chapter 4.6.

<u>Association of iron status with further health variables</u>
As there might be interactions between vitamin A and iron and as, for instance, low vitamin A and low iron stores were found to be associated (Hinderaker et al. 2002), Hb values were also checked for a relationship with retinol binding protein (RBP) values (see chapter 4.4). Yet, no association between the two variables could be detected for the given study population. The relationship between TfR and RBP was also not significant; further, there was no significant association between Hb and BMI within the study population.

<u>Multiple relations</u>
For regression analysis with Hb as the depend variable two different approaches were applied: first, the Hb of each season was used and respective seasonal production and consumption variables as predictors. Thus, differences between the seasons were to be detected. Second, the mean Hb across the three seasons was tested in different multiple regression models with, i.a., data on nutritional knowledge of participants and affiliation to food patterns.

Few significant associations existed for regression models for each season (Tables A43 and A44; Figure 4.3.1). When each model was controlled for district and age, no significant associations could be shown for Hb with consumption or production parameters. It was also tested to use the change in Hb from one to the next season as the dependent variable, however, these regression models were not significant.

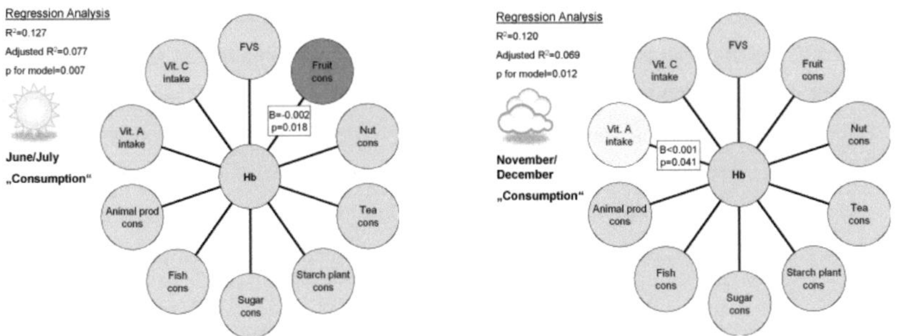

Figure 4.3.1   Results of multiple regression analysis "consumption model" during two differentseasons with Hb as dependent variable

When the mean Hb across three seasons was tested in different multiple regression models it was always controlled for further socio-economic variables. In the following, only those results from regression models are shown which demonstrated useful and significant results. High collinearity was found between DDS and FVS and, therefore, only DDS being more closely related to Hb was taken into account. The dummy variables for knowledge about the function of iron combined those participants who were indifferent in their answers ("1"), who named the correct function ("2"), or who named incorrect functions ("3"), while the fourth group of participants who gave no answer was excluded.

With a "consumption model" (Figure 4.3.2; Table A45), which was highly significant, 28% or adjusted 23% of variance could be explained. Besides the districts which were controlled for, only food pattern 5 ("animal products") and the knowledge about the function of iron "3" were significantly associated to Hb. Consequently, when the factor score, i.e. affiliation to food pattern 5, increased by one, the mean Hb of participants increased by 0.2. As discussed before, this is an expected association because iron can very easily be absorbed from foods of animal origin (Strain and Cashman 2002). Those participants within group 3 of knowledge about the function of iron were those who gave incorrect answers and did not know which function iron has in the human body. As this parameter was negatively associated with Hb it implies that no or poor nutritional knowledge can be associated with a poor iron status. While it was shown above that good education is not a guarantee for consuming a healthy and balanced diet, other studies could show that poor people, especially in developed countries, have less nutritional knowledge and are e.g. more prone to obesity (Mensink 2005; MRI 2008). While in developing countries the opposite is assumed, namely that poor people are less affected by obesity than wealthy individuals, this relationship can not be projected on micronutrient malnutrition, though both nutritional disorders could be changed to a certain extent with nutritional education. For the present study, the association between poor knowledge and poor iron status is seen as a hint that especially for the rural population in Tanzania nutritional knowledge is one important implement to independently choose the "right" foods and diets in a rapidly changing environment.

The predictor variables within the "production model" were, age, education, district, number of vegetables cultivated and number of vegetables collected. Additional, dummy variables for the results on the knowledge about the function of iron were included. The multiple regression model with the named predictors was highly significant (p<0.001) and explained about 27% (23% when adjusted) of variance.

Significant associations are highlighted in Figure 4.3.2 and Table A46: besides the districts Kongwa and Muheza, the number of vegetables collected was significantly but negatively related to Hb. This implies that, when the mean number of vegetables collected increases by one, the Hb value decreases by about 0.2; thus, women who collected more different vegetables had, on average, a poor iron status. The collection of vegetables was also negatively associated with wealth of participants during Mar/Apr (LR), meaning that those women with higher wealth status would collect vegetables from the wild only marginally. Though not on the basis of Hb but of TfR it could be shown that those women with low wealth, at the same time, had a poorer iron status during Mar/Apr (LR). While a close association between health and social factors is not a new finding (Gupta and Kumar 2007), it is suggested by means of the present data that the collection of wild vegetables serves as an indicator for poverty which, in turn, is associated to poor health, here in terms of iron deficiency. While the consumption of wild vegetables rich in micronutrients could imply good health, the collection of wild vegetables did not mean a high consumption of the same (chapter 4.1); further, there are multiple ways through which health is influenced by social and other factors (Gupta and Kumar 2007), so that vegetable cultivation and collection can not be viewed alone. While this relation is, i.a., controlled for districts, there is a significant difference between districts. This, in turn, suggests that the Hb value is influenced by further district-related values, e.g., malaria pressure, and it can be, therefore, not only traced back to the number and diversity of vegetables collected.

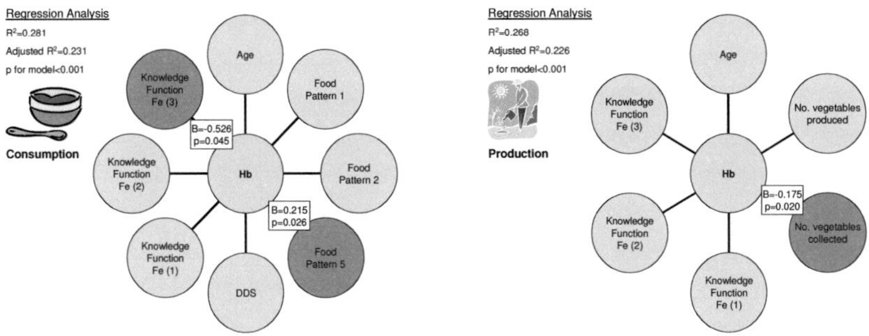

Figure 4.3.2   Results of multiple regression analysis "consumption model" and "production model" for mean Hb across three seasons as dependent variable

4.3.4   Conclusions

This chapter addressed the associations between iron status and dietary, health, as well as vegetable production characteristics of participants. The overall aim was to analyse the basis for improving micronutrient intake from natural foods and, consequently, to review the evidence that food-based strategies can meet the challenges of micronutrient malnutrition. In general, it was found that a high vegetable diversity in production and consumption alone does not guarantee good nutritional health in terms of iron status. Also the amount of vegetables that is actually consumed must be considered which was, however, not significantly associated to iron status for the study population. Further, the consumption of nutrients enhancing or inhibiting the iron intake and retention, different preparation methods through which nutrients can be lost or destroyed and bioavailability is affected, are of importance when analysing factors for iron deficiency. Finally, diseases influencing the iron status interfere with nutrition, and a major challenge in diagnosing anaemia is to distinguish between the anaemia of chronic disease and IDA in otherwise healthy subjects (Zimmermann 2008).

Main significant findings were that a poor iron status was associated with poor nutritional knowledge (Hb values) and with a low wealth status (TfR values). To measure TfR values of participants was a good addition next to the Hb value. The diagnostic value of TfR for IDA was found to be uncertain for participants from regions where malaria, inflammatory conditions and infections are endemic (Zimmermann 2008), which is the case for all three districts in Tanzania in terms of malaria. Also the Hb value as a measure for iron status need to be examined more carefully: to measure Hb is a widely used screening test for iron deficiency, yet, its specificity and sensitivity is low if used alone. For example, there are many other causes of anaemia other than iron deficiency, which must be taken into account as well as the fact that cut-off criteria can differ between ethnic groups (Zimmermann 2008). Still, during the surveyed season Mar/Apr (LR) Hb and TfR values were highly correlated and showed, thus, similar results which confirmed the findings regarding iron status of women.

As within the present study significant differences between districts within Tanzania were found, the location of districts and even villages can be an influencing factor on iron status. In the case of Singida villages are situated less close to town than in Kongwa and Muheza, yet, the town is much larger and more influential. A positive effect of this peri-urban setting with better access to education and health care can be assumed and, hence, contributing to a better health situation in general. On the other hand, the humid climate in Muheza district accompanied by a poor health system regarding malaria prevention will be responsible for the high rates of anaemia in this district.

To combat iron deficiency anaemia (IDA), a main challenge lies in the adverse interaction between malaria and iron administration. With the so-called Pemba trial an increase in serious adverse events among children receiving iron was shown; four suggestions are available as to possible means by which iron status might influence susceptibility to malaria (Prentice 2008). In general, iron lies at the center of the host-pathogen battle for nutrients, and it is suspected that extra iron might predispose to infection. Yet, even in areas with high malaria endemicity, iron appears beneficial in iron-deficient subjects and it is, therefore, only suggested to follow a cautious approach with either screening out iron-replete children or combining iron supplementation with effective disease control measures (Prentice 2008). *Vice versa*, the iron status of people living in malaria-endemic areas is highly influenced by this disease (Mamiro et al. 2005) and, consequently, it is difficult to analyse the exclusive link between nutrition and iron status, yet, a more comprehensive approach is needed.

4.4 Linking vitamin A status, food consumption and nutritional knowledge: "food taste versus nutritional knowledge and education"

An adequate provision of vitamin A is important for the human body, as a lack of this nutrient will affect the eyes. Nightblindness which is the inability to see at dusk and in dim light, is the first ocular sign of xerophthalmia with further symptoms including conjunctival xerosis, Bitot's spots, corneal xerosis, corneal ulceration/keratomalacia and xerophthalmic fundus (Sommer 1995). Besides having a role in vision, vitamin A also has a major role in the regulation of gene expression and tissue differentiation (Bender 2002) as well as in hearing and smelling (Sobeck et al. 2003). Vitamin A deficiency is a major public health problem in many areas of the world, and it is the result of two primary factors: persistent inadequate intake of vitamin A and a high frequency of infections (WHO1996).

In Tanzania, vitamin A deficiency (VAD) is still a public health problem, and it is classified to be most severe (Millstone and Lang 2003) and the prevalence of VAD (serum retinol < 0.7 mol/L) was estimated to be 36.6% of the population in 2000 (Mason et al. 2005). In Tanzania, the main source of vitamin A are foods of plant origin in the form of provitamin A carotenoids (Pepping 1988). While fruit consumption is seasonal depending on availability and with limited intake especially in semi-arid regions like Kongwa and Singida, vegetable consumption is common the whole year (Mulokozi 2004). Especially cooked green leafy vegetables constitute a common relish in rural households usually accompanying the starchy staple food as a side dish. In the following only the term 'vitamin A' is used without differentiating between the carotenes and retinol.

4.4.1 Vitamin A status of the study population

The vitamin A status of women was checked by different means. Yet, impression cytology for checking the long term situation of vitamin A status could not provide interpretable results because of a lack of morphological changes in conjunctival epithelium (see Focus C). Bitot spots and self-reported nightblindness were also no meaningful measures as only few participants were found to have or to report these clinical signs. Consequently, only retinol binding protein (RBP) measured in dried blood spots (DBS) was left for determining the vitamin A status. DBS were only taken during Mar/Apr (LR), thus, no seasonal but only a district comparison was possible.

In general, only 11% of participants had VAD, while 39% were classified as having marginal VAD and 51% had no VAD (Figure 3.4.1). Obviously, VAD described as a public health problem being "most severe" in Tanzania, is not correct for women in the surveyed areas of Tanzania. This explains the low rate of Bitot spots and self-reported nightblindness. Reasons for this could be a sufficient vitamin A supply through food and/or supplements.

Fruit and vegetable intake was below the recommended daily intake (300 g/d) for 32% of women. While the recommended daily intake for fruit and vegetables is 400 g/d (WHO 2003b), because of possible inaccuracies of assessment, all participants with a daily intake between 300 and 500 g/d were regarded as falling within the recommendation. With 49% of women lying within this ratio, and 19% eating even more than 500 g/d of fruit and vegetables, the study population had much higher intakes than the vegetable and fruit consumption per person and day as estimated by FAOSTAT for Tanzania (Figure 4.4). The FAOSTAT consumption data is, however, usually based on production data and, thus, most likely underestimating the real vegetable and fruit intake, as e.g. all vegetables and fruits collected from the wild and, most likely, from home-gardens will not have been taken into account. Consequently, the intake data measured within this study will be more realistic and suggests that the quantity of fruit and vegetable consumption is, at least, for two third of the surveyed women adequate.

Concerning the intake of vitamin A from all foods during the last 24-hour period, a share of 47% of participants were below the recommendation of 500 µg/d RE. However, the data on vitamin A intake must be viewed with caution because of the incomplete food data base in terms of nutrient contents. In fact, food frequency questionnaires, when compared with biochemical indicators, were found to provide valid information on intake for several micronutrients, however, not for vitamin A intake and retinol (Jacques et al. 1993).

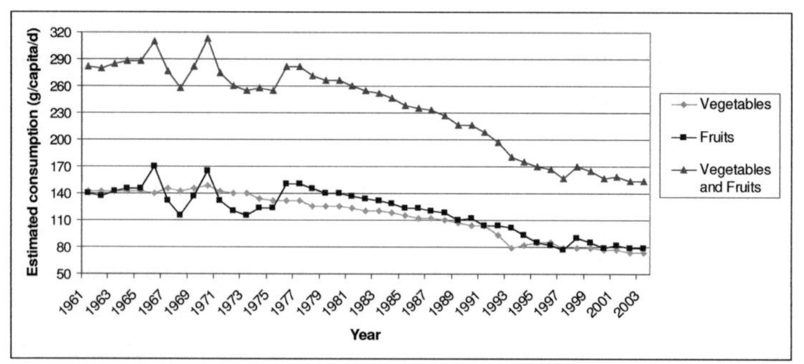

Figure 4.4   Estimated consumption of vegetables and fruits in Tanzania between 1961 and 2003 (Source: FAOSTAT 2009)

In all three districts, women who delivered their baby in a clinic received vitamin A supplements after delivery. In Kongwa, it was further explained that those who delivered at home but brought the new-born child for immunization also got vitamin A supplements for themselves. However, while vitamin A supplements were regularly given to children under five years ("Vitamin A day" twice a year in Tanzania), only few women within the present study took those supplements. During the Jun/Jul (DS) assessment, 18 out of 359 women (5%) took vitamin A supplements, during Nov/Dec (SR) 11 out of 291 women (4%), and 25 out of 298 women (8%) during Mar/Apr (LR). In general, supplementation strategies are a most straight forward approach to treat as well as prevent, e.g., xerophthalmia in endemic areas and to prevent nutritional blindness (Kuhnlein

1992). A major limitation of this strategy is that it is an expensive and only temporary solution and usually fails to sustain coverage; further challenges can be to fail to target those at risk and/or a low rate of coverage. With a strong focus on only one nutrient, the general poor nutritional status of the target population and their socio-economic conditions are usually not considered.

In contrast, dietary diversification has advantages in being a long-term intervention and is, therefore, more effective in terms of sustainability. Furthermore, natural food sources provide, next to specially desired nutrients, other nutritive and non-nutritive substances that can contribute to the prevention of illness (Kuhnlein 1992). They can improve dietary quality in general, and are consistent with the global need to lower the risk of overweight and chronic diseases (Allen 2008). A drawback of this intervention is that, as a gradual process, it takes time. Yet, as fortification and supplementation have even more limitations, the latter should not replace food-based strategies (Allen 2008), but rather complement them if necessary.

There were no significant differences regarding VAD among the three districts (Table 3.4.1). Still, more women with VAD were found in the semi-arid areas of Kongwa and Singida, where especially fruit consumption was seasonal. However, while vegetable diversity was much higher in Muheza district, the amount of vegetables consumed was higher in Kongwa and Singida (chapter 4.1).

4.4.2   Knowledge about vitamin A in nutrition

The knowledge of the interviewed women on nutritional aspects, surveyed during Nov/Dec (SR) (n=251), differed highly between whether they have heard about the term 'vitamin A' before and affirmed that they knew the function of vitamin A (Figure 3.4.13), and whether they were in fact able to name the function of vitamin A in the body correctly (Figure 3.4.14). Still, in comparison to the knowledge women had about iron, women were more knowledgeable about the correct function of vitamin A in the human body. Also regarding foods that contain vitamin A, women had more knowledge compared to their knowledge about iron-rich foods.

Clearly more women in Singida were knowledgeable compared to Kongwa and Muheza; this was probably due to the peri-urban setting of most villages in Singida district, assuming that access to knowledge is better in or close to a large town than in rural areas. Though villages in Singida district are situated less close to town than villages in the other two districts, Singida town itself is much larger than Muheza and especially Kongwa and influences the surrounding villages to a greater extent. In a study with female students in South Africa no differences between rural and urban backgrounds were found in terms of nutrition knowledge (Steyn et al. 2000); whereas adolescents in Cameroon had different food perceptions depending on the area – rural or urban – they lived in (Dapi et al. 2007).

Among the foods named as good for sight and eye health, vegetables were nearly always mentioned. However, while most participants named traditional vegetables (90%) only a few listed exotic vegetables (23%). While traditional vegetables are described in the literature to contain higher amounts of some micronutrients, e.g. vitamin A, as compared to their exotic counterparts (Chweya and Eyzaguirre, 1999; Weinberger and Msuya, 2004), this was obviously also known by the women participating in this study; at least, traditional vegetables were regarded as more healthy. Of course, it is not clear if women named traditional vegetables to be "good for sight and eye health" because they knew that they contained elevated amounts of vitamin A – according to the scientific understanding. However, they named these vegetables correctly, either according to a newly acquired modern knowledge or according to their traditional knowledge about what these foods are good for. Here, traditional and scientific knowledge seem to coincide, whereas it seems to be rather difficult to appreciate traditional knowledge from a scientific point of view. It would be of special interest to compare rural and urban areas: do people in rural areas have less modern but more traditional knowledge and is this, in terms of nutrition, an advantage or disadvantage?

The nutritional knowledge of women was further confronted with the VDS, however, without any significant relation (Table A48). From those participants who had a high VDS, only about 14% were knowledgeable about VARF, while 8% gave wrong and about 78% random answers – but they still consumed many different vegetables per week. Consequently, the consumption of a great variety of vegetables in the present study had definitely other causes than the knowledge about vitamin A rich vegetables or foods in general.

Overall, women who had some knowledge about vitamin A rich foods (VARF) were neither healthier nor did they consume a greater variety of vegetables than other women. Therefore, it could not be concluded that the higher the vegetable diversity consumed and the greater the nutritional knowledge the better the vitamin A status of women in the surveyed areas of Tanzania. Further hidden or confounding factors were decisive as, for example, taste, which was mentioned by about 53% of all participants as the most important criteria to choose a vegetable. This is obviously a stronger driving force for vegetable consumption, while nutritional aspects play a minor role. Also the convenience of preparation ("easy to prepare") was more important than health issues for women in Kongwa district (see also chapter 4.1). Still, other studies, e.g. in Nepal, found that women owning a kitchen-garden had more nutritional knowledge and consumed more micronutrient-rich fruits and vegetables (Jones et al. 2005); thus, nutritional knowledge and nutrition itself were found to be related to each other.

In general it is acknowledged that while the availability of (micro)nutrient-rich foods is one important precondition for a balanced diet, knowledge about the benefits from consuming this food, the amounts needed, and nutrient-preserving preparation methods is of equal concern. In fact, while more is known about physiological factors influencing the vitamin A status of an individual, less research focused on ecological, economic, and cultural factors which influence the intake of natural food sources of moderate and high vitamin A activity. Positive dietary change is initiated by education and choice, which has been given low priority in most vitamin A intervention programs so far (Kuhnlein 1992). Yet, as in the present study taste and availability were of high importance for vegetable consumption, the question arises to what extent knowledge is the key to a healthy diet and to what extent other factors affect dietary behaviour. For example, also the influence of certain policies and institutions which control the production, trade, distribution and marketing of food and, thus, its availability, is to be considered (see also chapter 4.2).

4.4.3 Vitamin A in associations and correlations

When correlated to socio-economic parameters, RBP values showed a significant correlation with wealth status (Figure 3.4.2). Thereby, to be affected especially by marginal VAD was rather a problem for poor people; yet, it was not clear if this was rather due to dietary, health or educational issues. An association between VAD and ethnic groups (Figure A19), though not significant, supported the fact that district differences did exist. Analogous to the districts, for the ethnic groups living in Muheza district, namely Bondei and Shambaa, no VAD was detected, while it was for all other tribes living in the semi-arid districts Kongwa and Singida.

Also the relationship among women with different education levels was not significant but interesting (Figure A20), as those women with higher education seemed to be more affected by VAD than those with no or low education. General education can probably not be equalised with nutrition education while at large, nutrition education is suggested to accompany always other measures such as home stead food production to combat VAD (de Pee and Bloem 2007). Because of the low number of participants in the high and low education groups in this study further interpretation of the data is refrained from.

Another interesting, though not significant, relationship was that of age and VAD (Figure A21). There was a trend from the low age group (16 to 25 years) with 20% of women having VAD, to 12% of women aged 26 to 35 years being affected and only 9% of the oldest group (36 to 45 years) having VAD. This suggests that older women (within the given range) had, in general, a better vitamin A status than their younger counterparts which was probably due to having more nutritional knowledge. While in the present study all participants cultivated vegetables in one way or the other, it was observed in all three districts that knowledge about vegetable processing techniques, especially of traditional types, was rather with older than with younger women (Keller 2004).

Only sugar consumption showed a significant negative correlation to RBP values, meaning that the higher the amount of sugar consumed, the lower was the RBP value and, thus, the worse was the vitamin A status. This relationship does not suggest that sugar has any influence on bioavailability, retention or intake of vitamin A, as sugar is even sometimes fortified with vitamin A (Pineda 1998). An increased sugar intake rather function as an indicator for an unhealthy or unbalanced diet, which goes along with a low vitamin A status among others. Consequently, to combat VAD in this particular case, sugar could be fortified with vitamin A, yet, with sugar being not a very healthy food vehicle which should be consumed in great amounts. A more healthy and sustainable solution would be to reduce the sugar intake and improve the diet in general and with this the vitamin A status of women.

Nutrient intakes as well as the food scores DDS and FVS were not correlated to RBP values. It was expected to confirm that a high nutritional diversity is desirable for a balanced diet (Burgess and Glasauer 2004) and, consequently, a precondition for nutritional health; yet, with RBP as the only measure for nutritional health, this was definitely not possible to show (see also chapter 4.5).

Though no bivariate significant correlation could be detected between VDS and RBP, this relationship was of special interest. In fact, participants without vitamin A deficiency were found in all VDS groups, consuming between one and twelve vegetables per week. They were completely equally distributed with 23 women having a medium VDS and 25 each having a low and high VDS, respectively. Women with marginal VAD and VAD were randomly distributed over the different VDS groups. Thereby, the distribution line of women with VAD over the different VDS groups was rather flat and not central but showing two peaks (Figure A22). This indicates that there must be a confounding factor, which breaks down the linear relationship and has a stronger influence on VAD than the diversity of vegetables consumed, e.g., the wealth status of women which showed a significant association to VAD.

Besides vegetable or, in general, dietary diversity, the amount of foods consumed is probably more crucial. However, when participants' fruit and vegetable intakes were plotted against their vitamin A status (Figure A23), also no significant relationship could be detected. In fact, the share of women with no VAD and consuming less than 300g of fruits and vegetables per day was even higher than those of women with VAD and marginal VAD taken together. While the accuracy of the 24h-recall can be put into question at this point, also the focus on the food groups fruits and vegetables might be too narrow.

Food patterns, generated from intake data of all three seasons, were also related to RBP values. In fact, one food pattern, namely No.5 "animal products", was positively associated to vitamin A status, meaning that those women consuming food according to this pattern also had high RBP values and, thus, most likely no VAD. This can be due to animal sources containing preformed vitamin A, which is much better absorbed by the human body than provitamin A from plant sources (Bender 2002). While vegetables and fruits are still the most important source for vitamin A in the researched areas in Tanzania, it is suggested that homestead food production can contribute to combating VAD, especially when it implies animal husbandry as it was also suggested by de Pee and Bloem (2007).

Obviously, it appears to be rather difficult to establish a straight link between only one food group and a single nutrient within human health. It would be more reasonable to establish food patterns for each participant and compare those with the nutritional health status (see chapter 4.6). Similarly, to examine diet-disease relations, dietary patterns are seen as a possible approach to consider overall eating patterns and not only single nutrients (Hu et al. 1999). Regarding vitamin A, it is especially suggested not only to investigate vegetables, but also (traditional) fruits and, for example, whether fruits rather complement or substitute vegetables. Other vitamin A rich foods, such as liver, milk or eggs were consumed very seldom in the study area and, thus, the main source for vitamin A was, in fact, dark green leafy vegetables and orange-fleshed vegetables and fruits.

The contribution of both traditional and exotic vegetables to nutritional health was also of interest. However, it was not possible to differentiate the influence of traditional or exotic vegetables on the vitamin A status, as only 22 women out of 145 consumed exotic vegetables during one surveyed week (tomatoes and onions excluded).

Moreover, relationships between the vitamin A status of women and further health variables were checked. For example, low vitamin A and low iron stores were found to be associated and iron deficiency was observed to decrease vitamin A mobilisation (Hinderaker et al. 2002); yet, in the present study no significant correlations between RBP and either Hb or TfR values were found. Furthermore, RBP values were also not associated with BMI. This was most likely due to VAD occurring only in few participants and severity in general was low. As only few bivariate associations between vitamin A status and further characteristics of participants were detected, it was refrained from conducting a multiple regression analysis as no meaningful results were expected.

4.4.4   Conclusions

In general, the prevalence of VAD in the present study was low as well as the severity. This is not consistent with the literature suggesting that VAD is still a public health problem in Tanzania. As the present study cohort is, of course, not representative for Tanzania, it is still suggested to scrutinise the vitamin A status of adults very carefully and to concentrate the combat of VAD mainly on children.

One important outcome of the present study was that traditional vegetables were regarded as more nutritious than exotic ones regarding vitamin A. Consequently, while traditional vegetables are still considered neglected and underutilised on a global scale, they continue to be used by the local population of Tanzania. While in the studied rural and peri-urban districts this publicity and utilisation enables the conservation of traditional vegetable diversity, especially for urban areas, it will be necessary to propagate these crops and their nutritional values to make sure that they will not perish but still can nourish future generations. This is even more important in the light of the rapid 'nutrition transition' (chapters 4.2 and 4.6) and the fact that people regard taste more important than health when choosing a vegetable and certainly also food in general.

It was not possible to measure a significant link between vitamin A status, food - especially vegetables - consumption and nutritional knowledge with the given study population. This might be due to the survey taking place during one season only, that vitamin A status as well as vegetable intake were similar for most participants, and that the severity of VAD was low. Studies which found relationships between vegetable consumption and vitamin A status measured different parameters over time: e.g. in Bangladesh, an increased intake of micronutrient-rich foods from homegardens resulted in a lower incidence of blindness in children (Yousuf and Islam 1994); or in the Philippines, vegetable prices were found to be positively related to morbidity rates (Bouis 1991).

While there exist different approaches to combat vitamin A deficiency, namely supplementation, fortification or dietary modification including diversification of the diet, this study attempted to identify a link between the latter approach and vitamin A status of women. However, by investigating only one nutrient against the background of insufficient data for vitamin A contents of food, unsolved problems regarding provitamin A conversion factors and bioavailability, as well as having data of only one biochemical indicator for assessing vitamin A status of women, no satisfying conclusions could be drawn. Therefore, the analysis of both dietary diversity and dietary patterns was emphasized more strongly (see chapters 4.5 and 4.6).

FOCUS C: Conjunctival Impression Cytology

Conjunctival Impression Cytology (CIC) is especially favoured for its use in developing countries as it is non-invasive and simple with no need for, e.g., cooling specimens. So far, this techniques was mainly used to study vitamin A status in children. Its feasibility to accurately characterise the risk of VAD in communities, though not in individuals, was confirmed by several studies (Coutsoudis et al. 1993; Fuchs et al. 1994; Pandit et al. 1998).

The idea behind this approach is to examine the morphology of epithelial cells obtained from the conjunctival surface on a piece of filter paper; through this, it is possible to assess whether changes have occurred that are associated with vitamin A deficiency. A normal impression of conjunctival cells will reveal sheets of small epithelial cells and an abundance of mucin-secreting goblet cells. In the presence of vitamin A deficiency, the epithelial cells are flattened and enlarged, and there is a marked reduction or absence of goblet cells. (WHO 1996)

To obtain an impression of the conjunctiva, two different methods can be applied, namely Conjungtival Impression Cytology (CIC) or Impression Cytology with Transfer (ICT). Both techniques require microscope slides, standard pore-size filter-paper, and a simple light microscope. For ICT, where the cells obtained on the filter-paper are immediately transferred to a slide and fixed, only a single staining solution is required. In contrary, for CIC more reagents and processing steps for fixing, staining and mounting specimens are required. Both chemicals and lab staff is rather expensive and was, therefore, a crucial factor for using ICT within the present study. However, with ICT difficulties can occur in obtaining an efficient transfer of cells of high quality from filter paper to slide. Consequently, while processing of specimen for ICT is simpler compared to CIC, the unreadable rate may be higher, as not the entire sample of cells is transferred and available for evaluation. (WHO 1996)

The latter point was a main problem in the current study, and several slides were found to contain too few material for interpretation. A further challenge was that there is no generally accepted reference standard for the classification of vitamin A status on the basis of impression cytology readings (WHO 1996). Furthermore, a standard set of reference slides or photographs for interpretation, against which unknown slides can be compared, was not available, which was the major factor why slides were not able to be interpreted. An alternative to reference slides is having two readers and accepting as definite only identical interpretations made by both (WHO 1996); however, this was also not given in the present study.

When slides from the present study were examined, it was also found that the epithelial cells, on which interpretation is based besides goblet cells, can also originate from other sources than the conjunctiva. The filter paper may have been also contaminated by epithelial cells, e.g., from the eye lid, which can not be distinguished from those from the conjunctiva and would lead to false interpretations.

In general, as this technique is non-invasive, relatively easy to perform and with minimal discomfort to the patient (Singh et al. 2005), it would be worthwhile to establish precise criteria and standards for interpreting results, and to introduce this technique into routine clinical practice in areas where VAD is of a major problem.

4.5   Dietary diversity score (DDS) and food variety score (FVS): „measuring dietary diversity and the association between nutritional diversity and nutritional health"

The food scores DDS and FVS are a qualitative measure that can serve as a proxy of the nutrient adequacy of the diet for individuals and can also reflect household access to a certain variety of foods (FAO 2007). In fact, it was already observed that the DDS represents very well the overall dietary quality of women in a poor rural African setting and can also be linked to their nutritional status (Savy et al. 2005). In general, an increase in individual DDS was found to be related to increased nutrient adequacy of the diet in several studies with different age groups (FAO 2007).

The scores can be created from data from 1-day dietary recalls or also more-day recalls to prevent the bias of missing out foods eaten less routinely or to avoid special days, which can be atypical in regard to food consumption. Yet, it was found that the DDS calculated from only a 1-day recall as compared to a 3-day recall was sufficient to predict the nutritional status of women (Savy et al. 2007). The DDS and FVS in the present study were created either from a 1-day recall (to compare the three seasons) or from all three recalls (mean across three seasons). The scores were grouped into terciles to distinguish between diets of low, medium and high diversity and variety.

4.5.1   DDS and FVS of the study population

The median DDS of 6.0 and the mean FVS of 8.3 of the study population (Tables 3.3.9 and 3.3.10) was low, suggesting an overall poor dietary quality. Yet, the scores were similar to those of other studies in the Middle East and Westafrica (Azadbakht et al. 2005; Savy et al. 2005). In a study from Kenya a DDS of 6 was considered as medium, however, only 12 instead of 14 food groups were used (ACF 2009). When dietary scores of different studies are compared, however, care has to be taken to check how food items and food groups were defined and, whether different cereal or vegetable varieties were counted as one or individually.

In general, about one third of participants had a monotonous diet with an alarmingly low DDS of only two to four food groups per day (Figure 3.3.22). These women consumed a very basic diet consisting of cereals and vegetables and usually oil or fat (mean for all three seasons and districts). Only few women of this group consumed foods from animal sources, such as eggs, meat and poultry, and also fruit consumption was fairly low (Figure 3.3.26). This is conform to findings from rural Burkina Faso, where women with a low DDS of only two or three food groups consumed mainly cereals, green leafy vegetables and condiments (Savy et al. 2005). In contrary, between 90% and 100% of women in the present study with a high DDS consumed, additionally to these three food groups, also grain legumes, nuts, fish and beverages, the latter usually being black tea (Figure 3.3.26). Those food groups, consumed not by everyone, were generally more expensive, such as foods from animal sources. Still, other food groups, like fruits, were probably less consumed due to availability, seasonality and nutritional knowledge, while starchy roots were presumably not consumed by every woman because of certain dietary habits or customs.

As could be expected, DDS and FVS differed both between the three districts and the three seasons, whereby the latter was less pronounced (Tables 3.3.9 and 3.3.10; Figures 3.3.23, 3.3.24, 3.3.28 and 3.3.29). Obviously, each district, with its particular ethnicities, culture, agro-ecology and nutritional habits, had a stronger influence on dietary diversity and food variety than season. One reason for Muheza having the highest DDS and FVS was probably the high plant diversity in this district (Keller et al. 2005). In fact, Muheza district involves parts of the Eastern Usambara Mountains that are, in turn, part of the Eastern Arc Mountains, which form together with other highlands the Eastern Afromontane Hotspot, one of 34 biodiversity hotspots in the world (Conservation International 2007). However, a high dietary diversity is not only high because of a great diversity of locally available and produced products. Because the various ethnic groups in Muheza possess different dietary customs, they will have a great share in the high dietary diversity of participants. Additionally, coastal trade influences food availability and usage, which can be traced back to the 5th to 7th century, when the Indian Ocean trade was already linked to a regional commerce (Iliffe1995). Reasons for a low dietary diversity, as it was consumed by women in Kongwa and Singida, can be due to the general low agricultural diversity in these districts and also because of only one or two ethnic groups dominating the districts with certain food habits (see also 'Associations of food scores with socio-economic variables').

Unlike the consumed vegetable diversity (VDS) that differed highly between the seasons, DDS and FVS seemed to be less dependent on seasonal changes. This applies especially to food items that are not or seldom home-produced, but can be bought all year round, such as tea, sugar, bread/cakes, oil/fat and also starchy staples. In this case, the differences observed between low and high dietary diversity typically refer to dietary changes that happen during the process of a 'nutrition transition' (Popkin 1999), and of which the early stage obviously already takes place in rural Tanzania (see also chapters 4.2 and 4.6). Still, vegetables were among those food groups forming the basic diet and being consumed by nearly every participant year round; here, the amount consumed must be taken further into account, which was done in chapter 4.1.

4.5.2   Dietary scores in associations and correlations

<u>Associations of food scores with socio-economic variables</u>
Both scores were strongly associated with the ethnic group of participants (Figures 3.3.30), which might have an influence in terms of dietary habits. This must be considered especially when proposing a high food variety for nutritional health in areas inhabiting many different ethnic groups. It was found in studies with migrants that ethnicity is an important factor influencing eating habits (IFAVA 2008).

Regarding the status within the household (Figures 3.3.31) being "only" the wife of the head of household was obviously going along with less food variety and, therefore, a minor food quality. Yet, that a poor household is equated with low dietary diversity does not agree with the fact that women-headed households are usually poorer than households headed by men (UNDP 1995). From the present results it is suggested that, besides wealth, it is also important how much power of decision the person responsible for family nutrition has got.

The relationship of occupation of participants to food scores (Figures 3.3.32) showed that women with an additional income from non-farming activities were able to eat a greater variety of foods, probably especially those that needed to be bought. Wealth itself was not significantly related to DDS or FVS, yet, it must be considered that the wealth indicator comprised of possessions of participants and not of income. Therefore, a certain occupational status can act as an indicator for additional income at this point, assuming that cash additional income from a business or work in any kind of service is higher than selling vegetable or agricultural products. The present finding, that those women with additional income also had a higher DDS and FVS, coincides with the findings in Burkina Faso, where a higher share (about 50%) of women with commercial incomes had a high FVS and DDS (Savy et al. 2005).

Religion of participants was only significantly associated with the FVS (Figure 3.3.33) with nearly 50% of Muslims having a high FVS, while only 25% of Christians consumed a high food variety. As many Muslims lived in Muheza district, influencing factors other than religion must be considered. In fact, in Burkina Faso it was the other way round with a larger share of Christians than Muslims having a high FVS (Savy et al. 2005).

<u>Associations of food scores with vegetable production variables</u>
The level of dietary diversity and food variety of study participants was also significantly associated to the number of vegetable types that women cultivated or collected, and if they bought and sold vegetables (Table 3.3.12). Those participants with a high DDS/FVS cultivated/collected more or less the same number of different vegetables year round (four to five). The women with low or medium DDS/FVS had more fluctuations in their vegetable cropping diversity (two to six), which can be an indicator for a more weather-dependent vegetable cropping. That seasonality influenced vegetable cropping and consumption to a great extent in the researched districts has already been shown in chapter 4.1. It is generally acknowledged that seasonal malnutrition in rural areas in Africa is widespread (SCN 1988) and is caused, to a great extent, by seasonal fluctuations in food availability and the ability to acquire food (Teokul et al. 1985).

Table 4.5   Median number of vegetables cultivated/collected by women in Tanzania in different DDS/FVS categories and during different seasons

	DDS				FVS			
	Low	Medium	High	p*	Low	Medium	High	p*
No of vegetables cultivated all year	3	4	4	<0.0001	3	4	4	<0.0001
No of vegetables cultivated Jun/Jul (DS)	2	3	5	<0.0001	2	4	5	<0.0001
No of vegetables cultivated Nov/Dec (SR)	2	3	4	<0.0001	2	3	4	<0.0001
No of vegetables cultivated Mar/Apr (LR)	5	6	4	<0.0001	5	5	4	0,003
No of vegetables collected all year	-	-	-	n.s.	-	-	-	n.s.
No of vegetables collected Jun/Jul (DS)	2	3	4	<0.0001	2	2	4	<0.0001
No of vegetables collected Nov/Dec (SR)	3	3	4	0,032	-	-	-	n.s.
No of vegetables collected Mar/Apr (LR)	5	5	3	<0.0001	5	5,0	3	<0.0001

* p for differences between categories according to K-W test; n.s.=not significant

While there were no significant correlations between the two scores and vegetable purchase or sales during Jun/Jul (DS), during Nov/Dec (SR) more women who sold their vegetables had a high DDS (58%) or high FVS (50%). Thus, not only those women who cultivated/collected vegetable diversity more consistently, but also those who sold their products, had a higher DDS and FVS. Consequently, not only the availability of agricultural diversity was associated with dietary diversity, but also marketing of agricultural diversity played a certain role.

There was also a trend that those women who bought vegetables in addition to those they produced on their own rather had a high DDS, while those who did not buy additional vegetables had more often a low DDS (data not shown). The purchase of additional vegetables can be considered as an indicator that women in general were able to buy food in addition to those they produced on their own. Self-produced vegetable/food diversity did not alone influence dietary diversity but the purchase of foods and all related factors, such as income and supply of foods, had a great impact on the diversity of foods consumed, even in rural environments. In general, women-controlled income has greater immediate benefits for the nutrition and well-being of all household members, especially children, than when income is controlled by men (FAO 1997). Moreover, the nearer women live to a town or economic centre, the more they will participate in the cash economy, e.g., by the sale of garden produce (Fratkin and Smith 1995) and both factors can contribute to an increase in dietary diversity.

<u>Associations of food scores with nutrition and health variables</u>

The intake of nutrients by women within the low DDS category were always lowest and by those having a high DDS always highest, except for iron intake (Figure A17). For the study population it could be demonstrated that a higher diversification of the diet was accompanied by an increased nutrient intake. Similarly, food scores were highly and positively correlated to the intake (g/d) of most food groups (Table A34). With an increase in food variety the intake amount of foods also increased. An exception were cereal (regarding DDS) and vegetable intake which were negatively correlated to food scores and, thus, decreasing with increasing food diversity. This is reasonable as cereals and vegetables formed the basic diet and were obviously substituted by other foods (in amount) when food diversity increased.

The Hb of women was negatively correlated to FVS and DDS for the overall data of one year (Figures 3.4.5 and 4.5.1) and also for each season (Table 3.4.5). The mean Hb was always significantly lower for women within the high DDS/FVS category (Table A35). This would suggest that a high dietary diversity alone may not generally be favourable for a healthy iron status which does not make sense on the first glance. In fact, a sufficient iron intake is not necessarily achieved by a great variety of foods but by foods with a high iron content of good bioavailability.

Yet, absorbtion of iron can be enhanced through other nutrients such as ascorbic acid (Strain and Cashman 2002), which would, in turn, suggest a high diversity of the "right" foods to be favourable for a good iron status. As some nutrients, such as phytates, can also decrease iron absorbtion (Strain and Cashman 2002), the type of foods consumed is much more crucial when considering iron status than food variety. Furthermore, a malaria infection can influence the iron status of a person and, therefore, a direct link between Hb and DDS or FVS was difficult to be drawn in the present study (see also chapter 4.3). TfR values were positively associated with the FVS as well as the DDS, showing, thus, the same result as the association between Hb and the food scores.

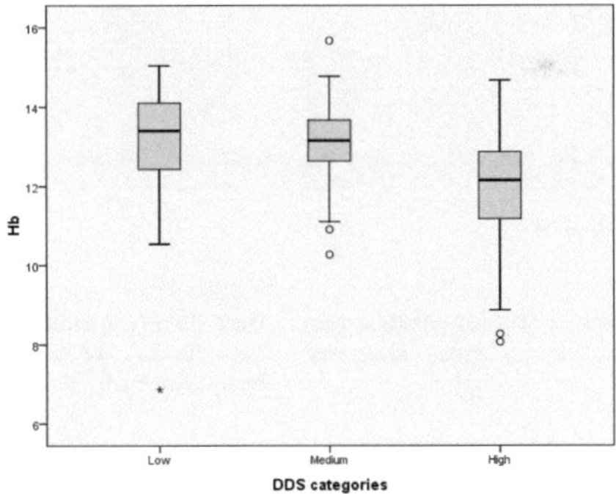

Figure 4.5.1
Hb values by DDS categories; n=189 women from three districts in Tanzania, mean across three seasons (ρ=-0.364, p<0.001)

BMI was positively correlated with both DDS (Figures 3.4.8 and 4.5.2) and FVS (Figures 3.4.7 and 4.5.3), suggesting that the greater the diversity of foods eaten the higher the BMI. In fact, the median BMI was significantly higher for women within the medium DDS/FVS category and again higher for those within the high DDS/FVS category (Table 3.4.9). As these median BMI values all range within the normal weight scope, it is difficult to suggest whether, through increased dietary diversity, health improved or deteriorated. A similar link between DDS/FVS and BMI was found in Burkina Faso where a more varied diet was considered to be associated with higher anthropometric indices reflecting a better nutritional status (Savy et al. 2005). This conclusion, of course, can only be applicable as long as the anthropometric indices do not pass the threshold indicating overweight or even obesity of participants. If the latter would be the case it must be questioned, which food types contribute to a higher dietary diversity and that a higher food diversity *per se* is not necessarily desirable. For example, with the 'nutrition transition', the access to cheap energy-dense foods is improved (Drewnowski 2000) and, in general, food diversity is increased, yet not necessarily standing for an improved, balanced and healthy diet. The 'nutrition transition' is, in fact, known to implicate the rapid rise of obesity (Hawkes 2006) (chapter 4.2) and it, thus, ,obviously improves dietary diversity but not quality. Therefore, it must be distinguished explicitly between different forms of increase in dietary diversity and food variety, as well as between quantity and quality of diversity.

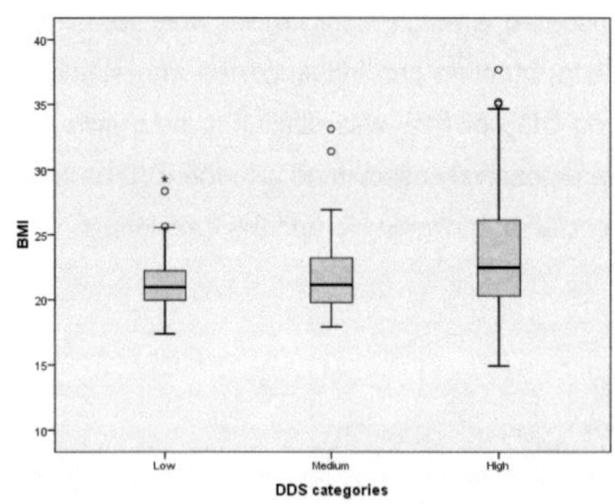

Figure 4.5.2
BMI by DDS categories; n=189 women from three districts in Tanzania, mean across three seasons (ρ=0.147, p=0.033)

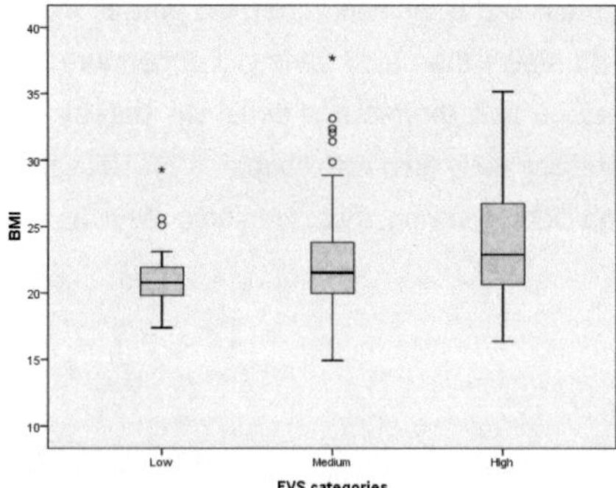

Figure 4.5.3
BMI by FVS categories; n=189 women from three districts in Tanzania, mean across three seasons (ρ=0.204, p=0.003)

<u>Multiple relations</u>

Both FVS and DDS were tested as dependent variables in multiple regression models for each season. Besides controlling for age and district, vegetable production parameters were chosen as predictors as well as BMI and Hb values, yet, the latter two predictors were tested in two different models because of the different composition of the study cohorts (Tables A49-A52). As results for both dietary scores were similar, only those for FVS (Figure 4.5.4) will be discussed in detail, while those for DDS are only summarised in Figure 4.5.5.

Regarding FVS, residuals of all models were normally distributed, all models were highly significant and explained between 38% and 42% of variance. In "model 1 (BMI)", significant correlations were found only during Nov/Dec (SR) (Figure 4.5.4), suggesting that when the mean number of collected vegetables increased by one, the mean FVS would decrease by 0.2. This was an interesting result as no bivariate association between these two variables was detected during Nov/Dec (SR) (chapter 3.3.3). While the influence of seasonality on this relationship was already discussed (chapter 4.5.2), it can be concluded that especially the collection of vegetables from the wild can stand as an indicator for less wealth, which, in turn, can be associated with a lower food variety. This coincides with earlier findings of different studies that, with higher income and wealth, food diversity is usually broadened (Clausen et al. 2005; Savy et al. 2005).

The BMI was more strongly correlated: with an increase of one in BMI the FVS increased by 0.1. As the link between FVS and BMI was found already as a bivariate relationship (Figure 4.5.3) and was discussed in detail above, the regression model mainly confirms this link even when it is controlled for age, district and vegetable production parameters.

From "model 2 (Hb)", it was learned that, during Nov/Dec (SR) (Figure 4.5.4), with an increase by one of the mean number of exotic vegetables cultivated, the mean FVS increased by 0.7. The number of cultivated exotic vegetables can stand for a certain degree of wealth, which is suggested to be positively associated to FVS and, thus, coincides with the findings within "model 1 (BMI)".

No significant correlation between FVS and Hb could be detected within the model. While Hb and FVS showed strong correlations when they were assessed in a bivariate way, the additional variables within the model obviously confounded this relationship; as hypothesised before, there are many other factors besides a diversified diet, which influences the iron status; here, especially the districts, with different prevalences of malaria, will affect the Hb level to a great extent (see also chapter 4.3).

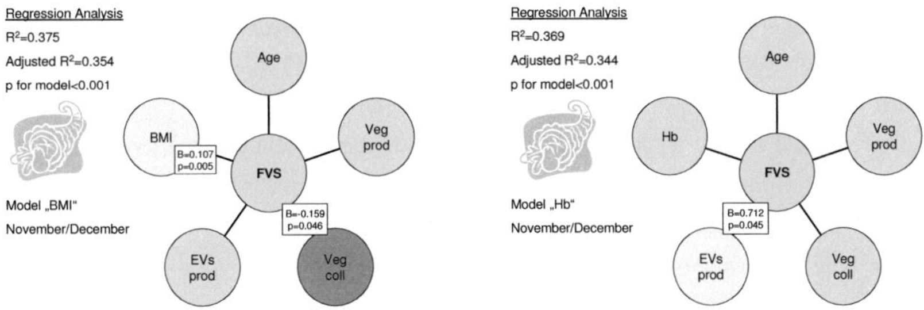

Figure 4.5.4   Results of two multiple regression analyses with FVS of women as dependent variable; n=210 for model "BMI"; n=185 for model "Hb"; data generated during three different seasons within three districts of Tanzania

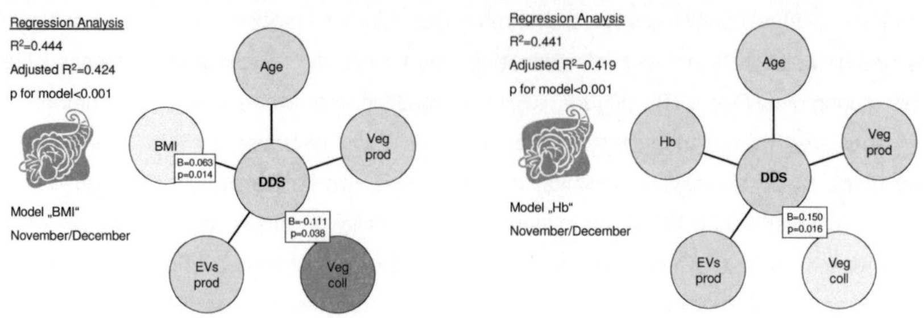

Figure 4.5.5   Results of two multiple regression analyses with DDS of women as dependent variable; n=210for model "BMI"; n=185 for model "Hb"; data generated during three different seasons within three districts of Tanzania

4.5.3   Conclusions

By means of DDS and FVS, minor seasonal and major regional differences of dietary habits of the study population in Tanzania could be shown. A main finding was that a higher diversification of the diet was associated with an increased intake of most nutrients and food groups. Another important fact regarding vegetable consumption was that the higher the DDS the more was the food group vegetables replaced by other food groups. While, in fact, many bivariate associations were found, the food scores were less related to other variables when analysed in multiple regression models. Still, the rather easy and quickly measured diversity scores are suggested to describe the dietary situation of the present study population in a meaningful way.

While no direct link between cropping diversity and dietary diversity could be shown, vegetable production functioned, to a certain extent, as an indicator for wealth (positively related to cultivation, negatively to collection). Additionally, a high wealth status was related to high food scores and, thus, to a more varied diet. Thus, an indirect link could be drawn between high cropping and dietary diversity. It is proposed that a study on nutritional health always needs to integrate food production. Contrariwise, nutritional concerns should always be integrated into agricultural research as it was also suggested by Bayani (2000).

While diversity should not be confused with quality, it was already shown that diversity scores clearly can reflect overall dietary quality (Hatloy et al. 1998; Torheim et al. 2004). As the scores are, moreover, linked to the nutritional status of women, here especially in terms of BMI, it is questioned if they could serve as proxies for dietary health. However, while the food scores account only for diversity of foods, the quantity of foods consumed definitely needs to be taken into account if addressing, e.g., the BMI. Furthermore, different forms of an increase in dietary diversity and food variety must be distinguished. A simple increase, for instance, in FVS could result from both a consumption of more different fruits and vegetables, and a consumption of more different convenience and fast foods of animal origin. A simple increase in food variety could also result from establishing and using a homegarden or from moving to town and being exposed to any type of 'nutrition transition' (see also chapters 4.2 and 4.6). Consequently, while usually a high nutritional diversity is desirable for a balanced diet (Burgess and Glasauer 2004), it does not necessarily result in a better health status. Likewise, a certain degree of diversity does not mean that the dietary quality will match people's particular needs (Brown et al. 2002). Therefore, dietary diversity scores should be enhanced as a tool for assessing dietary diversity together with dietary quality. This could be achieved through weighting of single food groups according to their importance for health, e.g. after the food pyramid or circle.

4.6   Dietary patterns and their association with nutritional health: "the strength and weaknesses of pattern analysis"

When investigating the complexity of human nutrition and its association with health, it is less meaningful to consider single nutrients alone. Nutrient-nutrient interactions, and complementary and synergistic mechanisms of the action of different phytochemicals also play important roles (Lampe 1999). Dietary patterns are regarded as one approach to consider overall eating habits instead of single nutrients (Hu et al. 1999).   These were also analysed for participants of the present study by applying two different approaches, i.e. principal component analysis (PCA) and cluster analysis.

4.6.1   Dietary patterns generated through principal component analysis

The five dietary patterns generated through PCA and their decisive food groups (Table 3.3.19) are summarised in Figure 4.6.1. The foods that characterise each pattern were, of course, not consumed exclusively but are important aspects of the patterns. The relationships between different variables and the factor scores of participants were investigated in order to characterise participants, which mostly followed the one or the other pattern, but also to find general relationships between one food pattern and other characteristics of the participants obtained within this study (see also chapters 4.2 for BMI and food patterns, 4.3 for iron status and food patterns, and 4.5 for diversity scores and food patterns).

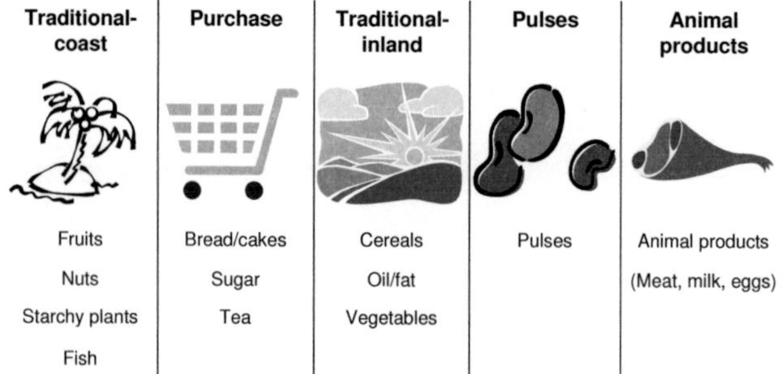

Figure 4.6.1   Typical food groups characterising five dietary patterns generated through PCA from data collected through three 24-h recalls (n=252) within three districts of Tanzania

Associations of dietary patterns with other variables

Exemplarily for all dietary patterns the "purchase" and the "animal products" pattern are presented to get an overview on their typical characteristics according to bivariate correlations. As summarised by figures 4.6.2 and 4.6.3 the "purchase" pattern had more negative associations than the "animal products" pattern. Further, the "purchase" pattern was much stronger correlated to some variables such as the FVS and DDS, which also show a significant relation to this pattern within multiple regression analysis (see below). In the following, some correlations found through bivariate analysis will be highlighted.

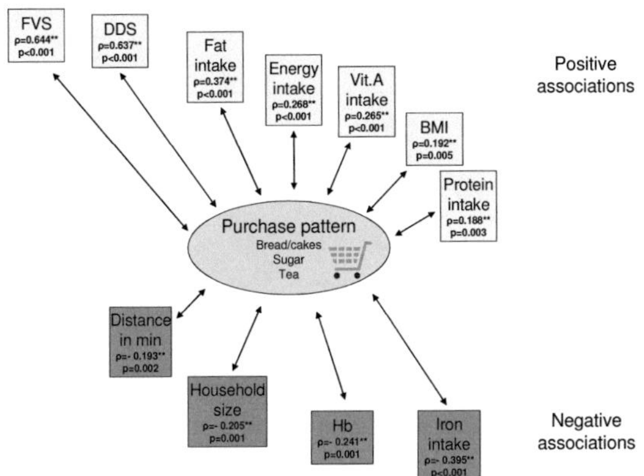

Figure 4.6.2   The purchase pattern and its positive and negative associations as determined through bivariate correlations

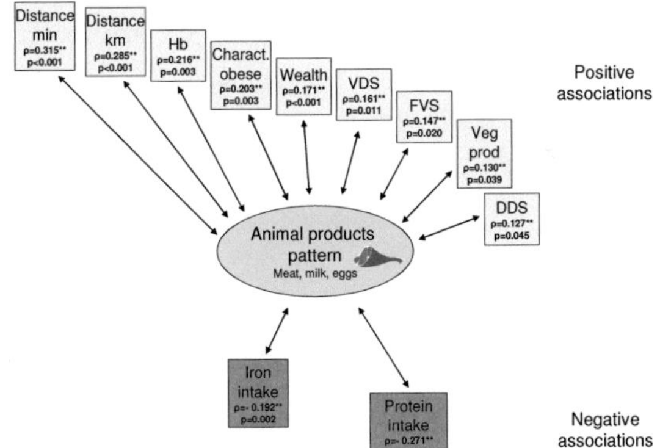

Figure 4.6.3   The animal products pattern and its positive and negative associations as determined through bivariate correlations

The districts were significantly associated with all five patterns (Table 3.3.22), from which can be inferred that also ethnic groups, which were highly correlated to districts, were associated with food patterns. Regarding the "traditional-inland" pattern (Table 4.6.1), there existed differences in food consumption between the "inland" districts Kongwa and Singida despite similar agro-ecological conditions.

Table 4.6.1   Share of participants (%) within the 5th quintile ("high consumption according to this pattern") of each food pattern (derived through PCA) in each of three districts, Tanzania

Food pattern	Kongwa	Muheza	Singida
Traditional-coast	10	82	8
Purchase	22	64	14
Traditional-inland	22	22	56
Pulses	16	42	42
Animal products	16	26	58

The "purchase" pattern was mainly consumed in Muheza (Table 4.6.1), while the 1st quintile of this pattern was dominated by participants from Singida (68%, data not shown), meaning that more than two thirds of women in this district did not follow this pattern. As the "purchase" pattern stands for an influence of the 'nutrition transition', the latter could be observed to a greater extent in Muheza district (see also chapter 4.2).

Overall, participants from Kongwa district were mostly found within the 2nd, 3rd or 4th quintile of all patterns. Also, the share of participants from Kongwa in the 1st quintile was always higher than the share of women within the 5th quintile (except for the "purchase" pattern). Thus, it was difficult to assign them to any of the food patterns in particular (Table 3.3.22).

Further, several significant bivariate correlations of socio-economic variables with food patterns were found (Table 3.3.23). For example, a high wealth status of participants was related to a high factor score for pattern 5, meaning that the food group "animal products" was consumed to a great extent by more wealthy participants (Figure 4.6.4). In fact, it was found in another study within Tanzania that an increase in income was associated with an increase in the quantities of meat consumed (Kaliba 2008); and it is acknowledged, in general, that the main determinant for per capita meat consumption is wealth (Speedy 2003).

The negative associations of patterns 2 and 3 with household size mean that the larger the household of a respondent, the less would she consume foods according to these two patterns (Figures 4.6.5 and 4.6.6). While for pattern 3 "traditional inland" the reason for this association can be found in other characteristics such as the districts, the second pattern "purchase" suggests that women living in smaller households can rather afford to purchase additional foods than those living in large households with many family members. This, however, is not conform with the finding that wealth levels are usually positively associated with sizes and compositions of households (Netting 1982). Yet, in the present study the opposite could be shown which is also reasonable, namely that large households are more poor households and can, thus, purchase less food.

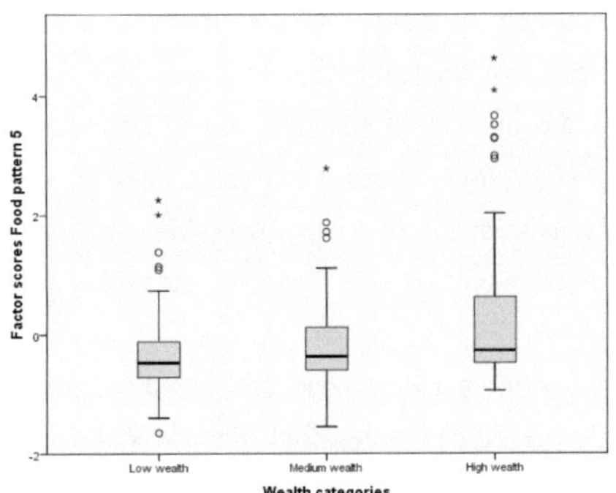

Figure 4.6.4
Association between three wealth categories and factor scores of food pattern 5 "animal products" (p<0.001)

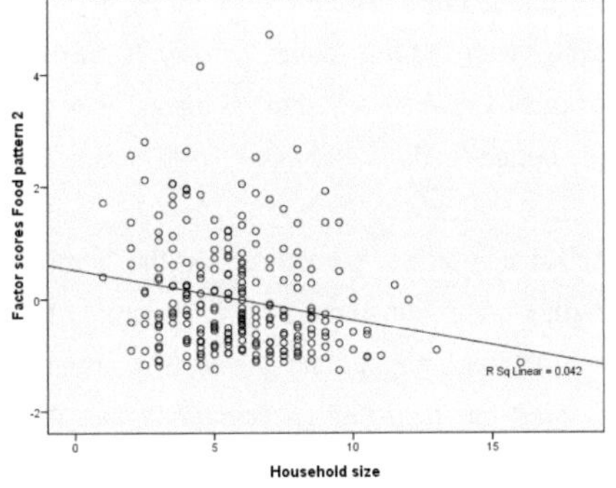

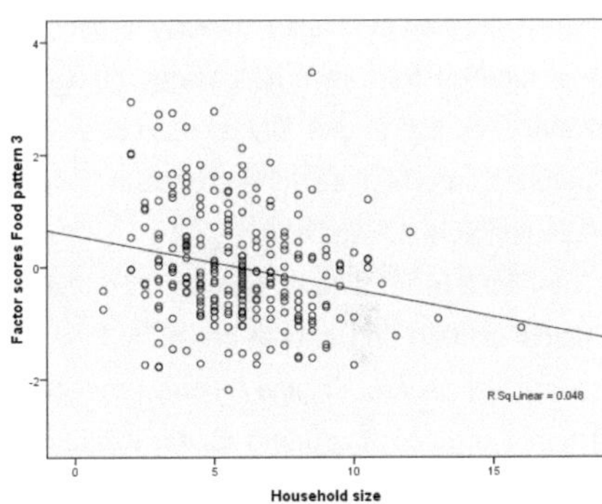

Figure 4.6.5
Association between household size and factor scores of food pattern 2 "purchase" (p=0.001)

Figure 4.6.6
Association between household size and factor scores of food pattern 3 "traditional-inland" (p=0.001)

To follow the "purchase" pattern makes especially sense for farmers with additional business/service as it can be assumed that they have additional cash income besides subsistence farming and, therefore, can also spend money to buy foods. However, many participants were not subsistence farmers alone but also sold part of their agricultural products and, therefore, were obviously able to follow the "purchase" pattern as well. Nearly one third of farmers with additional business/service consumed food according to the "pulses" pattern, while only 20% of mixed crop and livestock farmers did so, and even much less of crop farmers (Table 3.3.24). As stated earlier, pulses were said to replace vegetables in a meal if possible, meaning that those who could afford it (i.e. who had additional cash income) would rather go for pulses than for the everyday green leafy vegetables (see also chapter 4.1).

Regarding vegetable production variables, the "traditional coast" pattern, being significantly correlated to different vegetable production parameters, was favoured mainly by participants living in coastal Muheza. There, the number of seasons, during which vegetables were cropped, was highest as well as the cultivated and collected vegetable diversity overall. The "traditional-coast" pattern, however, was not characterised by a high vegetable consumption and, consequently, from these findings a direct relationship between vegetable cropping and consumption can not be detected.

The results found through bivariate correlations between food patterns and nutrition variables (Table 3.3.27) suggests that when the "traditional-inland" pattern was consumed, food diversity was generally low, while it was high when the "purchase" pattern was followed. Also when mean consumption values per quintile of each food pattern were calculated (Table 3.3.26), both FVS and DDS were considerably higher for women eating according to all patterns but the "traditional inland" pattern. If it is assumed that pattern 3 is traditional and pattern 2 influenced by the 'nutrition transition', then the higher diversity within pattern 2 is due to more different "new" foods; in general, the 'nutrition transition' is, besides others, associated with greater dietary variety (Drewnowski 2000). Yet, higher diversity as such does not necessarily implicate a "better" nutrition, as with the 'nutrition transition' also the access to cheap energy-dense foods is improved (Drewnowski 2000) (see chapters 4.2 and 4.5).

The intake of fat differed between the first and fifth quintile of the "purchase" and the "traditional inland" pattern and in both cases fat intake was higher when consumption was according to the respective pattern. For the "animal products" pattern, although being not significantly correlated to fat intake, a positive trend for fat intake was observed meaning that participants within the 5th quintile consumed significantly more fat than in the other quintiles. This makes sense as a higher intake of meat, dairy products and eggs can be associated with a higher intake in fat, especially saturated ones as well as cholesterol (Schmidhuber and Shetty 2004).

Vitamin A intake was significantly higher for participants eating according to pattern 1, which could be due to fruits playing a major role within this pattern, as well as pattern 2 for which no direct explanation is at hand (see chapter 4.4). Iron intake was, in turn, higher for women who did not follow food pattern 1 or 2. Interestingly, iron intake was also negatively and significantly correlated to pattern 5 and higher for participants from the first quintile of pattern 5, which were those consuming hardly any animal products. Yet, there was a significant trend for an increased iron intake the more the "traditional inland" pattern was followed (Table 3.3.26). This would suggest that while animal products are, of course, the best source of iron, dark green leafy vegetables as a main component within the "traditional inland" pattern contributed to an increased iron intake of women. Vegetables are, in fact, the major source of micronutrients for people using only vegetarian diets (Misra et al. 2008) (see also chapter 4.3).

Regarding health variables, the higher a participant's affiliation with the "purchase" pattern, the higher was her BMI (Figure 4.6.7). This is not surprising as this pattern was dominated by bread and cakes usually baked in oil, and by sugar and tea, thus, mainly high energy-dense foods contributing to a high dietary energy intake. Besides the intake, energy output and, thus, physical activity of a person need to be taken into account to tell something about energy imbalance and, consequently, about body weight (Caballero 2007), which has been discussed in chapter 4.2.

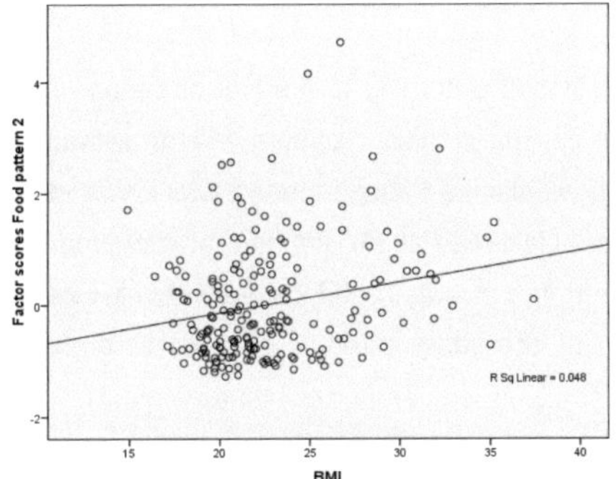

Figure 4.6.7
Association between BMI and factor scores of food pattern 2 "purchase" (n=210; p=0.005); data surveyed during three different seasons in three districts of Tanzania

When participants were grouped into quintiles, the same associations could be shown, e.g., that the median BMI was clearly higher for participants who ate according to the "purchase" pattern (Table 3.3.28). Also when the BMI was grouped into four categories, more obese and overweight women were found in the fifth quintile (high consumption according to this pattern) than in all other quintiles of the "purchase" pattern (Table 4.6.2). As overweight and obesity were, in fact, more prevalent than underweight in the researched districts (chapter 4.2), the recommended dietary pattern out of the five patterns discovered would be the "traditional-inland" pattern.

In addition, the number of positive characteristics named by participants for a corpulent person was significantly higher for those women consuming food mainly according to the "purchase" pattern (p=0.019), i.e., women who consumed many food items from the respective pattern thought rather positive about a corpulent or overweight person. Furthermore, for the "traditional-coast" pattern a significant and increasing trend was measured from the 1st to the 5th quintile (p=0.045), meaning that also women who favoured this pattern thought rather positive about a corpulent person. This can be related to the fact that women who ate according to this pattern mainly lived in coastal Muheza where more participants were overweight or even obese than in the other two districts (chapter 4.2).

Table 4.6.2   BMI of participants and quintiles of food pattern 2 "purchase" (p=0.038); data surveyed during three different seasons in three districts of Tanzania

	n	1st	2nd	3rd	4th	5th
Underweight	17	1 (5.9)	4 (23.5)	2 (11.8)	8 (47.1)	2 (11.8)
Normal weight	146	34 (23.3)	31 (21.2)	31 (21.2)	25 (17.1)	25 (17.1)
Overweight	34	7 (20.6)	6 (17.6)	6 (17.6)	4 (11.8)	11 (32.4)
Obese	13	0 (0.0)	1 (7.7)	3 (23.1)	5 (38.5)	4 (30.8)

10 cells (50%) had expected count less than 5; the minimum expected count is 2.60

Those women with eating habits according to the first or second pattern had a lower Hb, while those women following the "animal products" pattern had, on average, a higher Hb value (Figures 4.6.8 – 4.6.10). This can be explained through the fact that iron is especially easily absorbed from animal products (Strain and Cashman 2002); yet, the "animal products" pattern did not positively correlate with the measured iron intake. Therefore, the Hb-value as a measure for iron status as well as the determination of iron intake through 24-h recalls are examined in more detail (see chapter 4.3).

In accordance with these findings, the Hb value differed significantly between participants of the different quintiles of both the "traditional-coast" and the "purchase" pattern. For both patterns there was a decreasing trend from the 1st to the 5th quintile, meaning that the Hb was lower for those who followed these dietary patterns (Table 3.3.28). As this data only displays associations, the causation can not be explained and has to be treated especially careful in terms of Hb values. Both the "traditional-coast" and the "purchase" pattern were highly favoured by women from Muheza district. At the same time, it was found that malaria was much more prevalent in Muheza than in the other two districts, thus, having most likely certain consequences on the Hb value/the iron status of people living in Muheza (chapter 4.3). The lower Hb status could be partly due to the high consumption of black tea, which was a central food item in the traditional coastal diet. Black tea, if drunken directly before, after or with a meal can inhibit the bioavailability of iron in the food and, consequently, the intake of iron into the body (King and Burgess 1992); and in fact, iron intake was significantly lower for participants who followed the "'purchase" pattern (Table 3.3.28) as well as the "traditional-coast" pattern.

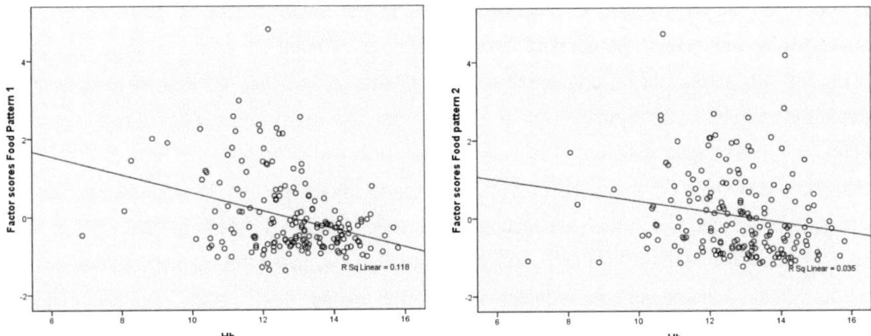

Figure 4.6.8
Association between Hb and factor scores of food pattern 1 "traditional-coast" (n=185; p<0.001); data surveyed during three different seasons in three districts of Tanzania

Figure 4.6.9
Association between Hb and factor scores of food pattern 2 "purchase" (n=185; p=0.001); data surveyed during three different seasons in three districts of Tanzania

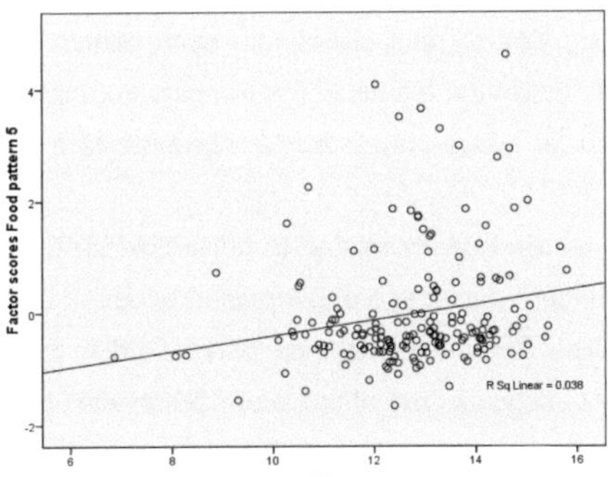

Figure 4.6.10
Association between Hb and factor scores of food pattern 1 "animal products" (n=185; p=0.003); data surveyed during three different seasons in three districts of Tanzania

RBP values did not show significant differences between quintiles of all five dietary patterns. Still, a significant positive trend was found from the 1st to the 5th quintile of the "animal products" pattern, suggesting that RBP was higher for those participants who followed this pattern. This could be due to the better availability of preformed vitamin A from foods of animal origin (Bender 2002).

While the intake of vitamin A was higher for those participants who followed either the "traditional-coast" or the "purchase" pattern, there was no relationship between these food patterns and the vitamin A status of a person according to the RBP value. This suggests that the vitamin A intake, as measured by 24-h recall, is probably difficult to be linked directly to RBP values. While bioavailability was taken into account through applying different conversion factors for different foods, still the amount of fat consumed with the vitamin A rich food as well as the mechanical preparation can influence vitamin A intake and retention to a great extent (Ahmed and Darnton-Hill 2004) (see also chapter 4.4).

Vitamin A intake was not associated with the "traditional-inland" pattern, in which vegetables were a major component, yet, it was associated with the first pattern in which the food group "fruits" loaded high. While many other factors will influence vitamin A intake this might be further evidence that for vitamin A intake not only vegetables – as it was focused by this study – but also fruits must be considered.

<u>Multiple relations</u>

While multiple regression analysis was used to determine whether the affiliation to a pattern influenced, e.g., a woman's BMI (chapter 4.2) or Hb (chapter 4.3), it was also calculated *vice versa* with regression analysis if the affiliation to a pattern itself is influenced by certain socio-economic, health or nutrition variables. The results for patterns 1 to 4 are shown in Figures 4.6.6 – 4.6.9.

Different variables were significantly associated with each pattern, and these were only two or three variables per pattern, unlike the results of the bivariate correlations. For pattern 1 "traditional coast", only vitamin A and fat intake played a role, however, to a minor extend as the B value suggests.

The "purchase" pattern was significantly related to household size: when the latter increased by one, the affiliation to the pattern (factor score) decreased by about 2%. Consequently, the smaller the household, the more likely a woman consumed food according to the "purchase" pattern. It can be argued that smaller households can afford more purchased food items as they are more wealthy compared to large households, which confirmed earlier results (Figure 4.6.5).

When a persons FVS increased by one, the affiliation to the "purchase" pattern increased by about 7%. This is conform with the assumption that, as soon as food items are not only self-produced but purchased from other sources, food variety increases. Food variety also increases if not only foods are purchased instead of being self-produced but also if sufficient money is available to be spent on non-staple food items (Torlesse et al. 2003) (see also chapter 4.1). The positive relation of the "purchase" pattern to fat intake suggests that with an increase in fat intake of 10 g/d the affiliation to this food pattern increases by 4%. This can be due to the dominating food group "bread/cakes" which refers to products fried or baked in oil. As it was mentioned before, pattern 2 revealed that the 'nutrition transition' already takes place in Tanzania (see also chapters 4.2 and 4.5).

The food group bread/cakes only became popular after cooking oil and wheat became available in sufficient and affordable quantities. In Tanzania, the production of, e.g., sunflower seeds increased from about 12,000 t in 1961 by more than double to about 28,000 t in 2006; similarly, overall oilcrop production nearly doubled during the same time period from about 72,000 t to about 138,000 t (FAOSTAT 2008). While the increase in dietary fat is part of a recent 'nutrition transition' in Tanzania, the consumption of black tea, usually in combination with large amounts of sugar, is not new but still part of a 'nutrition transition' if the latter is considered to have actually occurred over the past 400 years already, namely since the onset of colonial occupation (Raschke and Cheema 2008). Obviously, the 'nutrition transition' has reached all researched districts of Tanzania in terms of tea and fried breads and cakes, while the consumption of other food groups still differs between the districts and was rather local and traditional.

The associations between iron, protein and energy intake with pattern 3 "traditional inland" were only weak (Figure 4.6.13). Thereby, iron and protein intake were negatively associated which suggests that these nutrients were most likely insufficient in the "traditional-inland" pattern, consisting mainly of cereals and vegetables. Only the energy intake was positively related to this pattern which could be due to the oil or fat also being dominant within this pattern and usually being used to fry or cook the vegetables. Consequently, within one food pattern an increasing energy intake is accompanied by decreasing intakes of other nutrients which highlights again the importance of a holistic pattern approach instead of exploring single nutrients.

Pattern 4 "pulses" was associated with protein intake in a way that with an increase of protein intake by 10 g/d the average affiliation to the pattern increased by about 0.2. This is an expected relationship as pulses are a good source of protein (Young and Reeds 2002). Further, there is a strong relationship to the FVS: when the latter increased by one the average affiliation to pattern 4 increased by about 0.2. Obviously, the domination of pulses within this food pattern did not mean that food variety was low.

Within the models neither BMI nor Hb showed a relationship to any of the food patterns; similarly all four patterns were not associated to any vegetable production variable, thus, no direct connection was found between production and consumption. Still, this was the case when only vegetable production and consumption is considered (chapter 4.1).

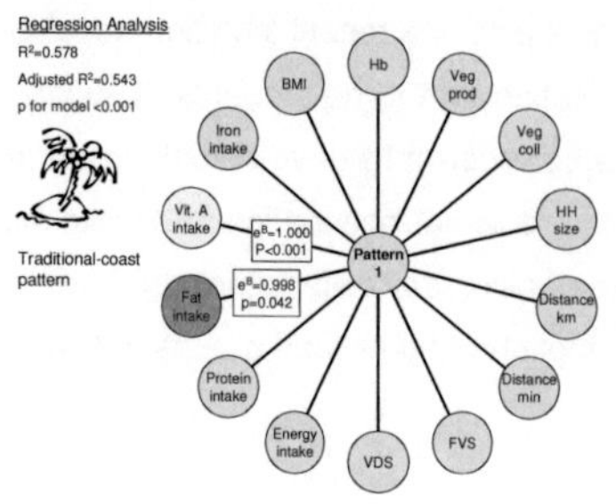

Figure 4.6.11
Results of multiple regression analysis with dietary pattern 1 as dependent variable; data surveyed during three different seasons in three districts of Tanzania

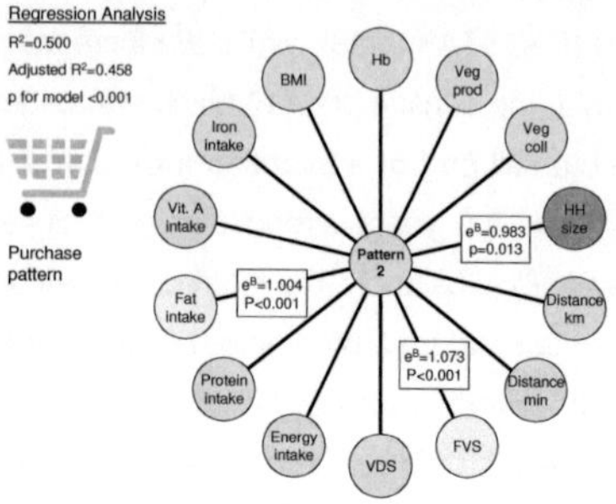

Figure 4.6.12
Results of multiple regression analysis with dietary pattern 2 as dependent variable; data surveyed during three different seasons in three districts of Tanzania

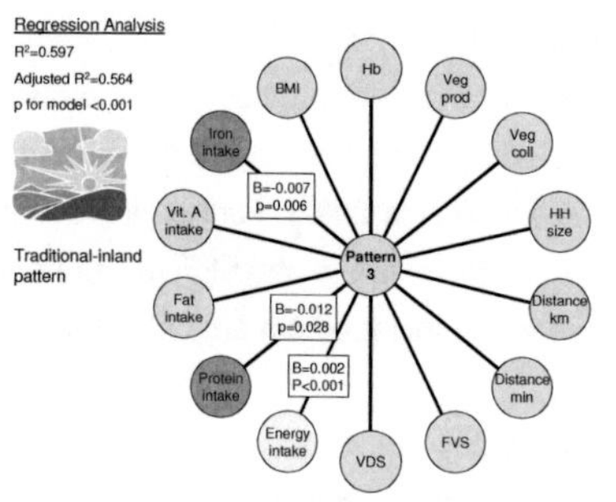

Figure 4.6.13
Results of multiple regression analysis with dietary pattern 3 as dependent variable; data surveyed during three different seasons in three districts of Tanzania

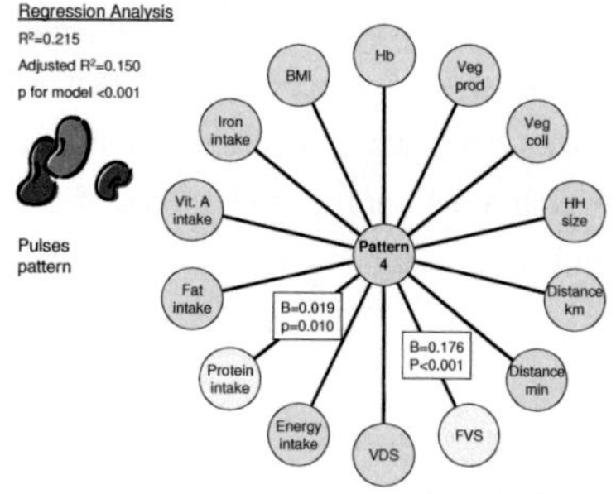

Figure 4.6.14
Results of multiple regression analysis with dietary pattern 4 as dependent variable; data surveyed during three different seasons in three districts of Tanzania

4.6.2 Dietary patterns generated through cluster analysis

Food consumption of study participants was clustered generating five food patterns (Table 3.3.21; Figure 3.3.43). The first cluster was clearly characterised by cereals and vegetables and, therefore, resembles the "traditional-inland" food pattern. Participants who followed the second cluster, consumed more cereals than vegetables but also pulses marked this dietary pattern. The third cluster was, besides cereals and vegetables, characterised by tea, while in the fourth cluster, besides cereals and tea, fruits were as important as vegetables. For participants within the fifth cluster, it seems as if tea was even substituting a meal, e.g. breakfast, which was quite common for some participants. In general, the two methods used, namely PCA and cluster analysis, did not result in exactly the same food patterns, yet, some similarities existed: cluster 1 equals pattern 3 "traditional-inland"; cluster 4 pattern 1 "traditional-coast" and cluster 5 pattern 2 "purchase" to a certain extent, while the other clusters and patterns were not possible to match.

<u>Associations of food clusters with other variables</u>

When different variables were related to the five food clusters, similar results as with food patterns derived through PCA were found. Therefore, in the following only new outcomes will be discussed. Regarding socio-economic characteristics (Table 3.3.29), namely the marital status, more single women were found in clusters 4 and 5, while married women rather ate food according to cluster 1. This coincides with the position of the participant within the household (hh) and confirms the relationship between food scores and status within the hh (chapter 4.5.2), namely that wives of the hh head consumed a rather simple and traditional diet (cluster 1), while women as the head of hh and single women consumed a more varied diet (clusters 4 and 5). This corresponds also to the finding that women with additional income from other activities than horticulture consumed food mainly according to clusters 5 as well as 3 and 4, while those being crop farmers only consumed rather a simple and traditional diet (Table 3.3.29). Position in the society as well as income parameters influenced the dietary pattern of women, yet, it must also be considered whether a dietary pattern could be linked to good nutritional health.

When the absolute number of vegetables cropped per woman was related to food patterns, there was a significant association (p=0.046), however, with too many cells (25%) having an expected count less than 5. About 50% of participants eating according to cluster 1 cropped only few vegetables (1-3). By contrast, most women within clusters 4 and 5 (74% and 72%, respectively) cropped a greater diversity of vegetables (4-9). Consequently, a high vegetable diversity in the field was accompanied by high dietary diversity on the plate, yet, not necessarily high vegetable diversity (see chapter 4.1).

The diversity of foods and food groups increased from cluster 1 to 5 (p<0.001 for both DDS and FVS). This means that participants eating according to clusters 4 and 5, which were comparable to the "traditional-coast" and "purchase" patterns generated with PCA, had a more diverse diet (Table 3.3.30).

The intake of vitamin A and iron differed significantly between clusters with the highest vitamin A intake achieved by participants within cluster 4, while the highest iron intakes were found in clusters 1 and 2 (Table 3.3.31). Like in dietary patterns derived through PCA, vitamin A and iron intake were opposed to each other, i.e., when the intake of one was high in a food cluster, the intake of the other was low in the same food cluster. This is rather unfortunate, as dietary deficiencies of both vitamin A and iron frequently coexist in developing countries (Mejia and Chew

1988), while, at the same time, an increased vitamin A intake may result in increased haemoglobin concentrations, at least in populations with low serum retinol levels (Suharno et al. 1993). Therefore, it is highly desirable to achieve sufficient intakes of both vitamin A and iron through one food pattern or cluster.

Regarding health variables (Table 3.3.32), similar to the food patterns generated by PCA, a low Hb was associated with those food patterns where black tea was a characteristic food item. A normal iron status was found in women who consumed the food pattern marked by vegetables. However, the connection between Hb and food consumption must be again treated carefully as malaria will interfere into this relationship (chapter 4.3).

4.6.3   Conclusions

There is no doubt that generating and using dietary patterns is highly reasonable in nutrition research and, particularly, in comparison to focusing on single nutrients (Schulze and Hoffmann 2006). This is especially so in the present study with insufficient data at hand about nutrient contents of processed local foods, nutrient bioavailability, intake and retention. Moreover, dietary patterns are considered a valuable tool in nutrition epidemiology for assessing the impact of dietary intakes on the risk of various diseases and mortality (Kant 2004).

In fact, both methods used in this study, PCA and cluster analysis, were found to work well in identifying major dietary patterns of a particular study population as stated by Hu et al. (2002). At the same time, dietary pattern research is not yet standardised, i.e., so far no standard procedures are used for generating such patterns. Consequently, comparison is difficult as it appears in the present study where two different types of analysis were applied but different results were gained to a certain extent. Besides cluster analysis and factor analysis or PCA, an additional tool for dietary pattern analysis, namely reduced rank regression (RRR), was recently introduced into nutritional epidemiology (Hoffmann et al. 2004). This method is a mix of exploratory and hypothesis-oriented approaches and, consequently, uses both prior knowledge and available dietary data of a study (Schulze and Hoffmann 2006). For the analysis of the present study the RRR was not applied due to lack of experience with this statistical method; yet, for further studies it is suggested to try this promising tool especially when dietary patterns should be linked to certain public health problems.

From the analysis of the present data from Tanzania and the applied methods, a number of conclusions can be drawn:

> No direct association between vegetable cropping diversity and vegetable consumption within the food patterns (generated through PCA) was established; rather, a high vegetable diversity in the field was accompanied by high dietary diversity on the plate in general, yet, not necessarily high vegetable diversity.

> Recommendations for a dietary pattern from the five patterns available (generated through PCA) would be in favour of the "traditional-inland" pattern. Participants within the 5th quintile of this pattern had a slightly lower BMI and were not inclined to overweight or obesity; they had further a higher Hb on average and, therefore, less or no iron deficiency. In addition, the median vitamin A intake of participants who followed this pattern was relatively higher. Still, food variety and dietary diversity was lowest for participants consuming food according to this pattern, suggesting that, before recommending it, this pattern needs to be adjusted according to sensible dietary guidelines.

> Concerning vitamin A intake, besides the "traditional-inland" pattern, also the patterns "traditional-coast" and "purchase" seemed to be of advantage, while for iron intake a positive trend was only seen within the "traditional-inland" pattern, which would again suggest this pattern as the favourite one among the five (generated through PCA).

> Both pattern analyses showed similar results in terms of a low Hb being associated with those food patterns where black tea was a characteristic food item, while a normal Hb was found with women who consumed the food patterns marked by vegetables.

> Similar results were also found regarding DDS and FVS, which were low for both participants within cluster 1 and those who followed pattern 3 "traditional-inland", and which were high for clusters 4 and 5 as well as patterns 1 "traditional-coast" and 2 "purchase".

Table 4.6.3 gives an overview of differences between associations of both patterns to different variables; while food patterns derived through cluster analysis could be only related as a group to further variables and were, several times, not related to a variable at all, dietary patterns derived through PCA could be viewed individually and, in most cases, at least one food pattern was related to a variable. Through this the difficulties in comparing results derived through the different methods become apparent, and it is concluded that though participants can be better allocated to food patterns when they are generated through cluster analysis, PCA is preferred as a tool in the present study, as more precise and sophisticated results were obtained and interpretability of results was more reasonable.

Table 4.6.3   Characteristics of participants (variables) and whether they were significantly associated to food patterns derived through PCA and cluster analysis

Dietary patterns (PCA) Association with pattern no.					Variables	Food patterns (Cluster) Association	
1	2	3	4	5		pos./neg.	no
					Socio-economics		
✓	✓	✓	✓	✓	Districts	✓	
					Ethnic group	✓	
				✓	Wealth		✓
	✓	✓			Household size		✓
			✓	✓	Distance (km and min)	✓	
	✓		✓		Occupation	✓	
			✓		Education		✓
		✓			Religion	✓	
					Status within household	✓	
					Marital status	✓	
					Vegetable production		
✓				✓	No. of vegetables cultivated	✓	
✓					No. of vegetables collected		✓
✓					No. of seasons vegetables were cultivated/collected		✓
					No. of seasons vegetables were sold	✓	
					Nutrition		
✓	✓	✓	✓	✓	FVS	✓	
✓	✓		✓	✓	DDS	✓	
				✓	VDS		✓
✓	✓	✓			Energy intake	✓	
✓	✓	✓	✓	✓	Protein intake	✓	
	✓	✓			Fat intake		✓
	✓	✓			Vitamin A intake	✓	
✓	✓	✓		✓	Iron intake	✓	
					Health		
	✓				BMI		✓
✓	✓			✓	Hb	✓	
				✓	No. of negative characteristics for a corpulent person		✓

In the introduction of this chapter, it was stated that exploratory approaches, as they were used in this study, do not necessarily represent optimal patterns and one has to carefully check if the generated patterns fit into common eating habits of the population (Hu et al., 2002). Therefore, it is important to know how many study participants followed a pattern, which is easy in terms of cluster-derived patterns, yet, for PCA-derived patterns only an estimation can be made (here: all participants with factor score >1); further it has to be kept in mind that one participant can theoretically follow several PCA-derived food patterns (Table 4.6.4). For dietary patterns generated through PCA, there were not many differences between the number of participants who followed a pattern, only less participants followed the "animal products" pattern number 5. This coincides with common eating habits in the researched rural areas, were still the consumption of animal products, especially meat, is less common than in urban areas (Njelekela et al. 2002).

Table 4.6.4   Number and share of participants (n=252) who followed a certain dietary pattern

	1	2	3	4	5
PCA pattern	41 (16.3%)	39 (15.5%)	44 (17.5%)	42 (16.7%)	33 (13.1%)
Cluster pattern	96 (38.1%)	29 (11.5%)	42 (16.7%)	40 (15.9%)	45 (17.9%)

PCA: number of participants out of 252 who had a factor score >1 for each pattern
Cluster: all 252 participants distributed among the five clusters

The distribution among participants within the cluster groups shows that a majority of more than one third followed a traditional food pattern determined by vegetables and cereals (cluster 1), while another 18% of participants consumed food according to the fifth cluster, which is comparable to the "purchase" pattern being characterised by tea and also the only one in which bread/cakes played a role. The latter gives a hint on the 'nutrition transition' which is on the rise according to different studies conducted in Tanzania (Njelekela 2002; Maletnlema 2004; Villamor 2006). While the 'nutrition transition' is often thought to be characterised by a sharp rise in meat and milk consumption, the spread of junk food and soft drinks, and rising food consumption away from home, this is only true for a later stage of the transition as it appears in wealthy countries. The early stages, however, are characterised by an increased use of cheap vegetable oils, i.a. palm and sunflower, which are rapidly integrated into local diets (Drewnowski 2000) and which, in fact, happens in rural Tanzania right now (see also chapters 4.2 and 4.5). It is suggested as a next step that the generated patterns should be compared to dietary guidelines, e.g., from FAO (Burgess and Glasauer 2004) or local institutions (Dietary guidelines for Morogoro and Iringa; TARP II – SUA Project 2004). From this comparison existing gaps between guidelines and reality could be identified, which could, in turn, help to find appropriate approaches and tools to overcome these gaps.

5   Overall conclusions and outlook

Within the results and discussion chapters findings of this study are presented topic-wise and several linkages between different parameters were tested, analysed and discussed. To conclude, first of all an overview of the most relevant links between vegetable production, vegetable and food consumption, and nutritional health in Tanzania shall be given. In addition, an outlook on a possible way forward and recommendations are presented. While the relationship between vegetable consumption and health is often the central topic of investigations, the link between production and consumption and, especially, the link between production and health seems to be less explored.

<u>Production and consumption</u>

Regarding the link between production and consumption, it can be concluded that there was a clear relationship between vegetable diversity produced and diversity consumed (vegetable diversity score (VDS)), while there was no clear link between production and

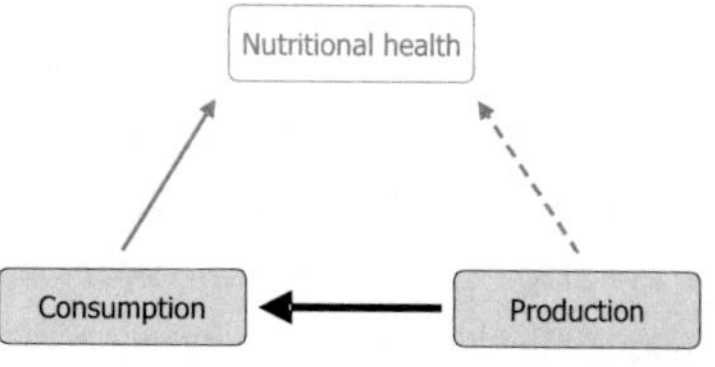

consumption quantity, except for single seasons. Influencing factors on consumed vegetable quantity are suggested to be seasonality; the purchase of additional vegetables (especially exotic) next to home-grown ones; and knowledge, attitudes and preferences of women regarding vegetable consumption (such as taste and convenience in preparation).

The VDS was significantly correlated to the number of vegetables produced and especially to the number of traditional vegetables collected. While this relationship could not be found for exotic vegetables, it is suggested that, as long as subsistence farming is dominant, and traditional vegetables are favoured and used, the diversity of vegetables in the field can very well influence vegetable diversity on the plate of women in rural Tanzania and, consequently, their nutritional health. Yet, similarly important is the preparation of foods, e.g., frying in oil, cooking for a long time, and a change in preparation techniques could have a lasting effect on nutritional health.

When analysing food patterns (generated through PCA), no direct association between vegetable cropping diversity and vegetable consumption was established; rather, a high vegetable diversity in the field was accompanied by high dietary diversity on the plate in general, yet, not necessarily high vegetable diversity. Still, for the study population in Tanzania it was found that food consumption was (still) influenced by production and that, therefore, food consumption issues should always integrate food production and *vice versa*.

<u>Consumption and nutritional health</u>
It could be shown that dietary pattern No. 2 "purchase" and food variety had an influence on the BMI of women. Additionally, through dietary pattern No. 2 the consumption of cheap vegetable oil was increased, which, in turn, was identified as part of the early stage of the 'nutrition transition'. Both food consumption and nutritional

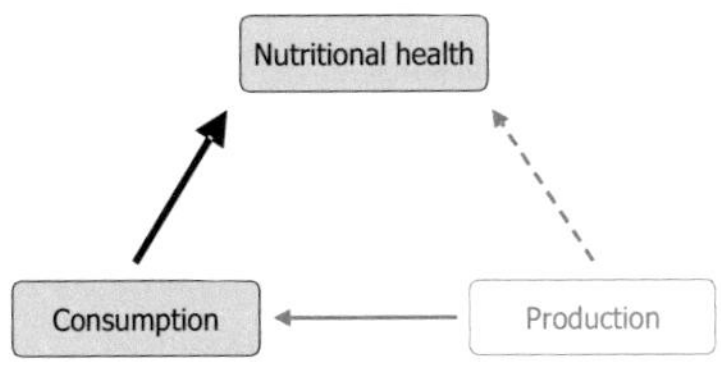

health data of the present study could show what was already found for other developing countries, namely, that the obesity epidemic is on the rise, even in rural, poor and underdeveloped regions, fuelled primarily by increased consumption of cheap oil, sugary drinks, animal products and a decrease in physical activity (Popkin 2007). DDS and FVS were linked to the BMI of women in a way that the higher the diversity of foods and food groups consumed the higher the BMI. Thus, a high dietary diversity is not *per se* a guarantee for a healthy diet, yet, food types and food groups that contribute to a high diversity are decisive.

Regarding iron in nutrition, it was found that vegetable diversity and vegetable amount consumed were not associated with the iron status of the study population. However, both pattern analyses (PCA and cluster) showed similar results in terms of a low Hb being associated with those food patterns where black tea was a characteristic food item, while a normal Hb was found with women who consumed the food patterns marked by vegetables. Though it is well known that the consumption of black tea influences bioavailability and intake of iron, participants of this study where not aware of this fact or at least did not act accordingly. In general, the investigation of iron nutritional status was highly affected by the occurrence of malaria.

With the given study population, it was not possible to identify significant links between vitamin A status, vegetable/food consumption and nutritional knowledge. However, participants who followed dietary pattern No. 3 "traditional-inland", had a higher vitamin A intake than other women; they had further a higher Hb on average and, therefore, less or no iron deficiency. In addition, those women had a slightly lower BMI and were not inclined to overweight or obesity. Obviously, linking food patterns to nutritional health was much more effective than linking single food or nutrient intake to the health of participants.

<u>Nutritional health and production</u>
The only production parameter associated significantly to BMI was the production of exotic vegetables, which was negatively related to BMI values. It was expected that the production of exotic vegetables was related to wealth and higher education and that it was, therefore,

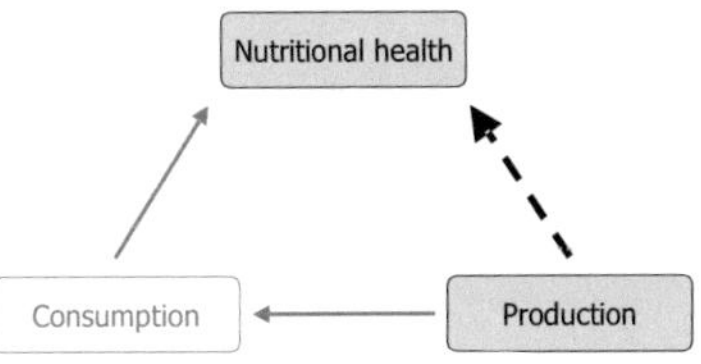

also related to a higher BMI, which would be a typical situation in developing countries (Monteiro et al. 2004; Villamor et al. 2006). However, wealth, which was associated with the cultivation of exotic vegetables, was not significantly related to BMI in this study. It was rather the other way round, suggesting that the cultivation of exotic vegetables was associated with a normal BMI and, thus, better health. Consequently, cultivating exotics in the respective districts in Tanzania would stand for better health in terms of BMI, whereas cultivating only traditional vegetables could be associated with a high BMI. The cultivation of exotic or traditional vegetables alone is, of course, not sufficient for determining or influencing BMI values, yet, this parameter may function as an indicator.

Regarding iron in nutrition, it was found that vegetable diversity in production was associated with Hb levels of participants, however, both positively and negatively during different seasons, thus, showing no clear direction. Furthermore, no relationship between vitamin A status of women and vegetable production was found. Consequently, it must be acknowledged that, from the present data, a direct conclusion from vegetable production to nutritional health can virtually not be drawn, and it is suggested to take the "step" in-between and analyse consumption, like in the nutrition security model (Figure 1.1). Still, vegetable or agricultural production should be definitely taken into account, when researching nutritional health, especially, when the investigated population consumes a great amount of self-produced foods.

Outlook, recommendations and arising research questions
While in the researched districts in Tanzania a clear link between consumption and health was found and a relationship between production and consumption, to a certain extent, a direct link between production and nutritional health could only partly be shown. Obviously, the focus on vegetable production seems to be not sufficient, but overall food production needs to be taken into account. Still, the link between agricultural production and health is emphasised, and both food-based and agricultural-based strategies are still seen as the most important approaches to improve nutritional status as well as household food and nutrition security, especially in rural areas (Khor 2008).

While usually a high nutritional diversity is desirable for a balanced diet (Burgess and Glasauer 2004), it does not necessarily result in a better health status, and a certain degree of diversity does not mean that the dietary quality will match people's particular needs (Brown et al. 2002). Similarly, in this study DDS and FVS could not clearly indicate a benefiting or a disadvantageous nutrition (e.g. the higher the better). Thus, the tool of food scores must be developed further in order to distinguish more precisely between desirable and non-desirable diversity; still, its easy and quick application and usage in the field should be maintained.

When investigating consumption and health, it is advised to refrain from analysing single nutrient intake in more detail, but rather concentrate on whole foods, food groups, dietary diversity and dietary patterns. Especially dietary intake in relation to diseases is highly complex, as foods contain many nutrients, and foods are eaten in combinations and, thus, it is difficult to credit effects to single dietary components (Schulze and Hoffmann 2006). A possible next step would be to assign certain dietary patterns to specific health effects (positive or negative) and, from this, to deduce dietary recommendations and dietary guidelines for a specific area or population in Tanzania.

The 'nutrition transition' is on the rise according to different studies conducted in Tanzania (Njelekela 2002; Maletnlema 2004; Villamor 2006). The early stage of the 'nutrition transition', characterised by an increased use of cheap vegetable oils, which are rapidly integrated into local diets (Drewnowski 2000), is already achieved in rural Tanzania. Following the early stage, the advanced stage of the 'nutrition transition' is well known and described (Popkin 1999). It must be questioned whether Tanzania is about to reach this stage sooner or later, how soon this could be and what could be done against it – or how to steer the transition into the right direction. Nutrition interventions should be, therefore, always accompanied by creating access to adequate public health services as well as to education.

In a nutshell, the main recommendations for future research are the following:

→ <u>Traditional vegetables</u>: If they shall be kept available for future generations, it is necessary to propagate these crops and make their nutritional values available, especially in urban areas.

→ <u>Exotic vegetables</u>: Check the influence of them or, in general, of exotic foods in production and consumption on nutritional health.

→ <u>"Muheza effect"</u>: Even if a great diversity is available by nature this does not guarantee its usage by the local population. The knowledge about this diversity, why it should be used and certain incentives are similarly important.

→ <u>Taste</u>: The latter was more important for participants when choosing a vegetable for consumption than health issues, suggesting that, next to nutritional knowledge, preferences of people explicitly need to be considered in both agriculture and nutrition research.

→ <u>Seasonality</u>: A link between vegetable production and consumption was only found during single seasons, therefore, seasonal change need to be always taken into account.

→ <u>Dietary diversity scores</u>: Enhance them as a tool for assessing dietary diversity together with dietary quality e.g. by weighing single foods according to their importance within the food circle or food pyramid.

→ <u>Dietary patterns</u>: determine typical dietary patterns of people as a measure for nutritional health and verify whether they are linked to a sound or poor health status. Modify existing patterns in order to create suitable, healthy and balanced dietary patterns i.a. for Tanzania.

→ <u>Dietary guidelines</u>: Eleborate them for the whole country of Tanzania, preferably district- or area-wise. When writing dietary guidelines in general, ethnic groups reflecting local customs and habits need to be focused on.

→ <u>Nutrition transition</u>: When investigating the trend of the 'nutrition transition' in Tanzania, there should be a particular focus on rural areas; besides studying eating habits, the nutritional knowledge of people, their attitudes and behaviour should be taken into account.

An improved nutrition and, consequently, improved health can only be achieved when the interactions among food diversity on the plate, people's nutritional knowledge, preferences, socio-economic factors and, finally, food diversity in the field are considered.

6 Summary

This cross-sectional sequential study investigated the link between vegetable diversity available ("production") and dietary diversity of women ("consumption") in three different districts of rural Tanzania. Furthermore, the relationship between the nutritional health status of participants and cropping and dietary diversity was analysed.

The study was carried out during three different seasons within one year (2006/2007) in 18 villages of three districts in north-eastern and central Tanzania including 252 women. The survey included an individual interview on vegetable production, food consumption (i.a. 24h-recall, 7d-recall on vegetables) and nutritional knowledge, and the measurement of body mass index (BMI), haemoglobin (Hb) for iron status and different parameters for vitamin A status. Besides studying single nutrients, food groups and health problems, also a more holistic view was taken on dietary diversity/food variety and dietary patterns and their relationships with nutritional status and vegetable production. Relationships were investigated both through bivariate correlations and multiple regression analysis; dietary diversity and food variety scores were calculated and dietary patterns were generated through principal component analysis (PCA) as well as cluster analysis.

Regarding the link between production and consumption, there was a clear relationship between vegetable diversity produced and diversity consumed, while this was not so clear between production and consumption quantity, except for single seasons. Influencing factors on consumed vegetable quantity were suggested to be seasonality; the purchase of additional vegetables (especially exotic) next to home-grown ones; and knowledge, attitudes and preferences of women regarding vegetable consumption. When analysing food patterns, no direct association between diversity of vegetable cropping and vegetable consumption was established. Yet, for the study population in Tanzania it was found that food consumption was (still) influenced by local production and that, therefore, food consumption issues should always integrate existing food production and *vice versa*.

The link established between food consumption and nutritional health data of the present study showed that the obesity epidemic is on the rise, even in rural, poor and underdeveloped regions of Tanzania. Furthermore, it was found that a high dietary diversity is not *per se* a guarantee for a healthy diet, yet, food types and food groups that contribute to a high diversity are decisive. A direct link between production and nutritional health could only partly be shown. Obviously, the focus on vegetable production seems to be not sufficient, but overall food production needs to be taken into account.

Further recommendations for future research are, i.a., to investigate the influence of exotic vegetables and, generally, exotic foods in production and consumption on nutritional health; to enhance dietary diversity scores as a tool for assessing dietary diversity together with dietary quality; to eleborate dietary guidelines for Tanzania, preferably district- or area-wise; to investigate the nutrition transition in Tanzania especially in rural areas with a focus on the nutritional knowledge of people, their attitudes, preferences (e.g. taste) and behaviour.

7   Zusammenfassung

Mit dieser sequentiellen Querschnittsstudie wurde die Verbindung zwischen der vorhandenen Gemüsevielfalt („Produktion") und der Nahrungsvielfalt von Frauen („Konsum") in drei verschiedenen Distrikten im ländlichen Tansania untersucht. Des Weiteren wurde die Beziehung zwischen dem Ernährungszustand der Teilnehmerinnen und deren Anbau- und Ernährungsvielfalt erforscht.

Die Studie wurde während drei verschiedener Jahreszeiten innerhalb eines Jahres (2006/2007) in 18 Dörfern der drei Distrikte in Nordost- und Zentraltansania mit insgesamt 252 Frauen durchgeführt. Die Erhebung beinhaltete ein Einzelinterview über Gemüseanbau, Nahrungsmittelkonsum (u.a. ein 24-Stunden-Recall und ein 7-Tage-Recall bzgl. Gemüse) und Ernährungswissen, sowie die Messung von Body Mass Index (BMI), Hämoglobin (Hb) für den Eisenstatus und verschiedene Faktoren für den Vitamin A Status. Neben der Untersuchung von einzelnen Nährstoffen, Nahrungsmittelgruppen und Gesundheitsproblemen wurde auch ein ganzheitlicher Ansatz verfolgt, um Nahrungsvielfalt sowie Ernährungsmuster und deren Verbindung zu Ernährungsstatus und Gemüseproduktion von Frauen zu prüfen. Beziehungen zwischen verschiedenen Variablen wurden sowohl durch bivariate Korrelationen als auch durch multiple Regressionen getestet. Werte für die Vielfalt an gegessenen Nahrungsmittelgruppen (DDS) sowie einzelnen Nahrungsmitteln (FVS) wurden errechnet und Ernährungsmuster wurden durch Hauptkomponentenanalyse (PCA) und Clusteranalyse gebildet.

Hinsichtlich der Verbindung zwischen „Produktion" und „Konsum" konnte eine eindeutige Beziehung zwischen der produzierten und konsumierten Gemüsevielfalt gefunden werden. Diese Beziehung war nicht vorhanden für produzierte und konsumierte Gemüsequantität, außer während einzelner Jahreszeiten. Die verzehrte Gemüsemenge wurde eher beeinflusst durch die Jahreszeit, den Zukauf von vor allem 'exotischen' Gemüse neben dem Selbstgezogenen, sowie Wissen, Einstellungen und Vorlieben der Frauen bezüglich Gemüsekonsum. Durch die Ernährungsmusterbildung konnten keine direkten Assoziationen zwischen Gemüsevielfalt im Anbau und Gemüsekonsum gefunden werden. Dennoch konnte für die Studienpopulation in Tansania aufgezeigt werden, dass die Ernährung (noch) durch die lokale Produktion beeinflusst wurde und dass deshalb Ernährungsfragen immer auch die aktuelle Lebensmittelproduktion mit einbeziehen und umgekehrt.

Die Beziehungen, die zwischen „Konsum" und Gesundheit gefunden wurden, zeigten, dass Adipositas sogar in ländlichen, armen und unterentwickelten Regionen Tansanias vermehrt vorkommt. Außerdem konnte dargelegt werden, dass eine hohe Vielfalt in der Ernährung nicht an sich eine Garantie für eine gesunde Ernährung ist, sondern dass die Nahrungsmittelart und Nahrungsmittelgruppen selbst, die zu dieser hohen Vielfalt beitragen, entscheidend sind. Eine direkte Verbindung zwischen „Produktion" und Gesundheit konnte nur teilweise aufgezeigt werden. Anscheinend war der Fokus auf Gemüseproduktion nicht ausreichend, sondern die gesamte Nahrungsmittelproduktion hätte in Betracht gezogen werden müssen

Weitere Empfehlungen für zukünftige Forschung beinhalten, u.a., den Einfluss von Anbau und Konsum von exotischen Gemüse und exotischen Lebensmitteln allgemein auf den Ernährungs- und Gesundheitszustand zu untersuchen; den Wert für die Vielfalt an gegessenen Nahrungsmittelgruppen (DDS) als ein Erhebungsinstrument für Nahrungsvielfalt und -qualität zu verbessern; einen Ernährungsleitfaden für Tansania zu erarbeiten, möglichst gebietsweise; die „nutrition transition" in Tansania vor allem in den ländlichen Regionen zu untersuchen mit dem Schwerpunkt auf Ernährungswissen, Einstellungen, Vorlieben (z.B. Geschmack) und Verhalten.

8   Acknowledgements

This study would not have been possible without the support and assistance of many people and institutions. I wish to sincerely thank Prof. Dr. Michael Krawinkel for his scientific guidance and for accepting me as an agricultural student at the Institute of Nutritional Sciences and thereby giving me the chance to carry out truly interdisciplinary research. I am deeply indebted to PD Dr. Brigitte Maass for her continuous advice and valuable criticism throughout all stages of the study as well as for linking my study to the ProNIVA project at AVRDC-RCA in Tanzania.

My special thanks goes to Prof. Dr. John Msuya, my 'local' supervisor in Tanzania, for his valuable advice and logistical support during the period of fieldwork. I extend sincere gratitude also to his family for their warm hospitality in Morogoro. I also extend great thanks to Dr. Detlev Virchow for many professional discussions and his constructive and logistical help at AVRDC-RCA, Arusha.

I am very grateful to my research team Adeltruda Massawe, Grene Tarimo, Mipanema Makame, Roseline Marealle, Ruth Mkopi, Mageni Pastori, Claude Kisinini and Mr Kimaro as well as to Ignaz Swai and Hassan Mndiga for their valuable support in organising the field research. Similarly, I am grateful to all staff at AVRDC-RCA who supported my stay and study in one or the other way. Thanks very much to Drs. Britta and Marc Swai for their cordial hospitality and assistance in Moshi.

I would like to acknowledge the prompt willingness, the excellent cooperation and contributions of all participating women who were interviewed and examined for this study and I am truly thankful to all horticultural extension workers in the different districts for their cooperation.

Many thanks to Patrick Maundu for sharing his great knowledge on indigenous vegetables and helping to create the vegetable picture folder. Dr. Jürgen Erhardt deserves many thanks for answering the never ending questions regarding Nutrisurvey, vitamin A and DBS analysis.

My sincere gratitude goes to the late Marion Mann and her colleagues, Dr. Stephan Schwarze, Dr. Gerrit Eichner and Alexander Kaever for their assistance with creating the questionnaires and performing the statistical analyses.

I wish to thank all my colleagues at both Giessen and Göttingen Universities for fruitful discussions and encouragement, especially Irmgard Jordan and Yujin Lee. Similarly, I am much obliged to the contributions and care in many ways of my husband, my parents, family and friends.

Last but not least I extend my thanks to BMZ/GTZ, DAAD and Eiselen Foundation for funding this study.

9 References

Abdulai A and Aubert D (2005) A cross-section analysis of household demand for food and nutrients in Tanzania. Agricultural Economics, 31(1):67-79

ACC/SCN (2000) Fourth Report on the World Nutrition Situation. ACC/SCN in collaboration with IFPRI. Geneva, Switzerland

ACF (2009) Mathare Sentinel Surveillance Report. ACF (Action Against Hunger) Kenya Mission, Nairobi, Kenya

Adedoyin SA and Taylor OA (2000) Homestead Nutrition Garden Project: An extension communication strategy for eliminating vitamin A deficiency (VAD) among poor households in Nigeria. Nigerian Journal of Nutritional Sciences, 21:34-40

Ahmed F and Darnton-Hill I (2004) Vitamin A Deficiency. In: Gibney MJ, Margetts BM, Kearney JM and Arab L (editors). Public Health Nutrition. The Nutrition Society Textbook Series. Blackwell Publishing, Oxford, UK, pp 192-215

Ali M and Tsou SCS (1997) Combating micronutrient deficiencies through vegetables – a neglected food frontier in Asia. Food Policy, 22:17–38

Allen LH (2008) To what extent can food-based approaches improve micronutrient status? Asia Pacific Journal of Clinical Nutrition, 17(S1):103-105

Anonymous (2003) Social conflict and water; lessons from north-east Tanzania. WaterAid, London, UK. <http://www.wateraid.org.uk/other/startdownload.asp?openType=forced&documentID =247> (accessed 12 August 2003)

Agho KE, Dibley MJ, D'Este C and Gibberd R (2008) Factors associated with haemoglobin concentration among Timor-Leste children aged 6-59 months. Journal of Health, Population and Nutrition, 26(2):200-209

Arkkola T, Uusitalo U, Kronberg-Kippilä C et al. (2007) Seven distinct dietary patterns identified among pregnant Finnish women – associations with nutrient intake and sociodemographic factors. Public Health Nutrition, 11(2):176-182

Azadbakht L, Mirmiran P and Azizi F (2005) Dietary diversity score is favorably associated with the metabolic syndrome in Tehranian adults. International Journal of Obesity, 29(11):1361-1367

Backhaus K, Erwachsen B, Pinke W and Weiber R (1994) Multivariate Analysemethoden. Eine anwendungsorientierte Einführung. Springer Verlag, Berlin, Germany

Baeten GH, Milner J, Corey P et al. (1979) Sources of variance in 24–hour recall data: implications for nutrient study design and interpretation. American Journal of Clinical Nutrition, 32:2546-2559

Bamia C, Trichopoulos D, Ferrari P et al. (2007) Dietary patterns and survival of older Europeans: The EPIC-Elderly Study (European Prospective Investigation into Cancer and Nutrition). Public Health Nutrition, 10(6):590-598

Barker DJ (1992) Fetal growth and adult disease. British Journal of Obstetrics and Gynaecology 99:275-6

Bayani EM (2000) Reducing micronutrient malnutrition: Policies, programmes, issues, and prospects—dietary diversification through food production and nutrition education. Food and Nutrition Bulletin, 21(4):521-526

Baynes RD (1996) Assessment of Iron Status. Clinical Biochemistry 29(3):209-215.

Bedigian D and Adetula OA (2004) Ceratotheca sesamoides Endl. Record from Protabase. Grubben, GJH and Denton, OA (editors). PROTA (Plant Resources of Tropical Africa / Ressources végétales de l'Afrique tropicale), Wageningen, Netherlands. <http://database.prota.org/search.htm> (accessed 20 August 2009)

Bellin-Sesay F (1995) How to combat food insecurity due to droughts? Agriculture and Rural Development, 2

Bender DA (2002) The Vitamins. In: Gibney MJ, Vorster HH and Kok FJ (editors). Introduction to Human Nutrition. The Nutrition Society Textbook Series. Blackwell Publishing, Oxford, UK, pp 125-176

Bioversity International (2009) Bioversity Homepage: Scientific Information – Nutrition. <http://www.bioversityinternational.org/scientific_information/themes/nutrition/overview.html> (accessed 26 November 2009)

Bouis H (1991) Food demand elasticities by income group by urban and rural populations for the Philippines. International Food Policy Research Institute, Washington DC, USA

Bouis H and Hunt J (1999) Linking Food and Nutrition Security: Past Lessons and Future Opportunities. Asian Development Review, 17(1,2):168-213

Brosius F (2006) SPSS 14. MITP, Redline GmbH, Heidelberg, Germany.

Brown KH, Peerson JM, Kimmons JE and Hotz C (2002) Options for achieving adequate intake from home-prepared complementary foods in low income countries. In: Black RE and Fleischer MK (editors). Public Health Issues in Infant and Child Nutrition. Nestle´ Nutrition Workshop Series, Pediatric Program, Vol. 48, Nestec, Vevey/Lipincott Williams and Wilkins, Philadelphia, USA

Bühl A (2006) SPSS 14. Einführung in die moderne Datenanalyse. 10. Auflage. Pearson Studium, München, Germany.

Bukenya-Ziraba R and Bonsu KO (2004) Solanum macrocarpon L. Record from Protabase. Grubben GJH and Denton OA (editors). PROTA (Plant Resources of Tropical Africa / Ressources végétales de l'Afrique tropicale), Wageningen, Netherlands. < http://database.prota.org/search.htm> (accessed 3 September 2009)

Burgess A and Glasauer P (2004) Family Nutrition Guide. FAO, Rome, Italy

Caballero B (2007) The Global Epidemic of Obesity: An Overview. Epidemiologic Reviews, 29:1-5

Castenmiller J and West C (1998) Bioavailability and Bioconversion of Carotenoids. Annual Review of Nutrition, 18:19- 38

Chen X, Ender P, Mitchell M and Wells C (2003) Regression with SPSS. <http://www.ats.ucla.edu/stat/spss/webbooks/reg/default.htm> (accessed 16 June 2008)

Chweya JA and Eyzaguirre PB (eds.) (1999) The Biodiversity of Traditional Leafy Vegetables. International Plant Genetic Resources Institute, Rome, Italy

Clausen T, Charlton KE, Gobotswang KSM et al. (2005) Predictors of food variety and dietary diversity among older persons in Botswana. Nutrition, 21:86-95

Cogill B (2003) Anthropometric Indicators Measurement Guide. Food and Nutrition Technical Assistance Project. Washington DC, USA

Conservation International (2007) Biodiversity Hotspots: Eastern Afromontane. <http://www.biodiversityhotspots.org/xp/Hotspots/afromontane/Pages/default.aspx> (accessed 13 September 2009)

Conway JM and Huffcutt AI (2003) A Review and Evaluation of Exploratory Factor Analysis Practices in Organizational Research. Organizational Research Methods, 6:147-168

Cook JD, Skikne BS and Baynes RD (1993) Serum transferrin receptor. Annual Review of Medicine 44:63-74

Coupland R (1956) East Africa and its invaders. Clarendon Press, Oxford, UK

Coutsoudis A, Mametja D, Jinabhai CC and Coovadia HM (1993) Vitamin A deficiency among children in a periurban South African settlement. American Journal of Clinical Nutrition, 57:904-907

Dapi LN, Omoloko C, Janlert U et al. (2007) "I eat to be happy, to be strong, and to live." perceptions of rural and urban adolescents in Cameroon, Africa. Journal of Nutrition Education and Behavior, 39(6):320-326

Darmon N and Drewnowski A (2008) Does social class predict diet quality? American Journal of Clinical Nutrition, 87(5):1107-1117

Delisle H (1990) Patterns of urban food consumption in developing countries: perspective from the 1980s. FAO, Rome, Italy

DeMaeyer EM (1989) Preventing and controlling iron deficiency anaemia through primary health care. A guide for health administrators and programme managers. Geneva, Switzerland: WHO

De Pee S and Bloem MW (2007) The bioavailability of (pro) vitamin A carotenoids and maximizing the contribution of homestead food production to combating vitamin A deficiency. International Journal for Vitamin and Nutrition Research, 77(3):182-192

DGE (2004) DGE-Ernährungskreis – Lebensmittelmengen. Beratungspraxis 05/2004, Deutsche Gesellschaft für Ernährung. <http://www.dge.de/modules.php?name=News&file=article& sid=415> (accessed 28 March 2008)

Dierßen K (1990) Einführung in die Pflanzensoziologie (Vegetationskunde). Wissenschaftliche Buchgesellschaft, Darmstadt

Doak CM, Adair LS, Bentley M et al. (2005) The dual burden household and the nutrition transition paradox. International Journal of Obesity 29, 129-136

Drewnowski A and Popkin BM (1997) The nutrition transition: new trends in the global diet. Nutrition Reviews, 55:31-43

Drewnowski A (2000) Nutrition transition and global dietary trends. Nutrition, 16(7/8):486-487

Eastwood MA (1999) Interaction of dietary antioxidants in vivo: how fruit and vegetables prevent disease? The Quartely Journal of Medicine, 92:527-30

Eckhardt CL, Torheim LE, Monterrubio E et al. (2008) The overlap of overweight and anaemia among women in three countries undergoing the nutrition transition. European Journal of Clinical Nutrition, 62:238-246

Erhardt JG (2004) NutriSurvey for Windows. <http:www.nutrisurvey.de> (accessed 29 August 2006)

Erhardt JG, Estes JE, Pfeiffer CM et al. (2004) Combined Measurement of Ferritin, Soluble Transferrin Receptor, Retinol Binding Protein, and C-Reactive Protein by an Inexpensive, Sensitive, and Simple Sandwich Enzyme-Linked Immunosorbent Assay Technique. The Journal of Nutrition, 134:3127-3132

FAO (1970) Food Composition Tables for Use in Africa. FAO, Rome, Italy

FAO (1997) Agriculture, food and nutrition for Africa. A resource book for teachers of agriculture. Food and Nutrition Division, FAO, Rome, Italy

FAO (2006) Spotlight 2006: Fighting hunger – and obesity. <http://www.fao.org/Ag/Magazine/0602sp1.htm> (accessed 30 Jun 2008)

FAO (2007) Guidelines for measuring household and individual dietary diversity. Version 2. FAO, Rome, Italy

FAOSTAT (2009) Food Balance Sheets. FAO, Rome, Italy <http://faostat.fao.org/site/368/default.aspx#ancor> (accessed 2 December 2009)

Ferro-Luzzi A, Scaccini C, Taffese S et al. (1990) Seasonal energy deficiency in Ethiopian rural women. European Journal of Clinical Nutrition, 44 Suppl 1:7-18

Field AP (2000) Discovering statistics using SPSS for Windows: advanced techniques for the beginner. Sage Publications, London, UK

Flegal KM, Graubard BI, Williamson DF and Gail MH (2005) Excess deaths associated with underweight, overweight and obesity. The Journal of the American Medical Association 293 (15), 1861-1867

Fondio L and Grubben GJH (2004) Corchorus olitorius L. Record from Protabase Grubben GJH and Denton OA (editors). PROTA (Plant Resources of Tropical Africa / Ressources végétales de l'Afrique tropicale), Wageningen, Netherlands. <http://database.prota.org/search.htm> (accessed 13 January 2009)

Fratkin E and Smith K (1995) Women's Changing Economic Roles and Pastoral Sedentarization: Varying Strategies in Alternative Rendille Communities. Human Ecology, 23(4):433-454

Friis H (2002) Micronutrients and infections: an introduction. In: Friis H, ed. Micronutrients and HIV infection. Boca Raton, Fl, USA: CRC:2–21

Frison E (2007) Agricultural biodiversity and traditional foods: important strategies for achieving freedom from hunger and malnutrition. SCN News 34

Fuchs GJ, Ausayakhun S, Ruckphaopunt S et al. (1994) Relationship between vitamin A deficiency, malnutrition and cunjunctival impression cytology. American Journal of Clinical Nutrition, 60:293-298

Futatsuka M, Arimatsu Y, Nagano M et al. (1985) A multiple regression analysis of hemoglobin values and iron status in Japanese farmers. Journal of Nutritional Science and Vitaminology (Tokyo), 31(6):573-584

Garner P and Gulmezoglu AM (2000) Prevention versus treatment for malaria in pregnant women. Cochrane Database of Systematic Revues, 4:Cd000169

Gould D, Kelly D, Goldstone L and Gammon J (2001) Examining the validity of pressure ulcer risk assessment scales: developing and using illustrated patient simulations to collect the data. Journal of Clinical Nursing, 10(5):697-706

Gupta R and Kumar P (2007) Social evils, poverty & health. Indian Journal of Medical Research, 126(4):279-288

Hart AD, Azubuike CU, Barimalaa IS and Achinewhu SC (2005) Vegetable consumption pattern of households in selected areas of the old rivers state in Nigeria. African Journal of Food, Agriculture and Nutritional Development (online), 5

Hatloy A, Torheim LE and Oshaug A (1998) Food variety – a good indicator of nutritional adequacy of the diet? A case study from an urban area in Mali, West Africa. European Journal of Clinical Nutrition, 52:891-898

Hawkes C (2006) Uneven dietary development: linking the policies and processes of globalization with the nutrition transition, obesity and diet-related chronic diseases. Globalization and Health, 2:4

Helen Keller International/Asia Pacific (2001) Homestead Food Production – A Strategy to Combat Malnutrition and Poverty. Jakarta, Indonesia: Helen Keller International

Hinderaker SG, Olsen BE, Lie RT et al. (2002) Anemia in pregnancy in rural Tanzania: associations with micronutrients status and infections. European Journal of Clinical Nutrition, 56:192-199

Hollenhorst M (2006) Einführung in die Statistik mit SPSS. Skript zur Veranstaltung des Hochschulrechenzentrums (HRZ) der Universität Gießen.

Hoffmann K, Schulze MB, Schienkiewitz A et al. (2004) Application of a New Statistical Method to Derive Dietary Patterns in Nutritional Epidemiology. American Journal of Epidemiology, 159(10):935-944

Hoffmeister M, Lyaruu IA and Krawinkel MB (2005) Assessment of Nutritional Intake, Body Mass Index and Glycimic Control in Patients with Type-2 Diabetes from Northern Tanzania. Annals of Nutrition & Metabolism, 49:64-68

Hogenkamp PS, Jerling JC, Hoekstra T et al. (2008) Association between consumption of black tea and iron status in adult Africans in the North West Province: the THUSA study. British Journal of Nutrition, 100(2):430-437

Holdsworth M, Delpeuch F, Landais E et al. (2006) Knowledge of dietary and bevaviour-related determinants of non-communicable disease in urban Senegalese women. Public Health Nutrition, 9:975–81

Hu FB, Rimm E, Smith-Warner SA et al. (1999) Reproducibility and validity of dietary patterns assessed with a food-frequency questionnaire. American Journal of Clinical Nutrition, 69:243-9

Hu FB (2002) Dietary pattern analysis: a new direction in nutritional epidemiology. Current Opinion in Lipidology, 13:3-9

Human N, Salim-Ur-Rehman, Anjum FM et al. (2007) Food fortification strategy – preventing iron deficiency anemia: a review. Critical Reviews in Food Science and Nutrition, 47(3):259-265

IFAVA (2008) Fruit and vegetable consumption among migrants. The IFAVA Scientific Newsletter No 29, International Fruit and Vegetable Alliance, Ottawa, Canada

Iliffe J (1995) Africans. The history of a continent. Cambridge University Press, Cambridge, UK

IOM – Institute of Medicine (2001) Dietary Reference Intakes for Vitamin A, Vitamin K, Arsenic, Boron, Chromium, Copper, Iodine, Iron, Manganese, Molybdenum, Nickel, Silicon, Vanadium and Zinc. National Academy Press, Washington, DC, pp.82-145 <http://www.nap.edu/openbook/0309072794/html/82.html> (accessed 20 March 2008)

IPGRI (2003) Rediscovering a forgotten treasure. IPGRI Public Awareness. Rome, Italy. http://www.ipgripa.grinfo.net/index.php?itemid=101 (accessed 10 March 2004)

Islam S (2007) Medicinal and nutritional qualities of sweetpotato tops and leaves. Coopeartive Extension Program, University of Arkansas at Pine Bluff. <http://www.uaex.edu/Other_Areas/publications/PDF/FSA-6135.pdf> (accessed 8 January 2009)

IVACG – International Vitamin A Consultative Group (2002) Conversion factors for vitamin A and carotinoids. <http://ivacg.ilsi.org/publications/pubslist.cfm?publicationid=220> (accessed 31 October 2007)

Jago R, Baranowski T, Baranowski JC et al. (2007) Distance to food stores & adolescent male fruit and vegetable consumption: mediation effects. International Journal of Behavioral Nutrition and Physical Activity, 13(4):35

Jansen PCM (2004) Sesamum angustifolium (Oliv.) Engl. Record from Protabase. Grubben GJH and Denton OA (editors). PROTA (Plant Resources of Tropical Africa / Ressources végétales de l'Afrique tropicale), Wageningen, Netherlands. <http://database.prota.org/search.htm> (accessed 10 June 2008)

Jones KM, Specio SE, Shrestha P et al. (2005) Nutrition knowledge and practices, and consumption of vitamin A-rich plants by rural Nepali participants in a kitchen-garden program. Food and Nutrition Bulletin, 26(2):198-208

Kaever A, Lingner T, Feussner K, Göbel C et al. (2009) MarVis: a Tool for Clustering and Visualization of Metabolic Biomarkers. BMC Bioinformatics, 10:92

Kaliba A (2008) Meat demand flecibilities for Tanzania: Implications for the choice of long-term investment. African Journal for Agricultural and Resource Economics, 2(2):208-221

Kant AK (2004) Dietary patterns and health outcomes. Journal of the American Diet Association, 104:615-635

Kayano M, Yoshiuchi K, Al-Adawi S et al. (2008) Eating attitudes and body dissatisfaction in adolescents: Cross-cultural study. Psychiatry and Clinical Neurosciences, 62(1):17-25

Keding GB und Jordan I (2008) Adipositas: Ein Public-Health-Problem in Tansania. Ernährung im Fokus, 8-10 2008, pp.372-377

Keller GB (2004) African nightshade, eggplant, spiderflower et al. – production and consumption of traditional vegetables in Tanzania from the farmers point of view. Masterarbeit im wissenschaftlichen Studiengang Agrarwissenschaften an der Georg-August Universität Göttingen, Fakultät für Agrawissenschaft

Keller GB, Mndiga H and Maass BL (2005) Diversity and genetic erosion of traditional vegetables in Tanzania from the farmer's point of view. Plant Genetic Resources, 3:400-413

Khor GL (2008) Food-based approaches to combat the double burden among the poor: challenges in the Asian context. Asia Pacific Journal of Clinical Nutrition, 17 Suppl 1:111-115

King FS and Burgess A (1992) Nutrition for Developing Countries. Oxford University Press, Oxford, UK

Koponen J (1988) People and production in late precolonial Tanzania. History and structures. Studia Historica 28. Gummerus Kirjapaino Oy, Helsinki, Finland

Krawinkel M (2006) Fehlernährung. In: Globalisierung – Gerechtigkeit – Gesundheit. Einführung in International Public Health. Razum O, Zeeb H and Laaser U (Hrsg). Verlag Hans Huber, Bern, Switzerland

Kuhnlein HV (1992) Food sources of vitamin A and provitamin A. Editorial introduction. Food and Nutrition Bulletin, 14(1):3-5

Kuhnlein HV, Erasmus B and Spigelski D (2009) Indigenous Peoples' food systems: the many dimensions of culture, diversity and environment for nutrition and health. FAO, Rome, Italy

Kyle UG, Bosaeus I, De Lorenzo AD et al. (2004) Bioelectrical impedance analysis – part I: review of principles and methods. Clinical Nutrition, 23(5):1226-1243

Lampe JW (1999) Health effects of vegetables and fruit: assessing mechanisms of action in human experimental studies. Am.J.Clin.Nutr., 70:475S-90S

Lang TA and Secic M (2006) How to report statistics in medicine: annotated guidelines for authors, editors, and reviewers. Second Edition. American College of Physicians, Philadelphia, USA

Lee CJ, Lawler GS and Panemangalore M (1987) Nutritional status of middle-aged and elderly females in Kentucky in two seasons: Part 2. Hematological parameters. Journal of the American College of Nutrition, 6(3):217-222

Lester RN and Seck A (2004) Solanum aethiopicum L. Record from Protabase. Grubben GJH and Denton OA (editors). PROTA (Plant Resources of Tropical Africa / Ressources végétales de l'Afrique tropicale), Wageningen, Netherlands <http://database.prota.org/search.htm> (accessed 15 January 2009)

Liebman M (1995) Polyculture Cropping Systems. In: Altieri MA, ed. Agroecology. The Science of Sustainable Agriculture. Second Edition. Westview Press, Colorado, USA

Lozán JL and Kausch H (1998) Angewandte Statistik für Naturwissenschaftler. Parey Buchverlag, Berlin, Germany

Lyimo MH, Nyagwegwe S and Mnkeni AP (1991) Investigations on the effect of traditional food processing, preservation and storage methods on vegetable nutrients: a case study in Tanzania. Plant Foods for Human Nutrition, 41:53-57

McIntyre BD, Bouldin DR, Urey GH, and Kizito F (2001) Modeling cropping strategies to improve human nutrition in Uganda. Agricultural Systems, 67:105–120

Maletnlema TN (2002) A Tanzanian perspective on the nutrition transition and its implications for health. Public Health Nutrition 5(1A):163-168

Malik VS, Schulze MB and Hu FB (2006) Intake of sugar-sweetened beverages and weight gain: a systematic review. American Journal of Clinical Nutrition, 84(2):274-288

Mamiro PS, Kolsteren P, Roberfroid D et al. (2005) Feeding practices and factors contributing to wasting, stunting, and iron-deficiency anaemia among 3-23-month old children in Kilosa district, rural Tanzania. Journal of Health, Population and Nutrition, 23(3):222-230

Mason J, Rivers J and Helwig C (editors) (2005) Recent trends in malnutrition in developing regions: Vitamin A deficiency, anemia, iodine deficiency, and child underweight. Special Section Reprint of the Food Nutrition Bulletin, 26:57-162

Massawe SN, Urassa EN, Mmari M et al. (1999) The   complexity of pregnancy anemia in Dar es Salaam. Gynecologic and Obstetric   Investigation, 47: 76-82

Massawe SN, Ronquist G, Nyström L et al. (2002) Iron Status and Iron Deficiency Anaemia in Adolescents in a Tanzanian Suburban Area. Gynecol Obstet Invest 54:137-144

Maxwell D, Levin C and Csete J (1998) Does urban agriculture help prevent malnutrition? Evidence from Kampala. Washington, DC, USA, Food Consumption and Nutrition Division (FCND). FCND Discussion Paper 45. <http://www.ifpri.org/divs/fcnd/dp/papers/dp45.pdf> (accessed 15 May 2008)

Maziya-Dixon B, Akinyele IO, Oguntona EB et al. (2004) Nigeria food Consumption and Nutrition Survey (2001-2003). International Institute for Tropical Agriculture (IITA), Ibadan, Nigeria

Mejia L A and Chew F (1988) Hematological effect of supplementing anemic children with vitamin A alone and in combination with iron. American Journal of Clinical Nutrition, 48:595-600

Mendez MA, Monteiro CA and Popkin BM (2005) Overweight exceeds underweight among women in most developing countries. American Journal of Clinical Nutrition, 81:714-721

Mensink GBM, Lampert T and Bergmann E (2005) Übergewicht und Adipositas in Deutschland 1984-2003. Bundesgesundheitsb-Gesundheitsforsch-Gesundheitsschutz 2005, 48:1348-1356

Messiaen CM and Rouamba A (2004) Allium cepa L. Record from Protabase. Grubben GJH and Denton OA (editors). PROTA (Plant Resources of Tropical Africa / Ressources végétales de l'Afrique tropicale), Wageningen, Netherlands <http://database.prota.org/search.htm> (accessed 15 January 2009)

Millstone E and Lang T (2003) The Atlas of Food. Who eats what, where and why. Earthscan Publications Ltd, London, UK

Misra S, Maikhuri RK, Kala CP et al. (2008) Wild leafy vegetables: a study of their subsistence dietetic support to the inhabitants of Nanda Devi Bioshphere Reserve, India. Journal of Ethnobiology and Ethnomedicine, 30(4):15

Monteiro CA, Conde WL, Lu B and Popkin BM (2004) Obesity and inequities in health in the developing world. International Journal of Obesity, advance online publication

MRI (Hg) (2008) Ergebnisbericht, Teil 1 – Nationale Verzehrsstudie II. Karlsruhe 2008. <http://www.was-esse-ich.de/uploads/media/NVS_II_Ergebnisbericht_Teil_1.pdf> (accessed 17 Jul 2008)

Müller O and Krawinkel MB (2005) Malnutrition and health in developing countries. Canadian Medical Association Journal, 173(3):279-286

Mulokozi G, Hedrén E and Svanberg U (2004) In vitro accessibility and intake of β-carotene from cooked green leafy vegetables and their estimated contribution to vitamin A requirements. Plant Foods for Human Nutrition, 59:1-9

Murphy SP and Allen LH (2003) Nutritional importance of animal source foods. Journal of Nutrition, 133(11 Suppl 2):3932S–5S

Netting RMC (1982) Some Home Truths on Household Size and Wealth. American Behavioral Scientist, 25(6):641-62

Njelekela M, Kuga S, Nara Y et al. (2002) Prevalence of obesity and dyslipidemia in middle-aged men and women in Tanzania, Africa: relationship with resting energy expenditure and dietary factors. Journal of Nutritional Science and Vitaminology, 48(5):352-358

Nyaruhucha CN, Achen JH, Msuya JM et al. (2003) Prevalence and awareness of obesity among people of different age groups in educational institutions in Morogoro, Tanzania. East African Medical Journal, 80(2):68-72

O'Dea JA (2008) Gender, ethnicity, culture and social class influences on childhood obesity among Australian schoolchildren: implications for treatment, prevention and community education. Health and Social Care in the Community, 16(3):282-290

Okeno JA, Chebet DK and Mathenge PW (2003) Status of indigenous vegetables in Kenya. Acta Horticulturae, 621:95-100

Owuor OB and Olaimer-Anyara E (2007) The value of leafy vegetables: an exploration of African folklore. African Journal of Food, Agriculture, Nutrition and Development, 7(3)

Pandit N, Rameshkumar K, Geethanjali S et al. (1998) Conjunctival impression cytology for detection of early vitamin A deficiency in children. Journal of Association of Physicians of India, 46(6):507-509

Patterson RE and Pietinen P (2004) Assessment of Nutritional Status in Individuals and Populations. In: Gibney MJ, Margetts BM, Kearney JM and Arab L (editors). Public Health Nutrition. The Nutrition Society Textbook Series. Blackwell Publishing, Oxford, UK, pp 66-82

Pepping F, Kavishe FP, Hacknitz EA and West C (1988) Prevalence of xerophthalmia in relation to nutrition and general health in preschool age children in three regions of Tanzania. Acta Paediatrica Scandinavica, 77:895–906

Pineda O (1998) Fortification of sugar with vitamin A. Food and Nutrition Bulletin, 19(2):131-136

Popkin BM (1994) The nutrition transition in low-income countries: an emerging crisis. Nutrition Reviews 52(9):285-98

Popkin BM (1999) Urbanization, Lifestyle Changes and the Nutrition Transition. World Development, 27(11):1905-1916

Popkin BM (2000) Urbanization and the Nutrition Transition. IFPRI 2020 Focus No. 03 – Brief 07. <http://www.ifpri.org/2020/focus/focus03/focus03_07.asp> (accessed 15 May 2008)

Popkin BM (2007) The world is fat. Scientific American. September 2007:60-67

Prentice AM (2008) Iron metabolism, malaria, and other infections: What is all the fuss about? Journal of Nutrition, 138:2537-2541

Raschke V and Cheema B (2008) Colonisation, the New World Order, and the eradication of traditional food habits in East Africa: historical perspective on the nutrition transition. Public Health Nutrition, 11(7):662-674

Ravelli GP, Stein ZA and Susser MW (1967) Obesity in young men after famine exposure in utero and early infancy. The New England Journal of Medicine, 295: 349-353

Renzaho A (2004) Fat, rich and beautiful: changing socio-cultural paradigms associated with obesity risk, nutritional status and refugee children from Sub-Saharan Africa. Health and Place, 10:105-113

Resnikoff S (1980) Malnutrition and weight excess: a paradoxical condition in Mauretanian Adrar. Medecine Tropicale, 40(4):419-423

Reynaud J, Brun T and Marek T (1989) Impact of a women's garden project on nutrition and income in Senegal. Arid Lands Newsletter

Rguibi M and Belahsen R (2006) Fattening practices among Moroccan Saharawi women. Eastern Mediterranean Health Journal, 12(5):619-624

Rigby P (1969) Cattle and Kinship among the Gogo. A Semi-pastoral Society of Central Tanzania. Cornell University Press, London, UK

Ruel MT (2001) Can food-based strategies help reduce vitamin A and iron deficiencies? A review of recent evidence. International Food Policy Research Institute, Washington DC, USA

Ruffo CK, Birnie A and Tengnäs B (2002) Edible Wild Plants of Tanzania. Technical Handbook No. 27. English Press, Nairobi, Kenya

Rush EC, Goedecke JH, Jennings C et al. (2007) BMI, fat and muscle differences in urban women of five ethnicities form two countries. International Journal of Obesity (2007) 31:1232-1239

Savy M, Martin-Prével Y, Sawadogo P et al. (2005) Use of variety/diversity scores for diet quality measurement: relation with nutritional status of women in a rural area in Burkina Faso. European Journal of Nutrition 59:703-716

Savy M, Martin-Prével Y, Traissac P and Delpeuch F (2007) Measuring dietary diversity in rural Burkina Faso: comparison of a 1-day and a 3-day dietary recall. Public Health Nutrition, 10(1):71-78

Schmidhuber J and Shetty P (2004) Nutrition transition, obesity & noncommunicable diseases: drivers, outlook and concerns. SCN News No 29, pp 13-19

Schmidt DR (1971) Comparative yields and composition of eight tropical leafy vegetables grown at two soil fertility levels. Agronomy Journal 63:546-550

Schulze MB, Manson JE, Ludwig DS et al. (2004) Sugar-sweetened beverages, weight gain, and incidence of type 2 diabetes in young and middle-aged women. Journal of the American Medical Association, 292(8):927-934

Schulze MB and Hoffmann K (2006) Methodological approaches to study dietary patterns in relation to risk of coronary heart disease and stroke. British Journal of Nutrition, 95:860-869

SCN (1988) First report on the world nutrition situation. SCN News No 2. United Nations Standing Committee on Nutrition, Geneva, Switzerland

SCN (2006) Diet-related Chronic Diseases and the Double Burden of Malnutrition in West Africa. SCN News No 33. United Nations Standing Committee on Nutrition, Geneva, Switzerland

Shaikh AR, Yaroch AL, Nebeling L et al. (2008) Psychosocial predictors of fruit and vegetable consumption in adults: a review of the literature. American Journal of Preventive Medicine, 34(6):535-543

Shi Z, Hu X, Yuan B et al. (2008) Vegetable-rich food pattern is related to obesity in China. International Journal of Obesity 32:975-984

Sichieri R (2002) Dietary patterns and their associations with obesity in the Brazilian city of Rio de Janeiro. Obesity Research, 10(1):42-48

Siemonsma JS and Kouamé C (2004) Abelmoschus esculentus (L.) Moench. Record from Protabase. Grubben, GJH and Denton, OA (editors). PROTA (Plant Resources of Tropical Africa / Ressources végétales de l'Afrique tropicale), Wageningen, Netherlands. < http://database.prota.org/search.htm> (accessed 16 September 2009)

Siervo M, Grey P, Nyan OA and Prentice AM (2006) A pilot study on body image, attractiveness and body size in Gamians living in an urban community. Eating and Weight Disorders, 11(2):100-109

Simondon KB, Ndiaye T, Dia M et al. (2007) Seasonal variations and trends in weight and arm circumference of non-pregnant rural Senegalese women, 1990-1997. European Journal of Clinical Nutrition, advance online publication, 30 May 2007, 1-8

Singh G, Kawatra A and Sehgal S (2001) Nutritional composition of selected green leafy vegetables, herbs and carrots. Plant Foods for Human Nutrition, 56:359-364

Singh R, Joseph A, Umapathy T et al. (2005) Impression cytology of the ocular surface. British Journal of Ophthalmology, 89:1655-1659.

SLE (2008) Market-driven development and poverty reduction: A value chain analysis of fresh vegetables in Kenya and Tanzania. SLE Publication Series (Centre for Advanced Training in Rural Development). Seminar für ländliche Entwicklung, Berlin, Germany

Smith FI and Eyzaguirre P (2007) African leafy vegetables: their role in the world health organization's global fruit and vegetables initiative. Afr. J. Food Agric. Nutr. Devel. (online) 7(3). <http://www.ajfand.net/Vol7No3.html> (accessed 11 September 2008)

Sobeck U, Fischer A and Biesalski H (2003). Uptake of vitamin A in buccal mucosal cells after topical application of retinyl palmitate: a randomised, placebo- controlled and double- blind trial. The British Journal of Nutrition 90:69- 74

Sommer A (1995a) Vitamin A deficiency and its consequences. A field guide to detection and control. Third edition. WHO, Geneva, Switzerland

Sommer A (1995b) Vitamin A status. In: WHO (ed.) Vitamin A deficiency and its consequences. A field guide to detection and control. Third Edition, Geneva, Switzerland

Speedy AW (2003) Global production and consumption of animal source foods. Journal of Nutrition, 133:4048S-4053S

Steinmetz KA and Potter JD (1996). Vegetables, fruit and cancer prevention: A review. J R Soc Health; 96:1027–39

Steyn NP, Senekal M, Brits S and Nel JH (2000) Urban and rural differences in dietary intake, weight status and nutrition knowledge of black female students. Asia Pacific Journal of Clinical Nutrition, 9:53-9

Strain JJ and Cashman KD (2002) Minerals and Trace Elements. In:Gibney MJ, Vorster HH and Kok FJ (editors). Introduction to Human Nutrition. The Nutrition Society Textbook Series. Blackwell Publishing, Oxford, UK, pp 177-224

Suharno D, West CE, Muhilal Karyadi D and Hautvast JGAJ (1993) Supplementation with vitamin A and iron for nutritional anaemia in pregnant women in West Java, Indonesia. Lancet 342: 1325–1328

Tanner M and Lukmanji Z (1987) Food consumption patterns in a rural Tanzanian community (Kikwawila village, Kilombero District, Morogoro Region) during lean and post-harvest season. Acta Tropica, 44(2):229-244

TARP II – SUA Project (2004) Dietary Guidelines for Morogoro and Iringa Region. The Food Security and Household Income for Smallholder Farmers in Tanzania Proejct. Sokoine University of Agriculture, Morogoro, Tanzania

Teokul W, Payne P and Dugdale A (1986) Seasonal variations in nutritional status in rural areas of developing countries: a review of the literature. Food and Nutrition Bulletin 8(4):8-10

Thurnham DI, McCabe GP, Norhtrop-Clewes CA and Nestel P (2003) Effects of subclinical infection on plasma retinol concentrations and assessment of prevalence of vitamin A deficiency: meta-analysis. The Lancet, 362:2052-2058

Tontisirin K, Nantel G and Bhattacharjee L (2002) Food-based strategies to meet the challenges of micronutrient malnutrition in the developing world. Proceedings of the Nutrition Society, 61(2):243-250

Torheim LE, Quattara F, Diarra MM et al. (2004) Nutrient adequacy and dietary diversity in rural Mali: association and determinants. European Journal of Nutrition 58:594-604

Torlesse H, Kiess L and Bloem MW (2003) Association of household rice expenditure with child nutritional status indicates a role for macroeconomic food policy in combating malnutrition. Journal of Nutrition, 133:1320-1325

Überla K (1972) Faktorenanalyse. 2. Aufl., Springer Verlag, Berlin, Germany

UNDP (1995) Human Development Report 1995. United Nations Development Programme. Oxford University Press, New York, USA

UNICEF (1998). The State of the World's Children 1998. Oxford University Press, New York, USA

UNICEF and WHO (1999) Prevention and Control of Iron Deficiency Anaemia in Women and Children. Gleason GR. Boston, Massachusets, USA, United Nations Children's Fund. Report of the UNICEF/WHO Regional Consultation. 3-4 February 1999 Geneva, Switzerland

USDA (2003) Tanzania agricultural seasons. Prodcution Estimates and Crop Assessment Division, US Department of Agriculture <http://151.121.3.140/pecad2/highlights/2003/03/tanzania/images/bimodal.htm> (accessed 29 January 2009)

USDHHS and USDA (2005) Dietary Guidelines for Americans, 6th Edition. US Department of Health and Human Services and US Department of Agriculture <http://www.health.gov/dietaryguidelines> (accessed 28 March 2008)

Vainio-Mattila K (2000) Wild vegetables used by the Sambaa in the Usambara Mountains, NE Tanzania. Annales Botanici Fennici, 37:57-67

van der Vossen HAM, Nono-Womdim R and Messiaen C-M (2004) Lycopersicon esculentum Mill. Record from Protabase. Grubben, G.J.H. & Denton, O.A. (editors). PROTA (Plant Resources of Tropical Africa / Ressources végétales de l'Afrique tropicale), Wageningen, Netherlands. < http://database.prota.org/search.htm> (accessed 7 August 2009)

Vijayaraghavan K (2004) Iron-deficiency Anemias. In: Gibney MJ, Margetts BM, Kearney JM and Arab L (editors). Public Health Nutrition. The Nutrition Society Textbook Series. Blackwell Publishing, Oxford, UK, pp 227-235

Villamor E, Msamanga G, Urassa W et al. (2006) Trends in obesity, underweight, and wasting among women attending prenatal clinics in urban Tanzania, 1995-2004. American Journal of Clinical Nutrition, 83:1387-1394

Visscher TLS, Seidell JC, Molarius A, van der Kuip D, Hofmann A, Witteman JCM (2001) A comparison of body mass index, waist-hip ratio and waist circumference as predictors of all-cause mortality among the elderly: the Rotterdam study. International Journal of Obesity 25, 1730-1735

Waterlow JC (1992) Protein energy malnutrition. London, GB: Edward Arnold.

Weinberger K and Msuya J (2004) Indigenous vegetables in Tanzania: Significance and Prospects. Technical Bulletin No. 31. AVRDC – The World Vegetable Center, Shanhua, Taiwan

Weinberger K and Swai I (2006) Consumption of traditional vegetables in central and north-eastern Tanzania. Ecology of Food and Nutrition, 45(2):87-103

Weinberger K (2007) Are indigenous vegetables underutilized crops? Some evidence from eastern Africa and south east Asia. Acta Horticulturae, 752:29-34

Weiß C (2002) Basiswissen medizinische Statistik. Springer Medizin Verlag, Heidelberg, Gemany

West CE, Pepping F and Temalilwa CR (1988) The composition of foods commonly eaten in East Africa. Wageningen Agricultural University, Wageningen, Netherlands

WHO (1996) Indicators for assessing Vitamin A Deficiency and their application in monitoring and evaluating intervention programmes. Micronutrient Series. World Health Organization, Geneva, Switzerland

WHO (1998) Obesity: Preventing and Managing the Global Epidemic. Report of a WHO Consultation on Obesity. WHO, Geneva, Switzerland

WHO (2000) Obesity: Preventing and managing the Global Epidemic. Report of a WHO Technical Consultation. WHO Technical Report Series 894. WHO, Geneva, Switzerland

WHO (2003a) Fact Sheet on Obesity and Overweight. <http://www.who.int/hpr/NPH/docs/gs_obesity.pdf> (accessed 21 Feb 2008)

WHO (2003b) WHO fruit and vegetable promotion initiative – report of the meeting. Geneva, Switzerland, 25-27 August 2003

WHO (2003c) Diet, nutrition and the prevention of chronic diseases. WHO Technical Report Series 916. WHO, Geneva, Switzerland

WHO (2004) Appropiate body-mass index for Asian populations and its implications for policy and intervention strategies, Lancet 363:157-63

WHO (2006) Mortality country fact sheet 2006. United Republic of Tanzania. World Health Organization, Geneva, Switzerland <http://www.who.int/whosis/mort/profiles/mort_afro_tza_tanzania.pdf> (accessed 21 January 2009)

WHO (2007) Prevention and control of noncommunicable diseases: implementation of the global strategy. Report by the Secretariat. World Health Organization, Geneva, Switzerland <http://apps.who.int/gb/ebwha/pdf_files/EB120/b120_22-en.pdf> (accessed 28 January 2008)

WHO (2008) Obesity and overweight - What are overweight and obesity? <http://www.who.int/mediacentre/factsheets/fs311/en/index.html> (accessed 17 Jul 2008)

Young VR and Reeds PJ (2002) Nutrition and Metabolism of Proteins and Amino Acids. In: Gibney MJ, Vorster HH and Kok FJ (editors). Introduction to Human Nutrition. The Nutrition Society Textbook Series. Blackwell Publishing, Oxford, UK, pp 46-68

Yousuf HKM and Islam NM (1994) Improvement of nightblindness situation through simple nutrition education intervention with the parents. Ecology of Food and Nutrition, 31(3-4):247-256

Zehnder A, Jeje B, Tanner M and Freyvogel TA (1987) Agricultural production in Kikwawila village, southeastern Tanzania. Acta Tropica, 44(2):245-260

Zimmermann MB (2007) Interactions between iron and vitamin A, riboflavin, copper, and zinc in the etiology of anemia. In: Kraemer K and Zimmermann MB (editors). Nutritional Anemia. Sight and Life press, Basel, pp 199-213

Zimmermann MB (2008) Methods to assess iron and iodine status. British Journal of Nutrition, 99, Suppl.3:S2-S9

10 Appendix

Table A1   Dietary pattern analysis: principal component analysis (PCA) with varimax rotation run for three, four and five components

Food group	Component 1	2	3
Tea	**0.702**	0.389	-0.264
Bread/cakes	**0.669**	0.230	-0.198
Oil/fat	**0.653**		0.467
Pulses	**0.557**	0.140	
Sugar	**0.535**	0.103	-0.163
Nuts	0.241	**0.752**	-0.106
Fruits	0.254	**0.657**	-0.119
Fish	0.258	**0.611**	
Starchy plants		**0.554**	-0.302
Animal prod	0.239	-0.450	-0.395
Cereals			**0.782**
Vegetables	-0.271	-0.163	**0.633**

Rotated Component Matrix

Extraction Method: Principal Component Analysis.

Rotation Method: Varimax with Kaiser Normalization.

Rotation converged in 6 iterations.

Food group	Component 1	2	3	4
Nuts	**0.762**	0.214		
Fruits	**0.662**	0.179		0.202
Fish	**0.623**	0.184	0.133	
Starchy plants	**0.559**	0.146	-0.304	
Animal prod	-0.439	0.365	-0.323	
Tea	0.419	**0.690**		0.260
Bread cakes	0.263	**0.688**		0.168
Sugar	0.137	**0.625**		
Cereals		-0.146	**0.797**	
Oil/fat		**0.572**	**0.595**	
Pulses	0.131	0.153	0.302	**0.811**
Vegetables	-0.138		0.483	**-0.724**

Rotated Component Matrix

Extraction Method: Principal Component Analysis.

Rotation Method: Varimax with Kaiser Normalization.

Rotation converged in 6 iterations.

Food group	Component 1	2	3	4	5
Fruits	**0.762**			0.196	
Nuts	**0.738**	0.244			-0.179
Starchy plants	**0.700**		-0.175		0.167
Fish	**0.531**	0.251	0.112		-0.288
Bread/cakes	0.209	**0.744**		0.156	
Sugar		**0.713**			
Tea	0.418	**0.703**		0.245	
Cereals		-0.245	**0.780**		-0.126
Oil/fat		0.453	**0.688**		0.104
Pulses	0.138	0.118	0.282	**0.827**	
Vegetables	-0.125	-0.192	**0.530**	**-0.691**	
Animal prod					**0.941**

Rotated Component Matrix

Extraction Method: Principal Component Analysis.

Rotation Method: Varimax with Kaiser Normalization.

Rotation converged in 5 iterations.

Table A2   Physical activity (PA) of women measured with visual analogue scale (VAS) in cm during two seasons and in three districts of Tanzania

	All districts		Kongwa		Muheza		Singida	
	Nov/Dec	Mar/Apr	Nov/Dec	Mar/Apr	Nov/Dec	Mar/Apr	Nov/Dec	Mar/Apr
n	251	252	71	72	76	76	104	104
Mean ±	6.3	6.2	6.2	5.9	7.0	6.5	5.8	6.1
SD	± 1.7	± 1.2	± 1.8	± 1.1	± 1.9	± 1.4	± 1.3	± 1.1
Median	6.0	5.8	6.2	5.7	7.6	6.9	5.6	5.7
(range)	(1.7-9.6)	(2.3-8.7)	(1.7-9.5)	(3.1-8.2)	(2.0-9.6)	(2.3-8.7)	(2.7-8.4)	(4.0-8.7)

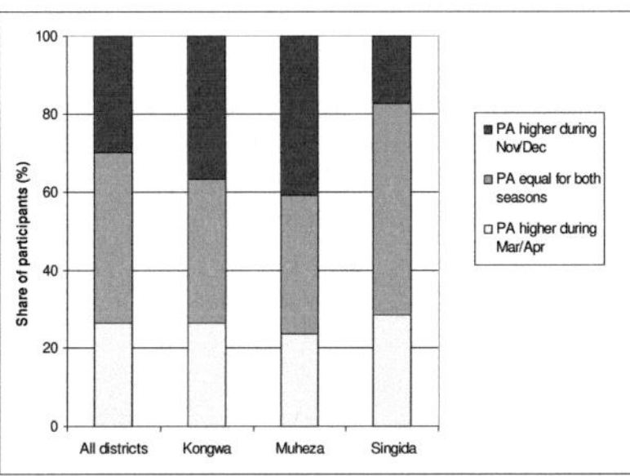

Figure A1   Seasonal change of physical activity (PA) of women in three districts of Tanzania

Table A3   Different socio-economic parameters of women in three districts of Tanzania

	n	Age	HH size	Distance (km)	Distance (min)
All districts	252				
Mean ± SD		33.3 (6.9)	5.8 (2.3)	28.3 (16.5)	109.0 (68.2)
Median (range)		34 (17-45)	6 (1-16)	33 (5-67)	120 (20-240)
Kongwa	72				
Mean ± SD		30.6 (7.0)	5.7 (2.3)	29.8 (23.0)	79.4 (54.9)
Median (range)		30 (17-44)	6 (1-13)	24 (8-67)	83 (20-150)
Muheza	76				
Mean ± SD		34.3 (7.5)	5.3 (2.4)	19.8 (15.6)	74.6 (63.8)
Median (range)		36 (18-45)	5 (1-16)	9 (5-51)	60 (30-240)
Singida	104				
Mean ± SD		34.4 (5.7)	6.2 (2.2)	33.5 (6.4)	154.6 (52.2)
Median (range)		35 (21-45)	6 (2-12)	35 (23-42)	150 (60-240)

Table A4   The pattern for classification of participants into different wealth categories

Parameters	Measured through	Cut-off points	
Possessions	Possession of radio, bicycle, mobile phone	0 or 1 = low wealth	0
		2 or 3 = high wealth	1
Housing	Setting of house: electricity, corrugated iron roof, brick/cement walls	0 or 1 = low wealth	0
		2 or 3 = high wealth	1
Livestock	Possession of chicken, goat, sheep, cow	0-2 (except cow) = low wealth	0
		3-4 (or cow only) = high wealth	1
Occupation	Type of main occupation	1;3;7-11* = low wealth	0
		2;4-6* = high wealth	1
Sale of vegetables	If own produced vegetables are sold	Vegetables not sold = low wealth	0
		Vegetables sold = high wealth	1

* For details see questionnaire

Table A5   The five <u>traditional</u> vegetables most often <u>cultivated</u> during three seasons in three different districts of Tanzania and share of participants (%) cultivating them

Jun/Jul (DS)	%	Nov/Dec (SR)	%	Mar/Apr (LR)	%
Kongwa					
Cowpea lvs (*Vigna unguiculata*)	69.4	Cowpea lvs (*Vigna unguiculata*)	63.9	Cowpea lvs (*Vigna unguiculata*)	97.2
Pumpkin lvs (*Cucurbita* sp.)	50.0	Pumpkin lvs (*Cucurbita* sp.)	58.3	Pumpkin lvs (*Cucurbita* sp.)	95.8
Sweet potato lvs (*Ipomoea batatas*)	34.7	Amaranth lvs (*Amaranthus* sp.)	27.8	Amaranth lvs (*Amaranthus* sp.)	88.9
Okra (*Abelmoschus esculentus*)	26.4	Sweet potato lvs (*Ipomoea batatas*)	27.8	Tree cassava lvs (*Manihot glaziovii*)	65.3
Amaranth lvs (*Amaranthus* sp.)	22.2	Okra (*Abelmoschus esculentus*)	13.9	Okra (*Abelmoschus esculentus*)	58.3
Muheza					
Pumpkin lvs (*Cucurbita* sp.)	86.8	Sweet potato lvs (*Ipomoea batatas*)	81.6	Sweet potato lvs (*Ipomoea batatas*)	81.6
Sweet potato lvs (*Ipomoea batatas*)	72.4	Pumpkin lvs (*Cucurbita* sp.)	80.3	Amaranth lvs (*Amaranthus* sp.)	72.4
Cowpea lvs (*Vigna unguiculata*)	64.5	Amaranth lvs (*Amaranthus* sp.)	77.6	Cassava lvs (*Manihot esculentus*)	42.1
Cassava lvs (*Manihot esculentus*)	63.2	Cowpea lvs (*Vigna unguiculata*)	56.6	Okra (*Abelmoschus esculentus*)	36.8
Amaranth lvs (*Amaranthus* sp.)	60.5	Cassava lvs (*Manihot esculentus*)	47.4	Tree cassava lvs (*Manihot glaziovii*)	36.8
Singida					
Sweet potato lvs (*Ipomoea batatas*)	60.6	Cassava lvs (*Manihot esculentus*)	61.5	Amaranth lvs (*Amaranthus* sp.)	79,8
Okra (*Abelmoschus esculentus*)	49.0	Sweet potato lvs (*Ipomoea batatas*)	41.3	Tree cassava lvs (*Manihot glaziovii*)	75.0
Pumpkin lvs (*Cucurbita* sp.)	44.2	Amaranth lvs (*Amaranthus* sp.)	35.6	Pumpkin lvs (*Cucurbita* sp.)	73.1
Amaranth lvs (*Amaranthus* sp.)	38.5	Pumpkin lvs (*Cucurbita* sp.)	31.7	Cowpea lvs (*Vigna unguiculata*)	68.3
Tree cassava lvs (*Manihot glaziovii*)	22.1	Wild cucumber (*Cucumis* sp.)	23.1	Sweet potato lvs (*Ipomoea batatas*)	68.3

Table A6   The five _traditional_ vegetables most often _collected_ during three seasons in three different districts of Tanzania and share of participants (%) collecting them

Jun/Jul (DS)		Nov/Dec (SR)		Mar/Apr (LR)	
Kongwa	%		%		%
False sesame (_Ceratotheca sesamoides_)	50.8	'Mhilile' (_Cleome_ sp.)	56.9	False sesame (_Ceratotheca sesamoides_)	94.4
'Mhilile' (_Cleome_ sp.)	35.8	Wild simsim (_Sesamum angustifolium_)	48.6	'Mhilile' (_Cleome_ sp.)	91.7
Amaranth lvs (_Amaranthus_ sp.)	23.3	Tree cassava lvs (_Manihot glaziovii_)	37.5	African spiderplant (_Cleome gynandra_)	84.7
African spiderplant (_Cleome gynandra_)	20.0	African spiderplant (_Cleome gynandra_)	36.1	Amaranth lvs (_Amaranthus_ sp.)	41.7
Tree cassava lvs (_Manihot glaziovii_)	13.3	Baobab tree lvs (_Adansonia digitata_)	31.9	Black jack (_Bidens pilosa_)	38.9
Muheza					
Bitter lettuce (_Launaea cornuta_)	61.3	Amaranth lvs (_Amaranthus_ sp.)	100.0	Bitter lettuce (_Launaea cornuta_)	72.4
Black jack (_Bidens pilosa_)	47.9	Bitter lettuce (_Launaea cornuta_)	93.4	Black jack (_Bidens pilosa_)	48.7
Amaranth lvs (_Amaranthus_ sp.)	33.6	Black jack (_Bidens pilosa_)	82.9	Beach morning glory (Ipomoea pescaprae)	38.2
Jute mallow (_Corchorus olitorius_)	29.4	African nightshade (_Solanum_ sp.)	60.5	Jute mallow (_Corchorus olitorius_)	15.8
African nightshade (_Solanum_ sp.)	27.7	Jute mallow (_Corchorus olitorius_)	59.2	Okra (Abelmoschus esculentus)	15.8
				Ngoswe	15.8
Singida					
Wild simsim (_Sesamum angustifolium_)	61.7	Amaranth lvs (_Amaranthus_ sp.)	100.0	Amaranth lvs (_Amaranthus_ sp.)	100.0
False sesame (_Ceratotheca sesamoides_)	45.0	False sesame (_Ceratotheca sesamoides_)	77.9	Wild simsim (_Sesamum angustifolium_)	94.2
Amaranth lvs (_Amaranthus_ sp.)	9.2	Wild cucumber (_Cucumis_ sp.)	54.8	False sesame (_Ceratotheca sesamoides_)	86.5
African spiderplant (_Cleome gynandra_)	5.0	African spiderplant (_Cleome gynandra_)	28.8	African spiderplant (_Cleome gynandra_)	80.8
Jute mallow (_Corchorus olitorius_)	3.3	Wild simsim (_Sesamum angustifolium_)	25.0	Wild cucumber (_Cucumis_ sp.)	76.0

Table A7   The five exotic vegetables most often cultivated during three seasons in three different districts of Tanzania and share of participants (%) cultivating them

Jun/Jul (DS)	%	Nov/Dec (SR)	%	Mar/Apr (LR)	%
Kongwa					
Tomato (*Lycopersicon esculentum*)	6.9	Chinese cabbage (*Brassica pekinensis*)	4.2	Swiss chard (*Beta vulgaris* subsp. *cicla*)	5.6
Chinese cabbage (*Brassica pekinensis*)	2.8	Swiss chard (*Beta vulgaris* subsp. *cicla*)	1.4	Tomato (*Lycopersicon esculentum*)	4.2
Common bean (*Phaseolus vulgaris*)	2.8	Tomato (*Lycopersicon esculentum*)	1.4	Chinese cabbage (*Brassica pekinensis*)	1.4
White cabbage (*Brassica oleracea* convar. *capitata*)	2.8				
Muheza					
Tomato (*Lycopersicon esculentum*)	19.7	Tomato (*Lycopersicon esculentum*)	11.8	Tomato (*Lycopersicon esculentum*)	2.6
Common bean (*Phaseolus vulgaris*)	17.1	Chinese cabbage (*Brassica pekinensis*)	5.3	Chinese cabbage (*Brassica pekinensis*)	1.3
Swiss chard (*Beta vulgaris* subsp. *cicla*)	14.5	Eggplant (*Solanum melongina*)	5.3		
Chinese cabbage (*Brassica pekinensis*)	6.6	Swiss chard (*Beta vulgaris* subsp. *cicla*)	3.9		
Eggplant (*Solanum melongina*)	2.6	White cabbage (*Brassica oleracea* convar. *capitata*)	1.3		
		Sweet pepper (*Capsicum annum*)	1.3		
Singida					
Tomato (*Lycopersicon esculentum*)	31.7	Chinese cabbage (*Brassica pekinensis*)	17.3	Tomato (*Lycopersicon esculentum*)	42.3
Chinese cabbage (*Brassica pekinensis*)	27.9	Tomato (*Lycopersicon esculentum*)	7.7	Chinese cabbage (*Brassica pekinensis*)	15.4
White cabbage (*Brassica oleracea* convar. *capitata*)	8.7	Swiss chard (*Beta vulgaris* subsp. *cicla*)	3.8	Onion (*Allium cepa*)	13.5
Swiss chard (*Beta vulgaris* subsp. *cicla*)	5.8	White cabbage (*Brassica oleracea* convar. *capitata*)	2.9	Swiss chard (*Beta vulgaris* subsp. *cicla*)	3.8
Eggplant (*Solanum melongina*)	5.8	Eggplant (*Solanum melongina*)	1.9	White cabbage (*Brassica oleracea* convar. *capitata*)	3.8
		Sweet pepper (*Capsicum annum*)	1.9		

Table A8   Vegetable species and number of women that cultivated/collected them in three districts of Tanzania during three different seasons

Scientific name	English name	Swahili name	Local name	Jun/Jul (DS)						Nov/Dec (SR)						Mar/Apr (LR)					
				Cultivated			Collected			Cultivated			Collected			Cultivated			Collected		
				Ko	Mu	Si	Ko	Mu	Si	Ko	Mu	Si	Ko	Mu	Si	Ko	Mu	Si	Ko	Mu	Si
Abelmoschus esculentus	Okra	Bamia		28	65	59	0	0	0	10	30	22	0	0	0	42	28	64	0	0	0
Adansonia digitata	Baobab (leaves)	Majani ya mbuyu	Ikuwi (Gogo)	1	0	0	0	0	0	0	0	0	23	0	0	0	0	0	1	0	0
Amaranthus dubius	Amaranth	Mchicha	Bwache (Sambaa), Chakaya, Lichakaya (Kaguru)	2	49	0	6	13	0	1	50	0	5	49	17	0	12	4	1	5	18
Amaranthus cruentus	Amaranthus cruentus	Mchicha		0	0	0	0	0	0	17	6	17	0	0	0	41	28	45	0	0	0
Amaranthus graecizans	Amaranth	Mchicha	Mugha (Nyaturu), Bwache (Bondei), Fwene (Gogo)	7	2	1	41	3	9	1	2	1	16	8	42	0	2	6	25	5	70
Amaranthus hybridus	Amaranth	Mchicha	Mugha (Nyaturu), Mafoene, Mjelu	16	16	38	4	1	1	0	1	0	1	20	1	5	0	1	2	0	43
Amaranthus hybridus ssp. cruentus	Amaranth	Mchicha		0	0	1	1	0	0	0	0	7	0	0	50	0	0	0	0	0	0
Amaranthus hypocondriacus	Amaranth	Mchicha		0	2	5	0	0	0	1	0	12	0	0	1	18	13	27	0	0	0
Amaranthus spinosus	Amaranth	Mchicha, Mchicha pori	Mbwache, Bwache dakaya, Bwache la miwa (Bondei), Bwache la pori (Muha)	0	1	1	1	36	1	0	0	0	0	12	0	0	0	0	0	1	6
Amaranthus viridis	Amaranth	Mchicha	Bwache (Bondei), Chakaya (Kongwa)	0	0	0	0	5	3	0	0	0	0	0	0	0	0	0	2	0	0
Amaranthus sp.	Amaranth	Mchicha	e.g. mchicha ngunuwe (Singida)	5	3	0	0	2	4	0	0	0	0	1	14	0	0	0	0	0	1
Basella alba	Malabar spinach	Nderema		0	3	10	0	12	0	0	1	0	0	11	0	0	0	0	0	7	0
Bidens biternata		Mutsungu-tsungu		0	0	0	0	0	0	0	0	0	0	0	0	0	0	0	0	0	0
Bidens pilosa	Black jack	Kishonanguo	Mpangalale/Mhangalale (Kigogo), Mbwembwe (Sambaa), Mbwimbiji (Zigua)	0	0	0	6	86	0	0	0	0	20	63	0	0	0	0	28	37	0

Scientific name	English name	Swahili name	Local name	Jun/Jul (DS)						Nov/Dec (SR)						Mar/Apr (LR)					
				Cultivated			Collected			Cultivated			Collected			Cultivated			Collected		
				Ko	Mu	Si	Ko	Mu	Si	Ko	Mu	Si	Ko	Mu	Si	Ko	Mu	Si	Ko	Mu	Si
Brassica carinata	Ethiopian kale/mustard	Sukuma wiki, Saro, Figili / Loshuu	Nyaulezi (Hehe, Nyaturu), Saladi (Ngoni)	6	11	1	1	2	0	2	5	0	0	0	0	15	1	11	0	0	0
Celosia argentea	Lagos spinach			0	0	0	0	0	0	0	0	0	0	0	0	0	0	0	0	0	0
Ceratotheca sesamoides	False sesame	Mlenda mbata	Ihonda, Mbara, Makhonda mbata (Nyaturu), Mgalu (fresh), Ilende (dried), Mlenda kigogo (Gogo), Ihombo (Kaguru)	3	0	2	103	1	61	0	0	0	35	0	26	0	0	3	68	0	90
Cleome gynandra	African spiderflower/ spiderplant	Mgagani	Moga (Nyaturu), Mwange (Sambaa), Mzimwe (Gogo), Mgange (Kaguru)	1	0	1	51	11	7	1	3	0	26	9	30	5	0	0	61	1	84
Colocasia esculenta	Cocoyam, Taro	Majani ya magimbi	Jimbi, Mayugwa (Sambaa), Gimbi (Bondei)	0	4	0	0	3	0	0	4	0	0	0	0	0	0	0	0	0	0
Corchorus olitorius	Jute mallow	Mlenda wima, Mlenda pori	Malimbe, Ihonda la ikuga, Montee (Nyaturu), Kibwando (Bondei), Limlikwi, Ighundaga (Kaguru), Munkhara (Singida)	0	1	1	1	52	5	0	1	0	1	45	1	0	0	0	18	0	8
Cucumis sp.	Wild cucumber	Matango pori	Mahanjo, Maimbe, Marimbe (Singida)	0	0	13	0	0	1	0	0	24	1	0	57	2	0	32	0	0	79
Cucumis sp.	Wild cucumber	Matango pori	Matanga-tumba (fruits not edible)	0	0	0	3	0	0	0	0	0	0	0	0	0	0	0	1	0	0
Cucumis sp.	Wild cucumber	Matango pori	Matanga (fruits edible)	0	0	0	2	0	0	0	0	0	2	0	0	0	0	0	2	0	0
Cucurbita spp.	Pumpkin leaves	Majani ya maboga	Mahukuma, Karubwagila (Nyaturu), Makamaboga (Mbena), Khoko (Sambaa), Muja (Kaguru)	68	100	51	8	0	0	42	61	33	0	1	0	69	24	76	0	0	0
Galinsoga parviflora	Gallant soldier	Majani ya sungura	Kachinja (Sambaa)	0	0	0	0	1	0	0	0	0	0	0	0	0	0	0	0	0	0

Scientific name	English name	Swahili name	Local name	Jun/Jul (DS)						Nov/Dec (SR)						Mar/Apr (LR)					
				Cultivated			Collected			Cultivated			Collected			Cultivated			Collected		
				Ko	Mu	Si	Ko	Mu	Si	Ko	Mu	Si	Ko	Mu	Si	Ko	Mu	Si	Ko	Mu	Si
Ipomoea batatas	Sweet potato leaves	Matembele	Matembea matoo (Nyaturu), Nkutu (Sambaa), Nijutu (Bondei)	34	83	72	0	0	0	20	62	43	0	1	0	33	62	71	0	0	0
Ipomoea sp.	Wild sweet potato leaves	Matembele, Matembele mwitu/pori	Ughulu, Chiwanda gulu							0	1	0	1	0	0	3	0	1	9	0	0
Ipomoea pescaprae	Beach morning glory	Majani ya mwaka	Talata?? (Muheza)	0	0	0	0	0	0	0	2	0	0	26	0	0	0	0	0	29	0
Lablab purpureus	Lablab	Fiwi		0	1	0	0	0	0	0	0	0	0	0	0	0	0	0	0	0	0
Launaea cornuta	Bitter lettuce	Mchunga	Ilisuka (Mbena), Mshunga (Sambaa), Sunga (Gogo, Kaguru)	0	0	0	5	111	0	0	0	0	3	71	0	0	0	0	19	55	0
Manihot esculenta	Cassava leaves	Kisamvu mhogo	Kayeva, Kisambu (Nyaturu), Pea (Bondei), Nyamattogo (Mbena), Kawea (Gogo)	4	76	10	0	0	2	0	35	8	0	0	0	0	32	5	0	0	0
Manihot glaziovii	Tree cassava leaves	Kisamvu mpira, Kisamvu mtii	Makaweya (Kongwa)	8	19	23	18	10	2	5	36	64	27	5	19	47	28	78	1	2	8
Moringa oleifera	Moringa leaves			0	0	0	0	0	0	0	0	0	0	0	0	0	0	0	0	0	0
Nicandra physaloides	(erect annual herb)	Mbirigi, Mbigili		0	0	0	0	0	0	0	0	0	0	0	0	0	0	0	0	0	0
Portulaca oleracea	Purslane	Tako la hasani		0	0	0	0	0	0	0	0	0	0	0	0	0	0	0	0	0	0
Rorippa nasturtium-aquaticum	Water cress	Saladi		0	0	0	0	0	0	0	0	0	0	0	0	0	0	0	0	0	0
Sesamum angustifolium	Wild simsim	Mlenda mwitu, mfuta, mlenda wima	Mahonda, Ikug'ha (Kinyaturu), Mlikwe (Kaguru), Nghonjera (Gogo)	0	0	2	4	0	84	0	0	0	0	0	81	0	0	0	2	0	98
Solanum aethiopicum	African eggplant	Ngogwe		0	0	0	0	0	0	1	8	5	0	0	0	7	1	10	0	0	0

Scientific name	English name	Swahili name	Local name	Jun/Jul (DS)						Nov/Dec (SR)						Mar/Apr (LR)					
				Cultivated			Collected			Cultivated			Collected			Cultivated			Collected		
				Ko	Mu	Si	Ko	Mu	Si	Ko	Mu	Si	Ko	Mu	Si	Ko	Mu	Si	Ko	Mu	Si
Solanum anguivi	African eggplant	Ngogwe		0	5	2	0	0	0	0	0	0	0	0	0	0	0	0	0	0	0
Solanum macrocarpon	African eggplant	Ngogwe	Uranangogwe (Nyaturu)	0	1	9	0	0	0	0	0	0	0	0	0	0	0	0	0	0	0
Solanum sp.	African eggplant	Ngogwe	Chimboto (Kuria)	0	1	0	0	0	0	0	0	0	0	0	0	0	0	0	0	0	0
Solanum sp.	African eggplant	Ngogwe	Tore (Muha)	0	1	0	0	0	0	0	0	0	0	0	0	0	0	0	0	0	0
Solanum sp.	African eggplant	Ngogwe	[Nyanya chungu (Mbena)]	2	8	4	0	0	0	0	0	0	0	0	0	0	0	0	0	0	0
Solanum americanum	African nightshade	Mnavu	Njuswa (Nyakusa), puche (Sambaa), Mhahi (Bena)	0	6	0	0	44	0	0	14	0	0	38	0	0	3	1	3	11	1
Solanum scabrum	African nightshade	Mnavu	Gimbankoko (Nyaturu), Chinswiga (Kuria), Njuswa (Nyakuswa)	0	22	3	0	2	2	1	9	2	0	0	0	22	5	10	0	0	0
Solanum villosum	African nightshade	Mnavu	Jimbo (Nyaturu)	1	1	1	0	3	0	0	0	0	2	2	1	5	1	14	6	1	64
Solanum sp.	African nightshade	Mnavu	Mnavu, Mnavu pori	0	1	0	1	4	0	0	4	0	1	6	0	0	0	0	0	0	0
Talinum sp.	Water leaf	Tonge, Pumbwiji, Mlenda		0	0	0	0	0	0	0	0	0	0	0	0	0	0	0	0	0	0
Vigna unguiculata	Cowpea (leaves)	(Majani ya) Kunde	Kusa, Sansa (Nyaturu), Shafa (Sambaa), Safwe (Gogo)	85	73	22	1	0	0	46	43	20	0	0	0	70	3	71	0	0	0
Vigna subterannea	Bambara groundnut	Njugu mawe	Pande (Nyaturu)	0	0	1	0	0	0	0	0	0	0	0	0	0	0	0	0	0	0
Tribulus terrestris	Caltrops, puncture vine		Mbigili, Mbigiri	0	0	0	0	1	0	0	0	0	0	0	0	0	0	0	0	0	0
Xanthosoma sagittifolium	Taro, Tannia	Majani ya magimbi	Iliyombo (Nyakusa)	0	1	0	0	0	0	0	1	0	0	1	0	0	7	0	0	5	0
			Bwete	0	0	0	0	1	0	0	0	0	0	0	0	0	0	0	0	0	0
			Chimua (Kongwa)	0	0	0	0	0	0	0	0	0	0	0	0	0	0	0	0	0	0
Ipomoea mombasana		Tembele mwitu	Chipali, Chiwanda-gulu	0	0	0	3	0	0	1	0	0	0	0	0	0	0	0	0	0	0

| Scientific name | English name | Swahili name | Local name | Jun/Jul (DS) | | | | | | Nov/Dec (SR) | | | | | | Mar/Apr (LR) | | | | | |
| | | | | Cultivated | | | Collected | | | Cultivated | | | Collected | | | Cultivated | | | Collected | | |
				Ko	Mu	Si	Ko	Mu	Si	Ko	Mu	Si	Ko	Mu	Si	Ko	Mu	Si	Ko	Mu	Si
			Dinohe (Kongwa)	0	0	0	0	0	0	0	0	0	0	0	0	0	0	0	1	0	0
			Ihundagwa	0	0	0	0	0	0	0	0	0	1	0	0	0	0	0	0	0	0
			Kayeva (Singida)	0	0	0	0	0	0	0	0	0	0	0	0	0	0	0	0	0	1
Corchorus sp.	Jute mallow		Kibwando	0	0	0	0	31	0	0	0	0	0	5	0	0	0	0	0	12	0
Platostema africanum (?)	(annual herb)		Kisogo	0	0	0	0	19	0	0	0	0	0	14	0	0	0	0	0	2	0
			Kisungu	0	0	0	0	1	0	0	0	0	0	1	0	0	0	0	0	1	0
			Kunduagundu (Muheza)	0	0	0	0	0	0	0	0	0	0	0	0	0	0	0	0	1	0
			Kungujulu	0	0	0	0	1	0	0	0	0	0	1	0	0	0	0	0	0	0
			Kuwi	0	0	0	0	0	0							0	0	0	1	0	0
			Kwake	0	0	0	0	1	0	0	0	0	0	0	0	0	0	0	0	0	0
		Mlenda (broad leaves)	Limana (Kaguru)	0	0	0	2	0	0	0	0	0	0	0	0	0	0	0	1	0	0
		like tembele	Liweja	0	0	0	0	0	0	0	0	0	1	0	0	0	0	0	0	0	0
			Mahon-dagwa, Manghun-daga	0	0	0	1	0	0	0	0	0	1	0	0	0	0	0	0	0	0
			Makasia	0	0	0	0	0	0	0	0	0	0	0	1	0	0	0	0	0	1
			Makawea	0	0	0	1	0	0	0	0	0	0	0	0	0	0	0	0	0	0
			Mhangalile (Kaguru)	0	0	0	1	0	0	0	0	0	0	0	0	0	0	0	0	0	0
Cleome sp.		Mhilile	Mhilile	1	0	0	71	0	0	0	0	0	41	0	0	0	0	0	66	0	0
		Mlenda	Mhumbulu	0	0	0	0	0	0	0	0	0	1	0	0	0	0	0	0	0	0
			Mkalifya	0	0	0	1	0	0	0	0	0	1	0	0	0	0	0	1	0	0
			Mkuku	0	0	0	0	0	0	0	0	1	0	0	13	0	0	0	0	0	2
			Mnyembeua	0	0	0	0	1	0	0	0	0	0	1	0	0	0	0	0	0	0
			Msangani	0	0	0	0	8	0	0	0	0	0	5	0	0	0	0	0	0	0
		Majani ya msufi	Msufi	0	0	0	0	1	0	0	0	0	0	0	0	0	0	0	0	0	0
			Mtanga	0	0	0	0	0	0	0	0	0	2	0	0	0	0	0	0	0	0
			Mtulu	0	0	0	0	0	0	0	0	0	2	0	0	0	0	0	0	0	0
			Ndelawari, Ndilawale (Singida)	0	0	0	0	0	0	0	0	0	0	0	1	0	0	0	0	0	1
			Ng'ongo	0	0	0	0	0	0	0	0	0	1	0	0	0	0	0	0	0	0

Scientific name	English name	Swahili name	Local name	Jun/Jul (DS)						Nov/Dec (SR)						Mar/Apr (LR)					
				Cultivated			Collected			Cultivated			Collected			Cultivated			Collected		
				Ko	Mu	Si	Ko	Mu	Si	Ko	Mu	Si	Ko	Mu	Si	Ko	Mu	Si	Ko	Mu	Si
Caylucea abyssinica			Ngoswe, Mghoswe, Mkoswe, Mngoswe	0	0	0	0	10	0	0	0	0	0	5	0	0	0	0	0	12	0
			Ngwiba							0	0	0	0	0	8	0	0	0	0	0	66
			Nkoso	0	0	0	0	1	0	0	0	0	0	0	0	0	0	0	0	0	0
			Nsonga, Songa (Singida)	0	0	0	0	0	0	0	0	0	0	0	0	0	0	0	0	0	2
			Nyangangati	0	1	0	0	0	0	0	0	0	0	0	0	0	0	0	0	0	0
			Nyangange	0	1	0	0	0	0	0	0	0	0	0	0	0	0	0	0	0	0
			Nyaulezi	1	0	0	0	0	0	0	0	0	0	0	0	0	0	0	0	0	0
	(legume)		Nzahe, Zahe	1	0	0	2	0	0	0	0	0	0	0	0	0	0	0	0	0	0
			Rongwe	0	0	0	0	0	0	0	0	0	0	0	0	0	0	0	0	0	0
			Sang'ala	0	0	0	0	0	0	0	0	0	1	0	0	0	0	0	0	0	0
Ipomoea sp. (?)		Tembele bwawa	Shambae, Smabae	0	0	0	0	18	0	0	0	0	0	11	0	0	0	0	0	5	0
			Talata, Tarata	0	0	0	0	32	0	0	0	0	0	1	0	0	0	0	0	0	0
			Tapa, Mtapa, Litapa	0	0	0	0	0	0	0	0	0	7	0	0	0	0	0	0	4	0
			Tebwa	0	0	0	0	0	0	0	0	0	0	0	0	0	0	0	0	0	0
			Tikini	0	0	0	0	25	0	0	0	0	0	15	0	0	0	0	0	6	0
			Zuma	0	0	0	0	2	0	0	0	0	0	6	0	0	0	0	0	2	0
Allium cepa	Onion	Vitunguu		0	0	1	0	0	0	0	0	0	0	0	0	0	0	14	0	0	0
Brassica oleracea convar. capitata	White cabbage	Kabichi		4	3	11	0	0	0	0	1	3	0	0	0	0	0	4	0	0	0
Brassica pekinensis	Chinese cabbage	Chainizi		4	9	34	0	0	0	3	4	18	0	0	0	1	1	16	0	0	0
Beta vulgaris subsp. cicla	Swiss chard	Spinachi		1	13	7	0	1	0	1	3	4	0	0	0	4	0	4	0	0	0
Lycopersicon esculentum	Tomato	Nyanya	(e.g. dumdum)	9	14	32	0	0	0	1	5	8	0	0	0	3	2	38	0	0	0
Lycopersicon esculentum	Cherry tomato (indigenous, wild)	Nyanya pori / kienyeji / gololi		2	7	6	0	1	0	0	4	0	0	0	0	0	0	6	2	1	28
Phaseolus vulgaris	Common bean	Maharage	Mahaage (Bondei)	4	22	5	0	0	0	0	0	0	0	0	0	0	0	0	0	0	0

Scientific name	English name	Swahili name	Local name	Jun/Jul (DS)						Nov/Dec (SR)						Mar/Apr (LR)					
				Cultivated			Collected			Cultivated			Collected			Cultivated			Collected		
				Ko	Mu	Si	Ko	Mu	Si	Ko	Mu	Si	Ko	Mu	Si	Ko	Mu	Si	Ko	Mu	Si
Capsicum annum	Sweet pepper	Pilipili hoho	Mhipii (Sambaa)	0	1	1	0	0	0	0	1	2	0	0	0	0	0	3	0	0	0
Capsicum baccatum (?)	Hot pepper	Pilipili mbuzi		0	0	0	0	0	0	0	0	0	0	0	0	0	0	1	0	0	0
Solanum melongina	Brinjal eggplant / Aubergine	Biringanya, Mabiringanya		0	4	6	0	0	0	0	4	2	0	0	0	0	0	2	0	0	0
Daucus carota	Carrot			0	0	0	0	0	0	0	0	0	0	0	0	0	0	0	0	0	0

Table A9   Share of participants (%) that sold vegetables in different ways (multiple answers permitted) during three different seasons in Tanzania

	Jun/Jul (DS)	Nov/Dec (SR)	Mar/Apr (LR)
n	65	24	33
Farm gate	63.1	29.2	48.5
Market	35.4	20.8	36.4
Walking from house to house	9.2	25.0	9.1
Others e.g. at homestead	9.2	41.7	30.3

Table A10   Number of participants that sold vegetables at different frequencies during one year in three districts of Tanzania

Jun/Jul (DS)

	n	< 4 times/ month	Once a week	Twice a week	Thrice a week	4-6 times a week	Every day
Kongwa	13	2	3	7	1	0	0
Muheza	27	2	11	9	4	0	1
Singida	29	0	11	13	4	0	1

Nov/Dec (SR)

	n	< 4 times/ month	Once a week	Twice a week	Thrice a week	4-6 times a week	Every day
Kongwa	4	0	3	1	0	0	0
Muheza	13	2	1	8	2	0	0
Singida	6	0	0	5	0	1	0

Mar/Apr (LR)

	n	< 4 times/ month	Once a week	Twice a week	Thrice a week	4-6 times a week	Every day
Kongwa	5	0	1	3	0	0	1
Muheza	13	3	3	5	2	0	0
Singida	15	0	7	4	4	0	0

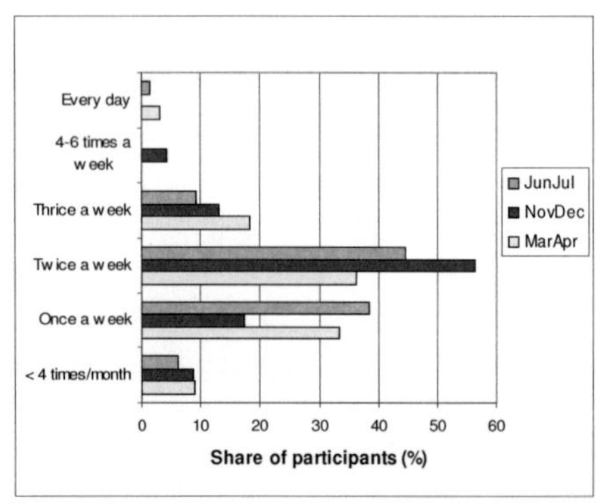

Figure A2
Share of participants that sold vegetables at different frequencies during one year in Tanzania (Jun/Jul (DS) n=65; Nov/Dec (SR) n=24; Mar/Apr (LR) n=33)

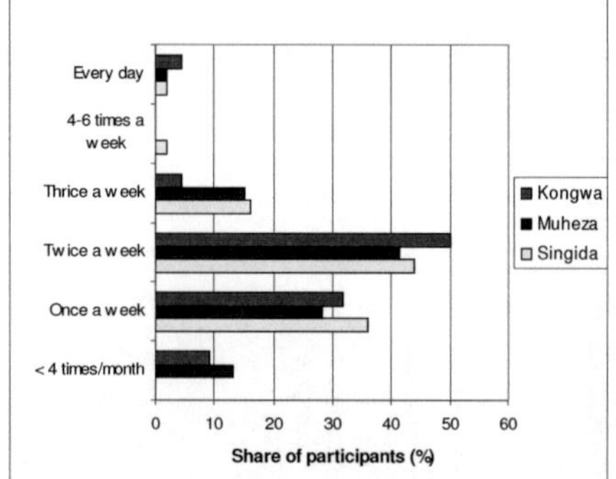

Figure A3
Share of participants that sold vegetables at different frequencies in three districts of Tanzania (Kongwa n=7; Muheza n=18; Singida n=17)

Table A11   Share of participants that purchased vegetables in different setups during two different seasons in Tanzania (multiple answers permitted; answers were not predetermined)

	Nov/Dec (SR)	Mar/Apr (LR)
n	**151**	**111**
Market	36.4	54.1
Farm gate/neighbours	15.9	16.2
Traders walking house to house	64.9	48.6
Shop/'Magenge'	1.3	5.4
Others	3.3	4.5

Table A12   Reasons for purchasing additional vegetables and share of participants (%) that stated these reasons (multiple answer permitted) during two different seasons in Tanzania

	Nov/Dec (SR)				Mar/Apr (LR)			
District	**All**	**Ko**	**Mu**	**Si**	**All**	**Ko**	**Mu**	**Si**
n	**151**	**54**	**48**	**49**	**111**	**26**	**51**	**34**
Not available in own garden/wild	39.7	31.5	**58.3**	30.6	**65.8**	**76.9**	37.3	**100.0**
Taste	**43.7**	**40.7**	50.0	38.8	**45.0**	19.2	**51.0**	41.2
Preparation/usage	9.9	11.1	2.1	16.3	3.6	0.0	3.9	2.9
Available in market/cheap	**43.7**	38.9	16.7	**51.0**	17.1	0.0	35.3	0.0
Nutrition/health	11.3	24.1	0.0	6.1	13.5	23.1	19.6	0.0

Highlighted: Highest share(s) within a column

Table A13   Number of participants that purchased vegetables at different frequencies during two seasons in three districts of Tanzania

Nov/Dec (SR)

	n	< 4 times/ month	Once a week	Twice a week	Thrice a week	4-6 times a week	Every day
Kongwa	55	6	14	14	11	10	0
Muheza	46	8	10	17	7	3	1
Singida	49	6	12	18	7	4	2

Mar/Apr (LR)

	n	< 4 times/ month	Once a week	Twice a week	Thrice a week	4-6 times a week	Every day
Kongwa	26	0	11	10	4	1	0
Muheza	50	0	10	25	13	2	0
Singida	33	0	8	13	6	6	2

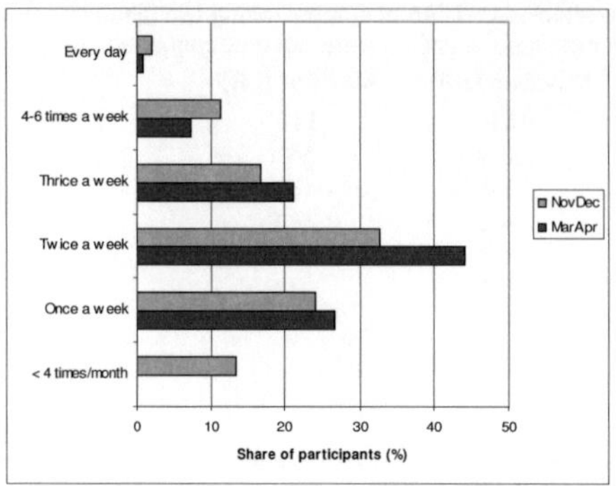

Figure A4
Share of participants that purchased vegetables at different frequencies during two different seasons in Tanzania (NovDec (SR) n=150; MarApr (LR) n=109)

Figure A5
Share of participants that purchased vegetables at different frequencies in three districts of Tanzania (Kongwa n=27; Muheza n=32; Singida n=27)

Table A14   Reasons for <u>not</u> cultivating vegetables that were cultivated during the previous season, and share of women (%) that stated the reason (multiple answer permitted; answers were not predetermined) during two different seasons in Tanzania

	Nov/Dec	Mar/Apr
n	107	79
Temperature too high/low	0.9	**67.1**
Water availability	**35.5**	**65.8**
Lack of input	**21.5**	1.3
Not in season	15.9	17.7
Flooding of garden	18.7	7.6
Pests+diseases	4.7	1.3
Plenty of other veg available	0.0	2.5
Others	8.4	5.1

Highlighted: The most often named two reasons per season

Table A15   Vegetables that were <u>not</u> cultivated during the current season because of different constraints, and share of women that named the vegetables (%) (multiple answer permitted) during two different seasons in Tanzania

Vegetable	Nov/Dec n=87	Mar/Apr n=66	Vegetable	Nov/Dec n=87	Mar/Apr n=66
African eggplant	9.2	7.6	Okra	**40.2**	10.6
African nightshade	3.4	3.0	Pumpkin fruit	0.0	1.5
Amaranth leaves	12.6	19.7	Pumpkin leaves	13.8	**22.7**
Chinese cabbage	9.2	15.2	Seeds (general)	2.3	0.0
Cowpea grain	0.0	18.2	Sweet pepper	1.1	1.5
Cowpea leaves	6.9	**22.7**	Sweet potato leaves	6.9	3.0
Eggplant	3.4	0.0	Swiss chard	12.6	7.6
Ethiopian kale	5.7	1.5	Tomato	**23.0**	4.5
			White cabbage	4.6	3.0

Highlighted: The most often named two vegetables per season

Table A16   Intake in gram per day of selected food groups by women of three districts in Tanzania; mean across three days during three different seasons

Food group	Kongwa (n=72)		Muheza (n=76)		Singida (n=104)	
	Mean ± SD	Median (Range)	Mean ± SD	Median (Range)	Mean ± SD	Median (Range)
Cereals	308 ± 115	299 (73-680)	283 ±114	294 (0 - 599)	370 ±115	368 (71 - 677)
Bread/cakes	12 ± 39	0 (0-333)	64 ±64	53 (0 - 327)	9 ±30	0 (0 - 167)
Fruits	22 ± 48	0 (0-292)	133 ±130	102 (0 - 700)	26 ±60	0 (0 - 333)
Vegetables	277 ± 134	264 (54-886)	208 ±96	194 (0 - 463)	351 ±151	342 (52 - 930)
Nuts	34 ± 60	15 (0-384)	104 ±92	88 (0 – 400)	1 ±4	0 (0 - 16)
Pulses	61 ± 75	52 (0-338)	92 ±92	67 (0 - 424)	67 ±71	50 (0 - 307)
Starchy plants	22 ± 59	0 (0-317)	108 ±123	77 (0 - 623)	34 ±62	0 (0 - 293)
Tea	128 ± 172	67 (0-933)	335 ±158	333 (0 - 767)	75 ±128	0 (0 - 733)
Oil/fat	17 ± 19	12 (0-96)	22 ±16	19 (0 -79)	19 ±12	17 (0 - 59)
Sugar	8 ± 10	5 (0-50)	12 ±13	9 (0 - 76)	5 ±12	3 (0 - 72)
Fish	7 ± 14	0 (0-80)	61 ±52	49 (0 - 230)	18 ±29	0 (0 - 137)
Animal prod	21 ± 43	0 (0-267)	26 ±53	0 (0 - 310)	42 ±62	10 (0 - 327)

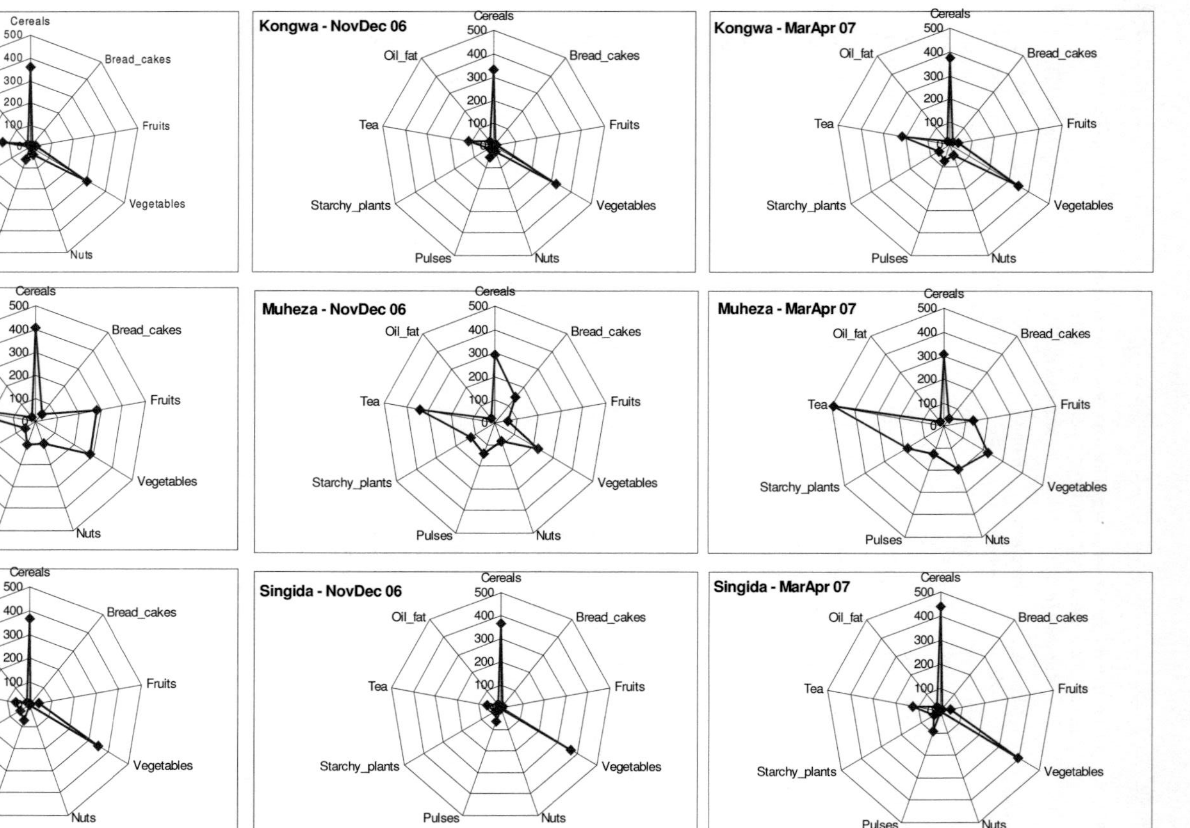

Figure A6
Mean intake of nine food groups in g/d of women from three districts in Tanzania during three seasons

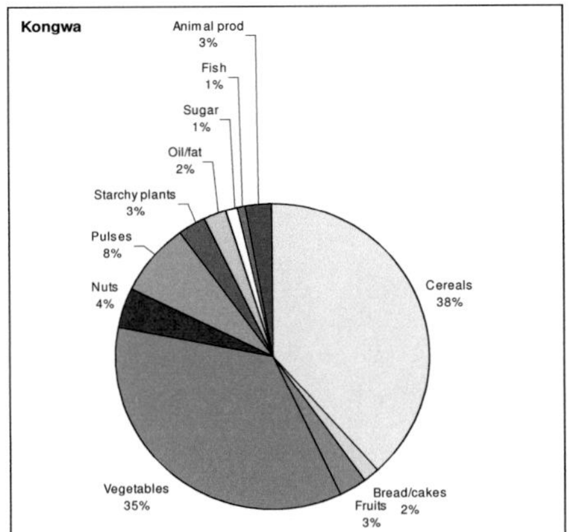

Figure A7
Share of eleven food groups consumed on average by women in Kongwa district (mean across three days; n=72)

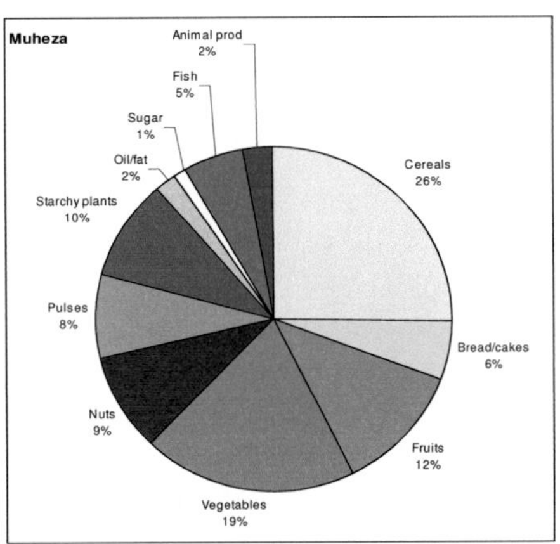

Figure A8
Share of eleven food groups consumed on average by women in Muheza district (mean across three days; n=76)

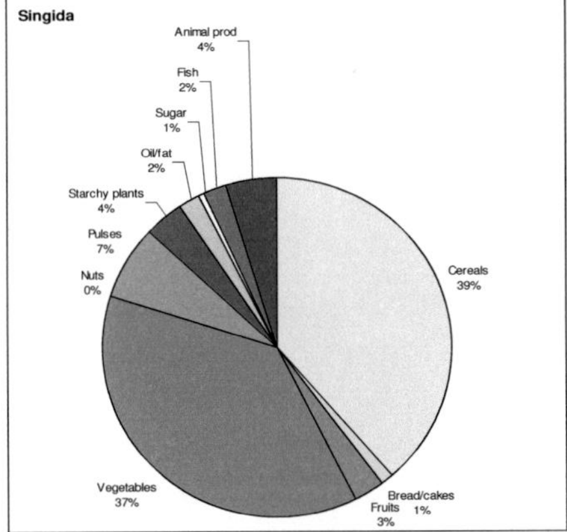

Figure A9
Share of eleven food groups consumed on average by women in Singida district (mean across three days; n=104)

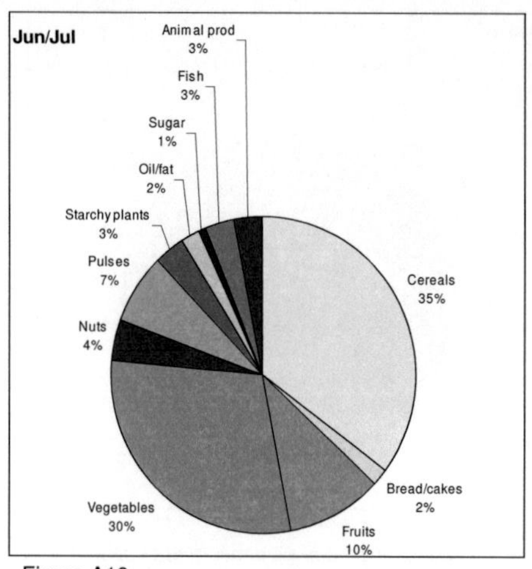

Figure A10
Share of eleven food groups consumed on average by women on one day during Jun/Jul (DS)

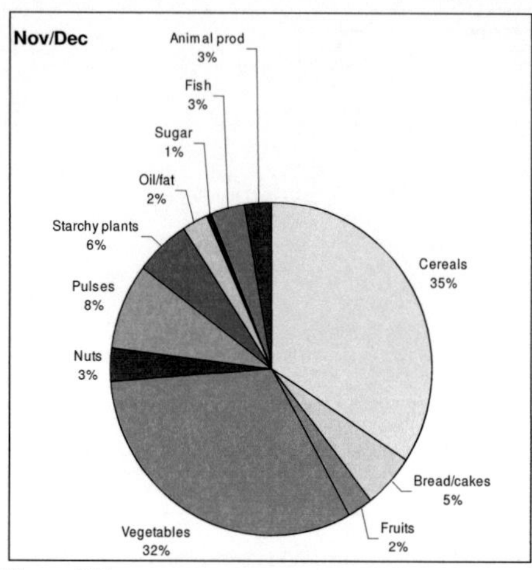

Figure A11
Share of eleven food groups consumed on average by women on one day during Nov/ Dec (SR)

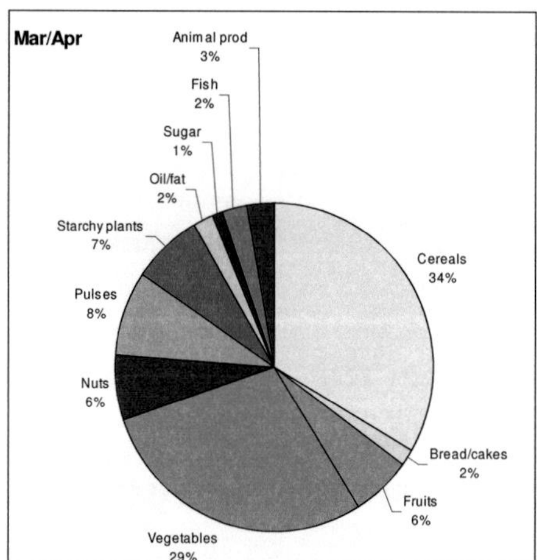

Figure A12
Share of eleven food groups consumed on average by women on one day during Mar/ Apr (LR)

Table A17   Share of participants that consumed five food groups with respect to their recommended daily allowance (RDA) during three different seasons in Tanzania

	Starchy staple food	Fruits and vegetables	Nuts and pulses	Animal products	Fats and oils
All seasons					
No consumption	0.0	0.0	8.8	29.2	4.0
Below	12.7	31.7	21.9	26.8	34.9
Within	53.6	48.8	42.2	34.4	38.5
Above	33.7	19.4	27.1	9.6	22.6
Jun/Jul (DS)					
No consumption	0.0	0.8	39.7	55.8	23.0
Below	21.4	39.7	10.3	7.2	29.8
Within	52.8	27.4	30.6	21.9	26.6
Above	25.8	32.1	19.4	15.1	20.6
Nov/Dec (SR)					
No consumption	0.4	0.8	41.8	60.2	17.1
Below	17.9	48.8	13.9	6.4	28.2
Within	54.0	33.3	24.3	16.3	29.8
Above	27.8	17.1	19.9	17.1	25.0
Mar/Apr (LR)					
No consumption	0.0	0.4	27.8	59.9	16.3
Below	15.1	38.9	11.9	6.3	29.0
Within	46.8	34.9	26.6	19.4	25.0
Above	38.1	25.8	33.7	14.3	29.8

Table A18   Share of participants that consumed five food groups with respect to their recommended daily allowance (RDA) in three districts of Tanzania (mean across three seasons)

	Starchy staple food	Fruits and vegetables	Nuts and pulses	Animal products	Fats and oils
Kongwa					
No consumption	0.0	0.0	1.4	52.8	11.1
Below	26.4	43.1	33.3	25.0	33.3
Within	50.0	44.4	50.0	18.1	36.1
Above	23.6	12.5	15.3	4.2	19.4
Muheza					
No consumption	0.0	0.0	0.0	1.3	0.0
Below	1.3	23.7	4.0	26.7	27.6
Within	46.1	56.6	30.7	57.3	42.1
Above	52.6	19.7	65.3	14.7	30.3
Singida					
No consumption	0.0	0.0	20.2	33.0	1.9
Below	11.5	28.8	26.9	28.2	37.5
Within	62.5	47.1	46.2	29.1	43.3
Above	26.0	24.0	6.7	9.7	17.3

Table A19   Correlation between intake of food groups and cultivation/collection of vegetables by women during three different seasons in Tanzania

Food group	No of vegetables cultivated		No of vegetables collected		No of TVs cult/coll		No of Evs cultivated	
	Correl. coeff. rho	p	Correl. coeff. rho	p	Correl. coeff. rho	p	Correl. coeff. rho	p
Jun/Jul (DS)								
Cereals	n.a.	n.s.	0.134	0.034	n.a.	n.s.	n.a.	n.s.
Bread and cakes	n.a.	n.s.	0.134	0.033	0.167	<0.001	n.a.	n.s.
Fruits	0.269	<0.001	0.413	<0.001	0.462	<0.001	n.a.	n.s.
Nuts	0.161	0.011	0.349	<0.001	0.362	<0.001	-0.163	0.010
Pulses	0.139	0.027	0.187	0.003	0.216	0.001	n.a.	n.s.
Tea	0.239	<0.001	0.411	<0.001	0.410	<0.001	n.a.	n.s.
Oil/fat	0.159	0.011	n.a.	n.s.	0.157	0.012	0.175	0.005
Sugar	0.144	0.022	0.365	<0.001	0.336	<0.001	n.a.	n.s.
Fish	0.134	0.034	0.167	0.008	0.180	0.004	n.a.	n.s.
Nov/Dec (SR)								
Cereals	n.a.	n.s.	-0.148	0.019	-0.140	0.027	n.a.	n.s.
Bread and cakes	0.241	<0.001	0.222	<0.001	0.301	<0.001	n.a.	n.s.
Fruits	0.135	0.032	0.206	0.001	0.176	0.005	n.a.	n.s.
Vegetables	n.a.	n.s.	n.a.	n.s.	n.a.	n.s.	-0.136	0.031
Nuts	0.126	0.046	n.a.	n.s.	n.a.	n.s.	n.a.	n.s.
Pulses	0.155	0.014	n.a.	n.s.	n.a.	n.s.	n.a.	n.s.
Starchy plants	0.218	0.001	0.245	<0.001	0.312	<0.001	n.a.	n.s.
Tea	0.327	<0.001	0.139	0.028	0.310	<0.001	n.a.	n.s.
Sugar	0.302	<0.001	0.134	0.034	0.277	<0.001	0.181	0.004
Fish	0.224	<0.001	0.127	0.045	0.246	<0.001	n.a.	n.s.
Mar/Apr (LR)								
Cereals	n.a.	n.s.	0.214	0.001	0.141	0.025	0.226	<0.001
Bread and cakes	-0.166	0.009	-0.215	0.001	-0.215	0.001	-0.167	0.008
Fruits	-0.182	0.004	-0.185	0.003	-0.231	<0.001	-0.158	0.013
Vegetables	0.291	<0.001	0.289	<0.001	0.357	<0.001	0.198	0.002
Nuts	-0.252	<0.001	-0.404	<0.001	-0.378	<0.001	-0.427	<0.001
Pulses	n.a.	n.s.	n.a.	n.s.	-0.166	0.008	n.a.	n.s.
Starchy plants	n.a.	n.s.	-0.167	0.005	-0.156	0.013	-0.168	0.007
Tea	-0.232	<0.001	-0.327	<0.001	-0.342	<0.001	-0.270	<0.001
Sugar	-0.141	0.025	-0.283	<0.001	-0.271	<0.001	-0.247	<0.001
Fish	-0.308	<0.001	-0.330	<0.001	-0.403	<0.001	-0.260	<0.001

Table A20 — Intake of macronutrients by study participants in three districts of Tanzania; mean across three seasons (outliers excluded for energy and fat)

Nutrient	Kongwa		Muheza		Singida	
	Mean ±SD	Median (Range)	Mean ±SD	Median (Range)	Mean ±SD	Median (Range)
Energy	1836 ±405	1777 (925-2811)	2053 ±328	2022 (1148-2688)	1694 ±322	1662 (922-2450)
Protein	60 ±14	58 (31-92)	66 ±15	65 (21-103)	51 ±12	50 (26-86)
Fat	40 ±19	37 (9-109)	40 ±17	37 (12-92)	35 ±14	34 (7-75)
Carbohydrate	308 ±64	310 (176-480)	359 ±64	356 (188-521)	305 ±59	305 (142-436)
Dietary fibre	32 ±9	31 (18-65)	33 ±7	33 (16-55)	43 ±16	39 (18-92)
PUFA	13 ±9	9,5 (2-54)	7 ±2	7 (3-17)	15 ±7	14 (3-39)
Cholesterol	9 ±13	4 (0-65)	41 ±38	27 (0-167)	20 ±28	9 (0-137)

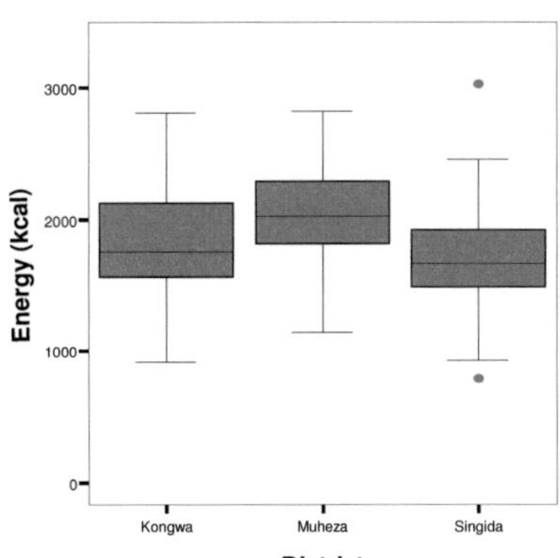

Figure A13
Energy intake of women within three districts of Tanzania (mean across three seasons)

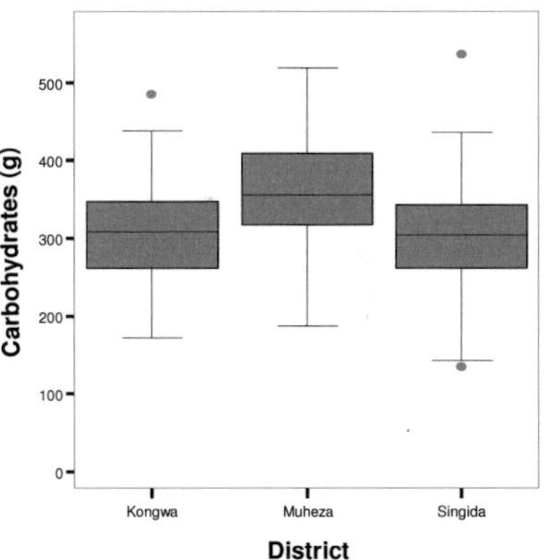

Figure A14
Carbohydrate intake of women within three districts of Tanzania (mean across three seasons)

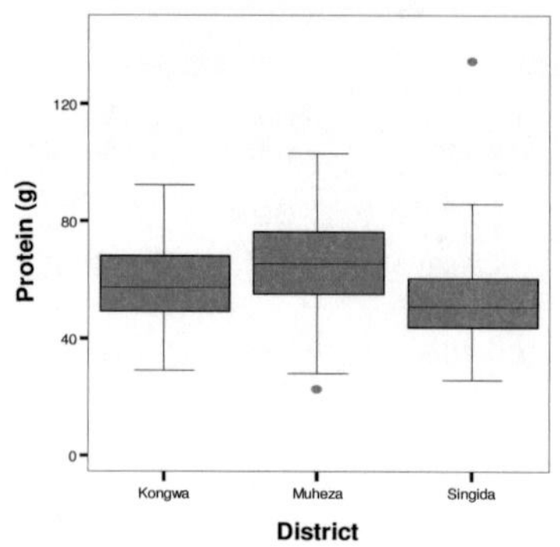

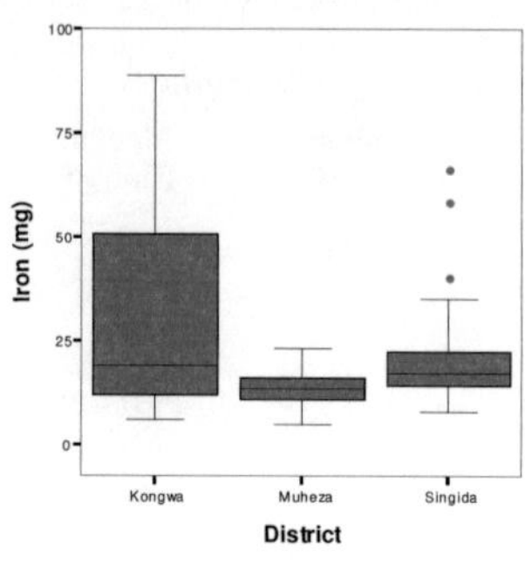

Figure A15
Protein intake of women within three districts of Tanzania (mean across three seasons)

Figure A16
Iron intake of women within three districts of Tanzania (mean across three seasons)

Table A21   Correlations between vegetable cultivation/collection and intake of eight nutrient groups by women of Tanzania (mean across three seasons)

Nutrient	No of vegetables cultivated		No of vegetables collected		No of TVs cult/coll		No of Evs cultivated	
	Correl. coeff. rho	p	Correl. coeff. rho	p	Correl. coeff. rho	p	Correl. coeff. rho	p
Energy	n.a.	n.s.	n.a.	n.s.	0.148	0.019	n.a.	n.s.
Protein	n.a.	n.s.	n.a.	n.s.	n.a.	n.s.	-0.133	0.034
Carbohydrates	0.132	0.036	0.129	0.040	0.167	0.008	n.a.	n.s.
Dietary fibre	n.a.	n.s.	n.a.	n.s.	n.a.	n.s.	0.268	<0.001
PUFA	n.a.	n.s.	n.a.	n.s.	n.a.	n.s.	0.161	0.010
Cholesterol	n.a.	n.s.	n.a.	n.s.	0.152	0.016	n.a.	n.s.
Vitamin A	n.a.	n.s.	n.a.	n.s.	n.a.	n.s.	-0.238	<0.001
Carotene	n.a.	n.s.	n.a.	n.s.	0.139	0.027	-0.323	<0.001
Vitamin E	n.a.	n.s.	n.a.	n.s.	n.a.	n.s.	0.260	<0.001
Vitamin B6	n.a.	n.s.	n.a.	n.s.	n.a.	n.s.	-0.169	0.007
Folic acid	-0.174	0.005	n.a.	n.s.	n.a.	n.s.	-0.237	<0.001
Vitamin C	n.a.	n.s.	0.162	0.010	0.136	0.031	n.a.	n.s.
Potassium	n.a.	n.s.	n.a.	n.s.	n.a.	n.s.	-0.233	<0.001
Calcium	-0.138	0.029	n.a.	n.s.	n.a.	n.s.	-0.248	<0.001
Phosphorus	n.a.	n.s.	n.a.	n.s.	n.a.	n.s.	0.145	0.021
Iron	-0.190	0.002	n.a.	n.s.	-0.210	0.001	n.a.	n.s.

Table A22   Uptake of vitamins by participants in three districts of Tanzania; mean across three seasons (outliers excluded for vitamin A)

Nutrient	Kongwa (n=57)		Muheza (n=67)		Singida (n=84)	
	Mean ± SD	Median (range)	Mean ± SD	Median (range)	Mean ± SD	Median (range)
Vitamin A	1143.8	1034	1441.6	1315	605.3	381
(µg/d)	±387.2	(235-3353)	±686.1	(458-3546)	±677.1	(89-3229)
Carotene	4.3	3	8.4	7	1.1	1
(mg/d)	±3.1	(0-12)	±4.5	(1-21)	±1.0	(0-6)
Vitamin C	127.8	112	148.9	139	133.3	96
(mg/d)	±74.8	(30-469)	±73.7	(29-398)	±120.3	(29-597)
Vitamin E	6,9	4	6.4	6	12.5	10
(mg/d)	±7.1	(0-40)	±2.3	(3-14)	±8.0	(1-39)
Vitamin B1	1,1	1	1.0	1	1.0	1
(mg/d)	±0.5	(1-4)	±0.2	(0-2)	±0.2	(0-2)
Vitamin B2	1,0	1	1.0	1	0.9	1
(mg/d)	±0.5	(0-2)	±0.2	(0-1)	±0.3	(0-2)
Vitamin B6	1,7	2	2.0	2	1.4	1
(mg/d)	±0.5	(1-3)	±0.6	(1-3)	±0.5	(1-2)
Folic Acid	496,2	452	392.6	372	318.7	300
(µg/d)	±229.4	(127-1149)	±138.2	(152-783)	±114.5	(132-708)

Table A23   Uptake of minerals by participants in three different districts of Tanzania; mean across three seasons (outliers excluded for iron)

Nutrient	Kongwa (n=35)		Muheza (n=60)		Singida (n=77)	
	Mean ± SD	Median (range)	Mean ± SD	Median (range)	Mean ± SD	Median (range)
Iron	17.2	13	13.4	14	18.4	17
(mg/d)	±10.1	(6-42)	±3.5	(5-23)	±6.9	(8-39)
Zinc	5.9	6	7.6	8	7.1	6
(mg/d)	±2.7	(2-17)	1.8	(3-12)	±3.2	(3-21)
Calcium	524.8	491	429.6	410	450.1	442
(mg/d)	±206.3	(212-945)	128.2	(222-873)	±156.7	(206-966)
Sodium	255.3	187	563.5	505	307.7	250
(mg/d)	±291.0	(32-1757)	280.9	(136-1285)	±206.1	(65-1086)
Potassium	2265.9	2267	2879.1	2863	2067.3	1999
(mg/d)	±910.6	(814-5594)	843.2	(1319-6347)	±633.6	(1123-4485)
Magnesium	320.3	281	341.6	322	328.6	297
(mg/d)	±178.7	(151-1163)	85.8	(189-619)	±123.7	(147-774)
Phosphorus	861.8	783	942.9	965	1023.8	1021
(mg/d)	371.5	(355-2366)	205.2	(330-1385)	±279.6	(481-2052)

Table A24   Correlation between vegetable cultivation/collection and intake of nutrients by women of Tanzania during three different seasons

Nutrient	No of vegetables cultivated		No of vegetables collected		No of TVs cult/coll		No of EVs cultivated	
	Correl. coeff. rho	p	Correl. coeff. rho	p	Correl. coeff. rho	p	Correl. coeff. rho	p
Jun/Jul (DS)								
Energy	0.141	0.025	0.317	<0.001	0.304	<0.001	n.a.	n.s.
Protein	n.a.	n.s.	0.242	<0.001	0.173	0.006	n.a.	n.s.
Carbohydrates	0.138	0.029	0.294	<0.001	0.296	<0.001	n.a.	n.s.
PUFA	n.a.	n.s.	-0.143	0.023	-0.129	0.040	0.156	0.013
Cholesterol	n.a.	n.s.	n.a.	n.s.	n.a.	n.s.	0.159	0.011
Vitamin A	0.202	0.001	0.280	<0.001	0.344	<0.001	n.a.	n.s.
Carotene	0.257	<0.001	0.455	<0.001	0.476	<0.001	n.a.	n.s.
Vitamin E	0.175	0.005	n.a.	n.s.	n.a.	n.s.	0.246	<0.001
Vitamin B1	0.141	0.025	0.234	<0.001	0.256	<0.001	n.a.	n.s.
Vitamin B2	n.a.	n.s.	n.a.	n.s.	n.a.	n.s.	-0.127	0.044
Vitamin B6	n.a.	n.s.	0.263	<0.001	0.237	<0.001	n.a.	n.s.
Folic acid	n.a.	n.s.	0.237	<0.001	0.202	0.001	-0.128	0.043
Vitamin C	0.153	0.015	0.240	<0.001	0.278	<0.001	n.a.	n.s.
Sodium	0.137	0.030	0.214	0.001	0.243	<0.001	n.a.	n.s.
Potassium	n.a.	n.s.	0.257	<0.001	0.233	<0.001	-0.152	0.016
Calcium	n.a.	n.s.	n.a.	n.s.	n.a.	n.s.	-0.189	0.003
Magnesium	n.a.	n.s.	0.144	0.022	0.147	0.020	n.a.	n.s.
Nov/Dec (SR)								
Energy	0.148	0.018	0.151	0.016	0.142	0.024	n.a.	n.s.
Carbohydrates	0.133	0.035	0.197	0.002	0.144	0.022	n.a.	n.s.
Dietary fibre	n.a.	n.s.	n.a.	n.s.	n.a.	n.s.	0.138	0.029
PUFA	n.a.	n.s.	-0.199	0.002	-0.197	0.002	n.a.	n.s.
Cholesterol	0.236	<0.001	n.a.	n.s.	0.228	<0.001	n.a.	n.s.
Vitamin A	n.a.	n.s.	0.198	0.002	0.142	0.024	n.a.	n.s.
Carotene	0.160	0.011	0.155	0.014	0.184	0.003	n.a.	n.s.
Vitamin B2	-0.154	0.015	n.a.	n.s.	n.a.	n.s.	n.a.	n.s.
Vitamin B6	n.a.	n.s.	0.197	0.002	n.a.	n.s.	n.a.	n.s.
Folic Acid	-0.180	0.004	n.a.	n.s.	-0.152	0.016	n.a.	n.s.
Vitamin C	-0.127	0.044	0.219	<0.001	n.a.	n.s.	n.a.	n.s.
Potassium	-0.133	0.035	0.153	0.015	n.a.	n.s.	n.a.	n.s.
Calcium	-0.201	0.001	n.a.	n.s.	-0.168	0.008	n.a.	n.s.
Iron	-0.216	0.001	n.a.	n.s.	-0.229	<0.001	n.a.	n.s.
Mar/Apr (LR)								
Protein	-0.202	0.001	-0.174	0.006	-0.198	0.002	-0.157	0.012
Dietary fibre	n.a.	n.s.	0.204	0.001	0.143	0.023	0.312	<0.001
PUFA	0.350	<0.001	0.301	<0.001	0.382	<0.001	0.268	<0.001
Cholesterol	-0.169	0.007	-0.226	<0.001	-0.224	<0.001	n.a.	n.s.
Vitamin A	-0.172	0.006	-0.261	<0.001	-0.260	<0.001	-0.266	<0.001
Carotene	-0.166	0.008	-0.337	<0.001	-0.294	<0.001	-0.357	<0.001
Vitamin E	0.283	<0.001	0.275	<0.001	0.316	<0.001	0.272	<0.001
Vitamin B6	-0.140	0.026	-0.168	0.008	-0.183	0.004	-0.156	0.013
Sodium	n.a.	n.s.	-0.194	0.002	-0.187	0.003	n.a.	n.s.
Potassium	n.a.	n.s.	-0.200	0.001	-0.191	0.002	n.a.	n.s.
Calcium	0.130	0.040	n.a.	n.s.	n.a.	n.s.	n.a.	n.s.
Phosphorus	n.a.	n.s.	n.a.	n.s.	n.a.	n.s.	0.198	0.002
Iron	0.167	0.008	0.188	0.003	0.190	0.002	0.311	<0.001

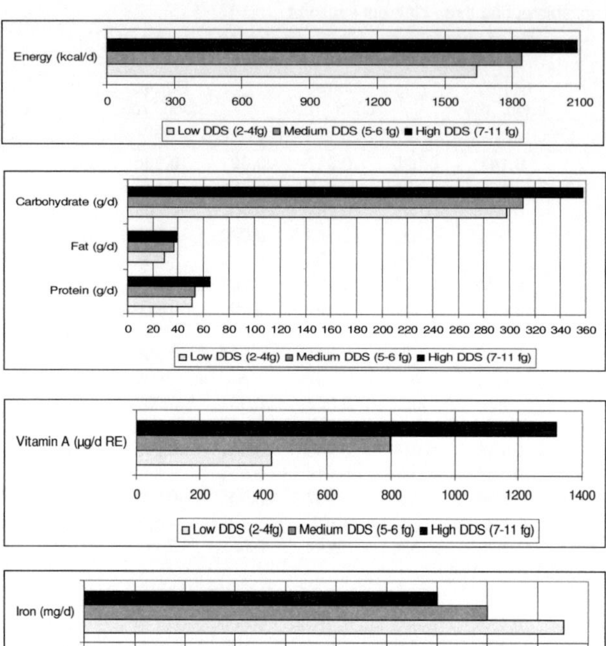

Figure A17   Mean (energy and carbohydrate) or median intake of participants as a function of the DDS categories; mean across three seasons; DDS=dietary diversity score, fg=food groups

Table A25   The five most important <u>traditional</u> vegetables consumed during three seasons in three different districts of Tanzania and percentage of women who consumed these at least once during one week

Jun/Jul (DS)	%	Nov/Dec (SR)	%	Mar/Apr (LR)	%
Kongwa					
Cowpea lvs (*Vigna unguiculata*)	70.8	False sesame (*Cerathoteca sesamoide*)	72.2	Pumpkin lvs (*Cucurbita* sp.)	80.6
False sesame (*Cerathoteca sesamoide*)	40.3	Amaranth lvs (*Amaranthus* sp.)	51.4	False sesame (*Cerathoteca sesamoide*)	73.6
Pumpkin lvs (*Cucurbita* sp.)	40.3	Cowpea lvs (*Vigna unguiculata*)	48.6	Amaranth lvs (*Amaranthus* sp.)	70.8
Wild simsim (*Sesamum angustifolium*)	31.9	Sweet potato lvs (*Ipomoea batatas*)	48.6	Mhilile (*Cleome* sp.)	69.4
Amaranth lvs (*Amaranthus* sp.)	29.2	Pumpkin lvs (*Cucurbita* sp.)	29.2	Cowpea leaves (*Vigna unguiculata*)	65.3
				Okra (*Abelmoschus esculentus*)	65.3
Muheza					
Pumpkin lvs (*Cucurbita* sp.)	65.3	Amaranth lvs (*Amaranthus* sp.)	76.3	Amaranth lvs (*Amaranthus* sp.)	80.3
Amaranth lvs (*Amaranthus* sp.)	54.7	Sweet potato lvs (*Ipomoea batatas*)	67.1	Sweet potato lvs (*Ipomoea batatas*)	65.8
Bitter lettuce(*Launaea cornuta*)	48.0	Bitter lettuce(*Launaea cornuta*)	57.9	Okra (*Abelmoschus esculentus*)	43.4
Sweet potato lvs (*Ipomoea batatas*)	42.7	Pumpkin lvs (*Cucurbita* sp.)	56.6	Pumpkin lvs (*Cucurbita* sp.)	38.2
Okra (*Abelmoschus esculentus*)	38.7	Okra (*Abelmoschus esculentus*)	50.0	Bitter lettuce(*Launaea cornuta*)	31.6
Singida					
Sweet potato lvs (*Ipomoea batatas*)	49.0	Amaranth lvs (*Amaranthus* sp.)	60.6	Amaranth lvs (*Amaranthus* sp.)	93.3
False sesame (*Cerathoteca sesamoide*)	48.1	False sesame (*Cerathoteca sesamoide*)	49.0	Okra (*Abelmoschus esculentus*)	76.9
Okra (*Abelmoschus esculentus*)	48.1	Wild simsim (*Sesamum angustifolium*)	44.2	False sesame (*Cerathoteca sesamoide*)	66.3
Amaranth lvs (*Amaranthus* sp.)	41.3	Sweet potato lvs (*Ipomoea batatas*)	38.5	Pumpkin lvs (*Cucurbita* sp.)	65.4
Wild simsim (*Sesamum angustifolium*)	40.4	Cassava lvs (*Manihot esculentus*)	25.0	Sweet potato lvs (*Ipomoea batatas*)	64.4

Table A26   The five most important exotic vegetables consumed during three seasons in three districts of Tanzania and percentage of women who consumed these at least once during one week

Jun/Jul (DS)	%	Nov/Dec (SR)	%	Mar/Apr (LR)	%
Kongwa					
Tomato	66.7	Tomato	80.6	Tomato	100.0
Onion	58.3	Onion	75.0	Onion	98.6
Chinese cabbage	11.1	Chinese cabbage	15.3	Chinese cabbage	4.2
White cabbage	1.4	White cabbage	5.6	White cabbage	1.4
				Cucumber, exotic	1.4
Muheza					
Onion	97.3	Tomato	96.1	Tomato	93.4
Tomato	93.3	Onion	94.7	Onion	89.5
Chinese cabbage	12.0	White cabbage	23.7	White cabbage	23.7
White cabbage	10.7	Chinese cabbage	9.2	Chinese cabbage	3.9
Swiss chard	2.7	Swiss chard	7.9	Carrot	2.6
Singida					
Tomato	100.0	Tomato	92.3	Tomato	99.0
Onion	99.0	Onion	90.4	Onion	98.1
Chinese cabbage	28.8	Chinese cabbage	41.3	Chinese cabbage	12.5
White cabbage	6.7	White cabbage	2.9	White cabbage	3.8
Sweet pepper	3.8	Sweet pepper	2.9	Swiss chard	3.8

Table A27   Traditional vegetable sources of consumed vegetables during three different seasons in Tanzania (according to a 7d-recall)

Jun/Jul (DS)	n	Purchased	Cultivated	Collected	Gift
Amaranth leaves	105	25.7	63.8	12.4	0.0
Bitter lettuce	36	0.0	36.1	63.9	0.0
Cassava leaves	33	6.1	87.9	6.1	0.0
Cowpea leaves	87	8.0	86.2	4.6	3.4
False sesame	79	2.5	15.2	82.3	1.3
Mhilile	18	0.0	0.0	100.0	0.0
Okra	91	33.0	67.0	0.0	0.0
Pumpkin leaves	115	8.7	86.1	0.9	4.3
Sweet potato leaves	99	9.1	87.9	1.0	3.0
Wild simsim	67	3.0	10.4	86.6	0.0
Nov/Dec (SR)					
Amaranth leaves	158	36.1	59.5	5.1	0.6
Bitter lettuce	46	0.0	89.1	13.0	0.0
Cassava leaves	58	5.2	89.7	1.7	3.4
Cowpea leaves	73	15.1	84.9	0.0	0.0
False sesame	106	7.5	49.1	46.2	0.0
Mhilile	28	7.1	50.0	42.9	3.6
Okra	60	51.7	48.3	0.0	0.0
Pumpkin leaves	76	18.4	80.3	1.3	0.0
Sweet potato leaves	126	22.2	75.4	1.6	0.8
Wild simsim	49	2.0	28.6	67.3	0.0
Mar/Apr (LR)					
Amaranth leaves	209	13.4	79.4	6.7	0.0
Bitter lettuce	25	0.0	64.0	36.0	0.0
Cassava leaves	42	2.4	92.9	4.8	0.0
Cowpea leaves	60	3.3	95.0	0.0	1.7
False sesame	122	0.0	63.1	36.1	0.8
Mhilile	50	0.0	60.0	40.0	0.0
Okra	160	26.3	73.1	0.6	0.0
Pumpkin leaves	155	3.9	96.1	0.6	0.0
Sweet potato leaves	160	6.9	89.4	3.8	0.0
Wild simsim	34	2.9	41.2	55.9	0.0

Table A28 Traditional vegetable sources of consumed vegetables in three different districts of Tanzania (according to a 7d-recall; mean across three seasons)

Kongwa	n	Purchased	Cultivated	Collected	Gift
Amaranth leaves	109	37.6	58.7	3.7	0.9
Bitter lettuce	3	0.0	66.7	33.3	0.0
Cassava leaves	25	4.0	88.0	4.0	4.0
Cowpea leaves	133	8.3	89.5	2.3	0.8
False sesame	134	1.5	49.3	48.5	0.7
Mhilile	96	2.1	45.8	52.1	1.0
Okra	67	25.4	74.6	0.0	0.0
Pumpkin leaves	108	11.1	86.1	0.9	1.9
Sweet potato leaves	94	30.9	68.1	1.1	0.0
Wild simsim	191	0.0	16.0	84.0	0.0
Muheza					
Amaranth leaves	160	24.4	73.1	2.5	0.0
Bitter lettuce	104	0.0	65.4	35.6	0.0
Cassava leaves	56	1.8	98.2	0.0	0.0
Cowpea leaves	62	8.1	90.3	0.0	1.6
False sesame	4	0.0	75.0	25.0	0.0
Mhilile	0	0.0	0.0	0.0	0.0
Okra	100	43.0	57.0	0.0	0.0
Pumpkin leaves	121	4.1	95.0	0.0	0.8
Sweet potato leaves	133	5.3	91.0	0.8	3.0
Wild simsim	225	0.0	33.3	66.7	0.0
Singida					
Amaranth leaves	203	15.8	71.9	13.3	0.5
Bitter lettuce	0	0.0	0.0	0.0	0.0
Cassava leaves	52	7.7	82.7	7.7	1.9
Cowpea leaves	25	16.0	76.0	4.0	4.0
False sesame	169	4.7	42.6	1.2	0.6
Mhilile	0	0.0	0.0	0.0	0.0
Okra	144	29.9	69.4	0.7	0.0
Pumpkin leaves	117	11.1	86.3	0.9	2.6
Sweet potato leaves	158	7.6	88.6	4.4	0.0
Wild simsim	122	3.3	25.4	72.1	0.0

Table A29 Exotic vegetable sources of consumed vegetables during three different seasons in Tanzania (according to a 7d-recall)

Jun/Jul (DS)	n	Purchased	Cultivated	Collected	Gift
Chinese cabbage	47	51.1	46.8	0.0	2.1
Onion	218	94.5	5.0	0.0	0.9
Sweet pepper	4	75.0	25.0	0.0	0.0
Swiss chard	2	0.0	100.0	0.0	0.0
Tomato	222	89.2	10.4	0.0	1.8
White cabbage	16	93.8	6.3	0.0	0.0
Nov/Dec (SR)					
Chinese cabbage	61	65.6	34.4	0.0	0.0
Onion	226	97.8	2.2	0.0	0.4
Sweet pepper	6	66.7	33.3	0.0	0.0
Swiss chard	9	66.7	22.2	0.0	11.1
Tomato	221	95.5	4.5	0.0	1.8
White cabbage	25	92.0	8.0	0.0	0.0
Mar/Apr (LR)					
Chinese cabbage	19	52.6	47.4	0.0	0.0
Onion	241	92.5	7.1	0.0	0.4
Sweet pepper	2	50.0	50.0	0.0	0.0
Swiss chard	4	25.0	75.0	0.0	0.0
Tomato	246	71.5	19.5	8.9	0.0
White cabbage	23	95.7	4.3	0.0	0.0

Table A30 Exotic vegetable sources of consumed vegetables in three different districts of Tanzania (according to a 7d-recall; mean across three seasons)

Kongwa	n	Purchased	Cultivated	Collected	Gift
Chinese cabbage	22	68.2	31.8	0.0	0.0
Onion	171	98.8	1.2	0.0	0.6
Sweet pepper	1	100.0	0.0	0.0	0.0
Swiss chard	1	100.0	0.0	0.0	0.0
Tomato	174	83.9	13.2	0.0	0.6
White cabbage	6	100.0	0.0	0.0	0.0
Muheza					
Chinese cabbage	19	68.4	31.6	0.0	0.0
Onion	228	99.1	0.9	0.0	0.0
Sweet pepper	3	66.7	33.3	0.0	0.0
Swiss chard	8	50.0	37.5	0.0	12.5
Tomato	214	99.1	0.9	0.0	0.5
White cabbage	44	93.2	6.8	0.0	0.0
Singida					
Chinese cabbage	86	53.5	45.3	0.0	1.2
Onion	301	89.7	9.6	0.0	1.0
Sweet pepper	8	62.5	37.5	0.0	0.0
Swiss chard	6	33.3	66.7	0.0	0.0
Tomato	301	75.4	18.6	5.6	2.0
White cabbage	14	92.9	7.1	0.0	0.0

Table A31 Share of women (%) within each dietary diversity score category for three districts of Tanzania and three different seasons

	Low DDS (2-4 food groups)	Medium DDS (5-6 food groups)	High DDS (7-11 food groups)
Kongwa	41.4	32.5	26.0
Muheza	0.4	13.6	86.0
Singida	53.5	34.3	12.2
Jun/Jul	34.1	27.4	38.5
Nov/Dec	36.7	29.5	33.9
Mar/Apr	31.3	25.8	42.9

Table A32 Share of women (%) within each food variety score category for three districts of Tanzania and three different seasons

	Low FVS (2-6 foods)	Medium FVS (7-9 foods)	High FVS (10-16 foods)
Kongwa	42.8	38.1	19.1
Muheza	1.3	26.3	72.4
Singida	38.5	49.7	11.9
Jun/Jul	17.0	51.0	32.0
Nov/Dec	19.9	49.7	30.5
Mar/Apr	10.7	51.5	37.8

Table A33 Share of women (%) within each vegetable diversity score category for three districts of Tanzania and three different seasons

	Low VDS (1-3 vegetables)	Medium VDS (4-5 vegetables)	High VDS (6-15 vegetables)
Kongwa	11.1	54.2	34.7
Muheza	23.7	44.7	31.6
Singida	17.3	64.4	18.3
Jun/Jul	44.6	35.9	19.5
Nov/Dec	39.4	32.7	27.9
Mar/Apr	17.7	30.2	52.1

Table A34 Correlation between food diversity scores and intake of food groups in g/d by women (n=251) from three districts of Tanzania (mean across three seasons)

Food group	FVS		DDS		VDS	
	Correl. coeff. rho	p	Correl. coeff. rho	p	Correl. coeff. rho	p
Cereals	n.a.	n.s.	-0.140	0.027	n.a.	n.s.
Bread/cakes	0.658	<0.001	0.649	<0.001	n.a.	n.s.
Fruits	0.627	<0.001	0.662	<0.001	0.127	0.044
Vegetables	-0.338	<0.001	-0.433	<0.001	n.a.	n.s.
Nuts	0.530	<0.001	0.612	<0.001	n.a.	n.s.
Pulses	0.452	<0.001	0.480	<0.001	n.a.	n.s.
Starchy plants	0.438	<0.001	0.482	<0.001	n.a.	n.s.
Tea	0.825	<0.001	0.855	<0.001	n.a.	n.s.
Oil/fat	0.474	<0.001	0.435	<0.001	n.a.	n.s.
Sugar	0.706	<0.001	0.729	<0.001	0.172	0.006
Fish	0.470	<0.001	0.500	<0.001	-0.151	0.017

Table A35   Different health parameters of women in Tanzania by dietary diversity score (DDS) and food variety score (FVS) categories during different seasons

	DDS				FVS			
	Low	**Medium**	**High**	**P***	**Low**	**Medium**	**High**	**P***
Hb* all year	13.2	13.1	12.0	<0.0001	13.1	13.2	12.1	0.004
Hb* Jun/Jul (DS)	13.9	13.5	12.8	<0.0001	13.5	13.5	12.9	0.004
Hb* Nob/Dec (SR)	12.7	13.1	11.9	0,001	12.6	12.8	12.0	0.004
Hb* Mar/Apr (LR)	12.9	13.1	11.7	<0.0001	12.8	12.9	11.6	<0.0001
TfR** Mar/Apr (LR)	5.3	6.0	6.8	0,008	5.4	5.8	7,2	0,023
BMI** all year	20,9	21,1	22,5	0,033	20,8	21,5	22,9	0,003

Hb=haemoglobin; TfR=transferrin receptor; BMI=body mass index
* Mean values; ** median values; *** p for differences between categories according to K-W test

Table A36   Minimum and maximum factor scores generated through PCA within quintiles of five dietary patterns of women in Tanzania

		Pattern 1		Pattern 2		Pattern 3		Pattern 4		Pattern 5	
Quintile	**n**	Min	Max	Min	Max	Min	Max	Min	Max	Min	Max
1	50	-1.62	-0.76	-1.27	-0.85	-2.19	-0.83	-2.84	-0.75	-1.65	-0.64
2	51	-0.76	-0.50	-0.84	-0.46	-0.81	-0.33	-0.74	-0.30	-0.64	-0.44
3	50	-0.50	-0.19	-0.46	-0.02	-0.31	0.08	-0.30	0.06	-0.44	-0.19
4	51	-0.18	0.70	-0.01	0.72	0.11	0.81	0.07	0.85	-0.19	0.54
5	50	0.70	4.79	0.74	4.71	0.84	3.46	0.88	3.43	0.54	4.62

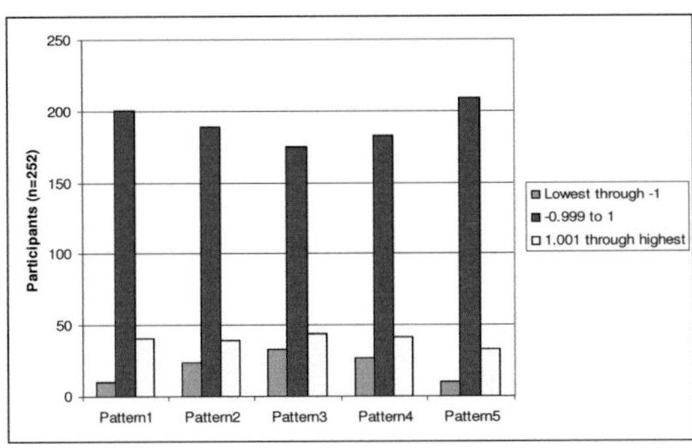

Figure A18   Number of participants within different categories of factor scores showing their affiliation to certain dietary patterns

Table A37   "General model" of multiple regression analysis with vegetable consumption quantity (g/d) as dependent variable; n=252 women from three districts of Tanzania

„General model"	Jun/Jul (DS) Residuals normal-distributed	Nov/Dec (SR) Residuals not normal-distributed	Mar/Apr (LR) Residuals normal-distributed
R Square	0,240	0,347	0.314
Adjusted R Square	0,212	0,314	0.279
Significance of model	<0,001	<0,001	<0.001
Predictor variables*			
Age	-3.002 0.075	0.650 0.681	-1.152 0.456
Kongwa	**-56.873** **0.045**	**-62.979** **0.022**	-7.285 0.812
Muheza	-31.119 0.354	-41.991 0.262	11.440 0.792
DDS	**-22.563** **0.006**	-26.737 0.001	**-35.330** **<0.001**
VDS	**17.500** **0.002**	9.688 0.055	**10.038** **0.040**
No. veg prod	VIF >10	14.714 0.163	-3.738 0.723
No. veg coll	VIF >10	33.974 0.002	3.793 0.707
No. TVs prod/coll	VIF >10	-20.996 0.007	4.596 0.583
No. EVs prod	-13.150 0.344	-11.836 0.620	21.176 0.202
Cereal cons	**0.363** **<0.001**	0.409 0.000	-0.040 0.497
Oil/fat cons	**1.940** **0.001**	2.470 0.000	**2.922** **<0.001**
Pulses cons	**-0.475** **<0.001**	-0.315 0.000	**-0.285** **0.001**

* Values shown for predictor variables are the B coefficient and the p-value

Table A38   "Consumption model" of multiple regression analysis with vegetable consumption (g/d) as dependent variable; n=252 women from three districts of Tanzania

"Consumption model"	Jun/Jul (DS) Residuals normal-distributed	Nov/Dec (SR) Residuals NOT normal-distributed	Mar/Apr (LR) Residuals normal-distributed
R Square	0.292	0.345	0.349
Adjusted R Square	0.250	0.306	0.311
Significance of model	0.000	0.000	0.000
Predictor variables*			
Age	-1.785 0.292	1.149 0.476	-1.044 0.492
Kongwa	**-79.099** **0.007**	-49.424 0.067	-36.419 0.168
Muheza	-50.811 0.198	-28.960 0.484	-62.178 0.128
DDS	**-26.719** **0.004**	-26.570 0.004	**-40.758** **0.000**
VDS	**17.441** **0.002**	10.306 0.049	**10.930** **0.019**
Cereal cons	**0.346** **0.000**	0.388 0.000	-0.040 0.494
Bread/cake cons	-0.099 0.657	-0.088 0.557	-0.180 0.329
Fruit cons	-0.028 0.665	-0.147 0.381	**0.162** **0.041**
Nut cons	**0.452** **0.001**	0.370 0.009	**0.244** **0.017**
Pulse cons	**-0.592** **0.000**	-0.324 0.000	**-0.230** **0.007**
Starchy plant cons	-0.061 0.559	0.085 0.387	0.129 0.070
Tea cons	0.052 0.441	-0.081 0.299	-0.064 0.186
Oil/fat cons	**2.529** **0.000**	2.546 0.000	**3.307** **0.000**
Fish cons	-0.453 0.032	-0.194 0.290	-0.138 0.501

* Values shown for predictor variables are the B coefficient and the p-value

Table A39 "Production model" of multiple regression analysis with vegetable consumption quantity (g/d) as dependent variable; n=252 women from three districts of Tanzania

"Production model"	Jun/Jul (DS) Residuals <u>not</u> normal-distributed	Nov/Dec (SR) Residuals <u>not</u> normal-distributed	Mar/Apr (LR) Residuals normal-distributed
R Square	0.040	0,121	0,164
Adjusted R Square	24	0,096	0,139
Significance of model	0.041	0,000	0,000
Predictor variables*			
Age	-3.650 0.051	-0.015 0.993	-1.832 0.273
Kongwa	-70.671 0.032	-51.072 0.096	-37.598 0.241
Muheza	-71.155 0.017	-151.866 0.000	**-122.608** **0.001**
No. veg prod	VIF >10	12.137 0.314	-1.019 0.928
No. veg coll	VIF >10	29.763 0.015	1.689 0.876
No. TVs prod/coll	VIF >10	-15.594 0.081	6.687 0.461
No. EVs prod	-12.332 0.424	-11.023 0.685	7.223 0.683

* Values shown for predictor variables are the B coefficient and the p-value

Table A40 Sørensen coefficient for three pairs of districts and three pairs of seasons, Tanzania, regarding vegetable types consumed and produced (cultivated/collected)

	Kongwa/ Muheza	Kongwa/ Singida	Muheza/ Singida	Jun/Jul / Nov/Dec	Jun/Jul / Mar/Apr	Nov/Dec / Mar/Apr
Consumption	79.5	83.6	87.9	86.8	87.5	86.8
Production	57.9	68.0	66.7	79.4	78.7	89.1

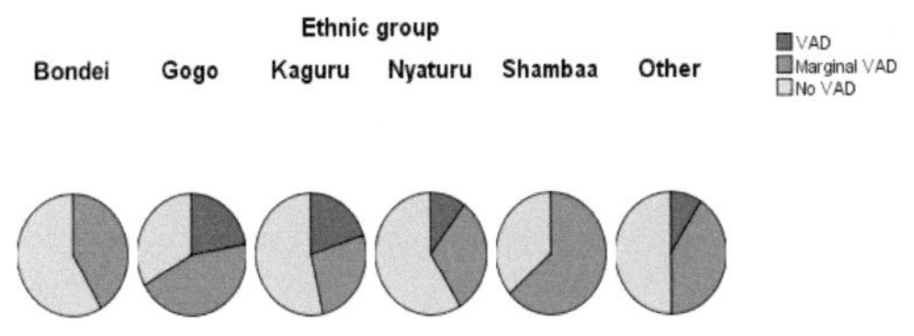

Figure A19   Association between ethnic group and vitamin A deficiency of women in Tanzania (p=0.234)

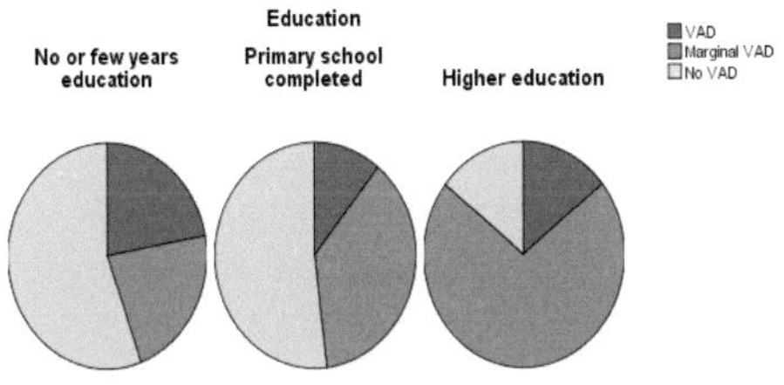

Figure A20   Association between education and vitamin A deficiency of women in Tanzania (p=0.234)

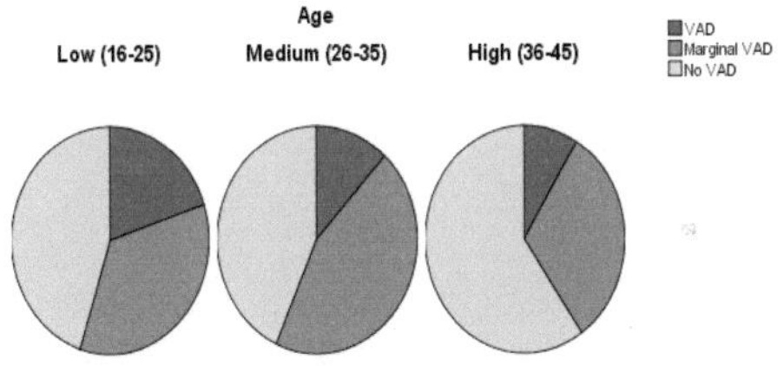

Figure A21   Association between age and vitamin A deficiency of women in Tanzania (p=0.307)

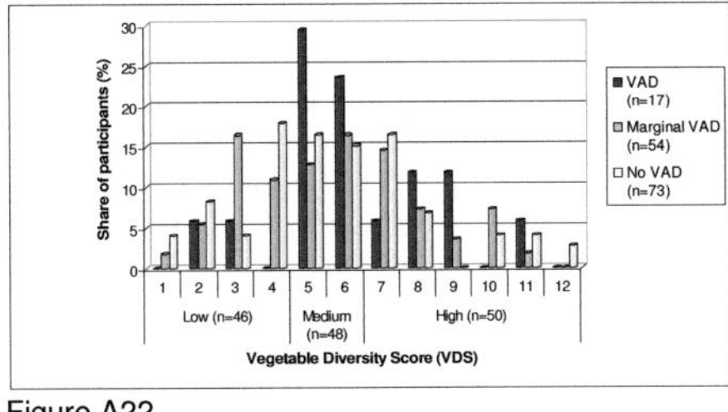

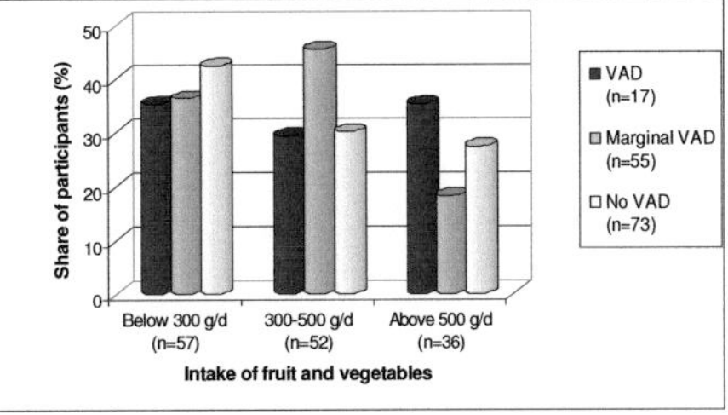

Figure A22
Association between vegetable diversity score (VDS) and vitamin A status (n=144; p=0.257); share of participants within vitamin A status

Figure A23
Association between the intake of fruit/vegetables and vitamin A status (n=145; p=0.334); share of participants within vitamin A status

Table A41 Share of women from three districts in Tanzania being underweight, overweight or obese and the ratio of overweight/obese to underweight (mean across three seasons)

	n	% Underweight BMI<18.5	% Overweight BMI 25-29.9	% Obese BMI≥30	% Overweight and obese BMI≥25	Overweight and obese : Underweight
All districts	210	7.1	15.7	6.2	21.9	3.1
Kongwa	52	1.9	19.2	3.8	23.0	12.1
Muheza	69	8.7	14.5	10.1	24.6	2.8
Singida	89	9.0	14.6	4.5	19.1	2.1

Table A42 Results of multiple regression analysis with ln(BMI) as dependent variable; n=210 women from three districts of Tanzania, mean across three seasons if applicable

	Unstandardized Coefficients		Standardized Coefficients			
	B	Std. Error	Beta	t	Sig.	e^B
(Constant)	2.898	0.079		36.486	0.000	
Age	0.003	0.002	0.121	1.748	0.082	1.003
Low education	0.068	0.041	0.112	1.642	0.102	1.070
High education	-0.092	0.072	-0.086	-1.277	0.203	0.912
Kongwa	0.020	0.029	0.053	0.692	0.490	1.020
Muheza	-0.032	0.034	-0.093	-0.935	0.351	0.969
Pattern2	0.022	0.013	0.139	1.682	0.094	1.022
FVS	0.014	0.007	0.204	2.008	0.046	1.014

Table A43 "Production model" of multiple regression analysis with Hb as dependent variable; n=252 women from three districts of Tanzania

„Production model"	Jun/Jul (DS)		Nov/Dec (SR)		Mar/Apr (LR)	
R Square	0.086		0.081		0.073	
Adjusted R Square	0.066		0.061		0.052	
Sig. of model	0.003		0.004		0.009	
	B	Sig.	B	Sig.	B	Sig.
No. veg prod	-0.116	0.652	-0.099	0.330	-0.068	0.572
No. veg coll	-0.197	0.443	**-0.220**	**0.040**	-0.051	0.663
No. TVs prod/coll	0.006	0.980	0.041	0.583	0.110	0.274
No. EVs prod	0.412	0.199	0.433	0.062	**0.413**	**0.014**

Table A44 "Consumption model" of multiple regression analysis with Hb as dependent variable; n=252 women from three districts of Tanzania

„Consumption model"	Jun/Jul (DS)		Nov/Dec (SR)		Mar/Apr (LR)	
R Square	0.127		0.120		0.191	
Adjusted R Square	0.077		0.069		0.144	
Sig. of model	0.007		0.012		<0.001	
	B	Sig.	B	Sig.	B	Sig.
FVS	0.058	0.381	0.004	0.944	-0.122	0.094
Fruit cons	**-0.002**	**0.018**	-0.001	0.469	-0.-001	0.411
Nut cons	<0.001	0.898	-0.002	0.140	-0.002	0.107
Tea cons	-0.001	0.135	-0.001	0.298	<0.001	0.666
Starchy plant cons	0.001	0.397	-0.002	0.087	-0.001	0.308
Sugar cons	-0.007	0.287	0.017	0.211	-0.007	0.102
Fish cons	-0.004	0.081	-0.001	0.541	-0.004	0.163
Animal prod cons	0.001	0.361	<0.001	0.928	0.001	0.302
Vitamin A intake	<0.001	0.149	**<0.001**	**0.041**	<0.001	0.364
Vitamin C intake	0.001	0.568	<0.001	0.894	0.001	0.204

Table A45   "Consumption model" of multiple regression analysis with mean Hb across three seasons as dependent variable; n=252 women from three districts in Tanzania

	Unstandardized Coefficients		Standardized Coefficients	t	Sig.
	B	Std. Error	Beta		
Age	-0.009	0.014	-0.044	-0.652	0.515
Food Pattern 1	-0.282	0.167	-0.190	-1.694	0.092
Food Pattern 2	-0.056	0.138	-0.042	-0.409	0.683
Food Pattern 5	0.215	0.099	0.160	2.242	0.026
DDS	-0.029	0.098	-0.039	-0.298	0.766
Knowledge function of iron (1)	0.724	0.449	0.107	1.612	0.109
Knowledge function of iron (2)	-0.115	0.267	-0.031	-0.430	0.668
Knowledge function of iron (3)	-0.526	0.260	-0.140	-2.022	0.045
Education - low	-0.116	0.366	-0.021	-0.317	0.751
Education - high	-1.047	0.609	-0.117	-1.720	0.087
Kongwa	-0.699	0.255	-0.215	-2.745	0.007
Muheza	-0.909	0.415	-0.291	-2.192	0.030

Table A46   "Production" Model of multiple regression analysis with mean Hb across three seasons as dependent variable; n=252 women from three districts of Tanzania

	Unstandardized Coefficients		Standardized Coefficients	t	Sig.
	B	Std. Error	Beta		
Age	-0.009	0.014	-0.043	-0.630	0.530
No. of vegetables cultivated	0.051	0.062	0.059	0.828	0.409
No. of vegetables collected	-0.175	0.075	-0.164	-2.341	0.020
Knowledge function of iron (1)	0.848	0.451	0.126	1.880	0.062
Knowledge function of iron (2)	0.006	0.266	0.002	0.024	0.981
Knowledge function of iron (3)	-0.419	0.259	-0.112	-1.617	0.108
Education - low	-0.099	0.367	-0.018	-0.269	0.788
Education - high	-1.161	0.608	-0.130	-1.908	0.058
Kongwa	-0.751	0.244	-0.231	-3.082	0.002
Muheza	-1.493	0.239	-0.477	-3.594	0.000

Table A47   Interdependencies between vitamin A status categories as measured by retinol binding protein (RBP) during one season (Mar/Apr) and knowledge of women on vitamin A rich foods (VARF) in Tanzania

Knowledge on VARF		No vitamin A deficiency	Marginal vitamin A deficiency	Vitamin A deficiency
Mainly correct answers (no.)	37			
Share within "knowledge" (%)		40.5	40.5	18.9
Share within "RBP" (%)		20.5	27.3	41.2
Random answers (no.)	93			
Share within "knowledge" (%)		55.9	34.4	9.7
Share within "RBP" (%)		71.2	58.2	52.9
Mainly wrong answers (no.)	15			
Share within "knowledge" (%)		40.0	53.3	6.7
Share within "RBP" (%)		8.2	14.5	5.9
Total	145	73	55	17

Table A48   Interdependencies between vegetable diversity score (VDS) categories and knowledge of women on vitamin A rich foods (VARF) in Tanzania

			VDS	
Knowledge on VARF		**Low**	**Medium**	**High**
Mainly correct answers (no.)	36			
Share within "knowledge" (%)		30.6	50.0	19.4
Share within "VDS" (%)		23.4	37.5	14.3
Random answers (no.)	93			
Share within "knowledge" (%)		32.3	26.9	40.9
Share within "VDS" (%)		63.8	52.1	77.6
Mainly wrong answers (no.)	15			
Share within "knowledge" (%)		40.0	33.3	26.7
Share within "VDS" (%)		12.8	10.4	8.2
Total	144	47	48	49

Table A49   "Model 1 (BMI)" of multiple regression analysis with food variety score (FVS) as dependent variable; n=252 women from three districts in Tanzania

"Model 1 (BMI)"	**Jun/Jul (DS)** Residuals normal distributed		**Nov/Dec (SR)** Residuals normal distributed		**Mar/Apr (LR)** Residuals normal distributed	
R Square	0,376		0,375		0,417	
Adjusted R Square	0,361		0,354		0,396	
Significance of model	<0,001		<0,001		<0,001	
	B	Sign.	B	Sign.	B	Sign.
Age	0.006	0.797	<0,001	0.989	-0.013	0.543
Kongwa	-0.336	0.441	0.243	0.546	0.616	0.125
Muheza	**3.477**	**<0.001**	**3.504**	**<0.001**	**3.969**	**<0.001**
No. veg prod	VIF>10		0.017	0.846	0.084	0.271
No. veg coll	VIF>10		**-0.159**	**0.046**	0.045	0.591
No. TVs prod/coll	VIF>10		VIF>10		VIF>10	
No. EVs prod	-0.080	0.675	0.560	0.072	0.053	0.789
BMI	0.075	0.073	**0.107**	**0.005**	0.058	0.107

Table A50   "Model 2 (Hb)" of multiple regression analysis with food variety score (FVS) as dependent variable; n=252 women from three districts in Tanzania

"Model 2 (Hb)"	**Jun/Jul (DS)** Residuals normal distributed		**Nov/Dec (SR)** Residuals normal distributed		**Mar/Apr (LR)** Residuals normal distributed	
R Square	0,393		0,369		0,42	
Adjusted R Square	0,376		0,344		0,399	
Significance of model	<0,001		<0,001		<0,001	
	B	Sign.	B	Sign.	B	Sign.
Age	0.009	0.715	0.021	0.396	-0.014	0.512
Kongwa	-0.180	0.693	0.596	0.172	**1.056**	**0.012**
Muheza	**3.804**	**<0.001**	**3.781**	**<0.001**	**4.085**	**<0.001**
No. veg prod	VIF>10		0.014	0.888	0.030	0.701
No. veg coll	VIF>10		-0.200	0.027	0.081	0.360
No. TVs prod/coll	VIF>10		VIF>10		VIF>10	
No. EVs prod	0.089	0.675	**0.712**	**0.045**	0.168	0.414
Hb	0.065	0.523	-0.149	0.220	-0.110	0.260

Table A51   "Model 1 (BMI)" of multiple regression analysis with dietary diversity score (DDS) as dependent variable; n=252 women from three districts in Tanzania

„Model 1 (BMI)"	Jun/Jul (DS) Residuals normal distributed		Nov/Dec (SR) Residuals normal distributed		Mar/Apr (LR) Residuals normal distributed	
R Square	0.407		0,444		0,514	
Adjusted R Square	0.393		0,424		0,497	
Significance of model	<0.001		<0.001		<0.001	
	B	Sign.	B	Sign.	B	Sign.
Age	0.010	0.550	0.008	0.590	-0.012	0.448
Kongwa	0.187	0.534	0.274	0.313	0.777	0.010
Muheza	**2.771**	**<0.001**	**2.899**	**<0.001**	**3.574**	**0.000**
No. veg prod	VIF>10		-0.018	0.757	0.050	0.377
No. veg coll	VIF>10		**-0.111**	**0.038**	0.007	0.905
No. TVs prod/coll	VIF>10		VIF>10		VIF>10	
No. EVs prod	-0.064	0.626	0.218	0.298	0.072	0.625
BMI	0.018	0.529	**0.063**	**0.014**	0.029	0.278

Table A52   "Model 2 (Hb)" of multiple regression analysis with dietary diversity score (DDS) as dependent variable; n=252 women from three districts in Tanzania

„Model 2 (Hb)"	Jun/Jul (DS) Residuals normal distributed		Nov/Dec (SR) Residuals normal distributed		Mar/Apr (LR) Residuals normal distributed	
R Square	0.434		0,441		0,514	
Adjusted R Square	0.418		0,419		0,495	
Significance of model	<0.001		<0.001		<0.001	
	B	Sign.	B	Sign.	B	Sign.
Age	0.009	0.590	0.018	0.286	-0.013	0.430
Kongwa	0.319	0.324	0.478	0.111	**1.029**	**0.001**
Muheza	**3.034**	**<0.001**	**3.200**	**<0.001**	**3.678**	**<0.001**
No. veg prod	VIF>10		-0.035	0.604	0.005	0.932
No. veg coll	VIF>10		**0.150**	**0.016**	0.032	0.631
No. TVs prod/coll	VIF>10		VIF>10		VIF>10	
No. EVs prod	0.038	0.800	0.381	0.116	0.147	0.348
Hb	-0.008	0.916	-0.091	0.273	-0.054	0.464

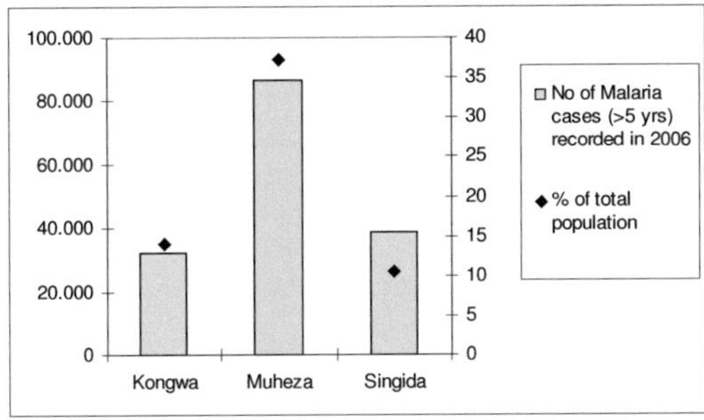

Figure A24
Malaria cases in 2006 in three districts of Tanzania; Source: District Medical Officers/Medical Records of the respective districts

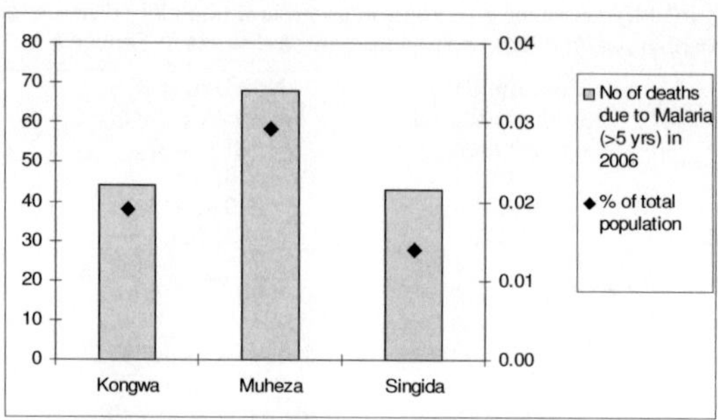

Figure A25
Death due to malaria in 2006 in three districts of Tanzania;
Source: District Medical Officers/Medical Records of the
respective districts

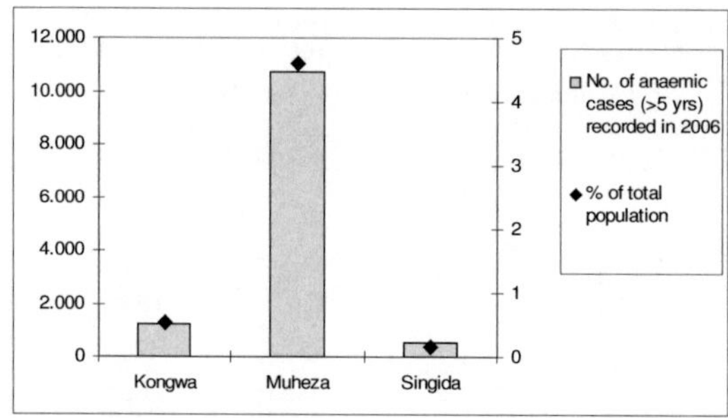

Figure A26
Anaemic cases in 2006 in three districts of Tanzania; Source:
District Medical Officers/Medical Records of the respective
districts

Table A53   Measures in connection with iron deficiency and malaria in the three districts researched in Tanzania as stated by the district medical officers or malaria coordinator of the respective district in 2007

Measure	Kongwa	Muheza	Singida
Iron supplementation	All pregnant women who come to the clinic get routinely iron supplements; Children, non-pregnant women and men get iron supplements after checking, in worse cases also blood transfusion	All pregnant women who attend ante-natal clinic get routinely iron supplements	All pregnant women who come to the hospital/health facility get routinely iron supplements
Malaria treatment	For complicated malaria Quinine; for uncomplicated malaria is A.L.U. (Artemether Lumefantrine) / *Artemisia annua*	For complicated malaria is A.L.U. (only)	Before 2007: "SP", Amodiaquin, Quinine; Since 2007: A.L.U.
Payment exemption	Pregnant women, children < 5 yrs and older people	Pregnant mothers, children and older people	Patients have to pay only a contribution: 5000 TZS per year per family to the Community Health Fund (CHF); then get medicinal treatment (e.g. for Malaria) for free
Malaria campaigns	National malaria day each year on 25th May (one to two weeks moving around in villages)	Different (uncoordinated) campaigns from different donors, e.g. in Amani region a campaign from Europe with treated nets	When A.L.U. was introduced: information campaign in each ward
	Tanzanian National Voucher System (TNVS): programme for all pregnant mothers and children < 1 yr to get an insecticide-treated net at low cost	-	TNVS: Tanzanian National Voucher Scheme, vouchers for pregnant mothers to buy mosquito nets at low cost
	Malaria Coordinator of school health in each primary and secondary school arranges for a continuing process for almost the whole year of information/ education on malaria	-	Once a year: seminar/training for school teachers through Village Health Workers/Committee

Linking Vegetable Diversity and Nutrition Security
Household Survey Tanzania June/July 2006

No. of clinical test

1. Basic data

Date of survey (dd/mm/yyyy)

Name of enumerator ___________________________________

Village _________________________ Ward _______________ District ____________

2. Respondents personal information

Respondent Name _______________________________ Age RAGE

Which tribe do you belong to? RTRI

Arusha	1	Kaguru	5	Nyiramba	9
Bondei	2	Meru	6	Pare	10
Chagga	3	Nguu / Zigua	7	Shambaa	11
Gogo	4	Nyaturu	8	Other (specify)	12

Are you the head of household? If not, what is your relation to the head of household? RHOU

Head	1	Sister	4	Other relative	7
Wife	2	Mother	5	Non relative	8
Companion	3	Daughter	6		

What is your marital status? RMAR

Single	1	Widowed	3	Separated	5
Married	2	Divorced	4		

What is your main occupation? ROCC

Crop farmer	1	Artisan	8
Livestock farmer	2	Handcrafter	9
Crop and livestock farmer	3	Student	10
Farm labour service	4	Housewife	11
Non-farm casual service	5	None	12
Business/Private service	6	Other	13
Formal employment	7		

What is your highest level of education? REDU

No formal education	1	Few years secondary	4	College/Univ.	7
Few years primary	2	Completed secondary only	5	Other	8
Completed primary only	3	High School	6		

Which religion do you belong to? RREL

Christian	1	Muslim	2	Traditionalist	3	Others	4

3. Household data

How many people belong to your household? *Read categories*

Male adults (15 or older) HMAD	Male children (5-14 years) HMCH	Male children (< 5 years) HMUF
Female adults (15 or older) HFAD	Female children (5-14 years) HFCH	Female children (< 5 years) HFUF

Note: Please remember to include respondent into the number of people belonging to the household

Do you own any of the following in your household / house? *Read categories*

Possessions	No	Yes	Livestock	No	Yes	How many?
Radio HRAD	0	1	Chicken HCHI	0	1	
Bicycle HBIC	0	1	Goat HGOA	0	1	
Mobile phone HMOB	0	1	Sheep HSHE	0	1	
Housing			Cow HCOW	0	1	
Electricity HELE	0	1				
Corrugated iron roof HCOR	0	1				
Brick/cement wall HBRI	0	1				

4. Vegetable consumption

What are the three most important criteria that you consider while choosing a vegetable?
Rank - do not tick but place "1", "2", "3" into boxes CCRI1, CCRI2, CCRI3

Taste	1	Fruit/leaf colour	5
Price	2	Fruit/leaf size	6
Availability	3	Other (specify)	7
Good for health	4		

____

What are the three most important factors influencing your decisions to consume many
<u>different</u> types of vegetables? *Rank - do not tick but place "1", "2", "3" into boxes* CDIV1, CDIV2, CDIV3

Different tastes of different vegetables	1
Prices fluctuate – always buy the cheapest vegetables	2
Availability – supply changes frequently (e.g. seasonality, quantities available)	3
A diverse diet is good for health	4
Vegetable diversity is high in own garden	5
Certain vegetables need to be cooked/ mixed with others and certain foods need to be consumed with certain vegetables	6
Availability of cash	7
Others (specify)____	8
I do not consume many different vegetables (less than 3 different per week)	9

5. Vegetable production

Which vegetables (exotic and traditional) do you <u>produce</u> at the moment in your homegarden for own (household) consumption? PPRO

No.	Local name	Kiswahili name	English name	Scientific name
1				
2				
3				
4				
5				
6				
7				
8				
9				
10				

Which naturally growing vegetables do you <u>gather</u> at the moment for your own (household) consumption? PGAT

No.	Local name	Kiswahili name	English name	Scientific name
1				
2				
3				
4				
5				
6				
7				
8				
9				
10				

Do you sell any of your own produced vegetables at the moment (this time of the year)? PSEL

No ☐ 0   Yes ☐ 1

PSWE If yes, how many times a week? ☐ times
PSMO If yes, how many times a month? ☐ times

PSHO If yes, how do you normally sell?   Farm gate ☐ 1   Market ☐ 2   Others ☐ 3

6. 24h-recall

Which <u>food and drinks</u> and how much of it did you (*alone* <u>not</u> *your household*) consume yesterday? Was the food item processed and where did you get it from?

Food item	Quantity - all foods - (piece, handful, cup, table spoon, etc.) (*show three containers - small, medium, large*)	Source - all foods - 1 = purchased 2 = produced 3 = collected 4 = gift 5 = other	Processing - vegetable only - (raw or dried, steamed, cooked, fried + for how many minutes) *e.g. steamed (10 min)*
Breakfast			
Lunch			
Dinner			
Snacks			

7. Seven-day recall of vegetables
(*Do not tick but place number for Source:* 1 = purchased; 2 = produced; 3 = collected; 4 = gift; 5 = other)

Vegetables	On which days of the last week did you consume a certain vegetable and where did you get it from? (*Use key above for source*)							No. of days	Source
	Mon	Tue	Wen	Thu	Fri	Sat	Sun		
Leafy vegetables									
African nightshade									
African spiderflower									
Amaranth leaves									
Baobab tree leaves									
Bitter lettuce / Hair lettuce									
Black jack									
Cassava leaves									
Chinese cabbage									
Cowpea leaves									
Crotalaria									
Ethiopian kale									
False sesame (Mlenda mbata)									
Gallant soldier (Galinsoga parviflora)									
Hyacinth bean leaves									
Jute mallow									
Knobwood (*Zanthoxylum chalybeum*)									
Malabar spinach (*Basella alba*)									
Moringa leaves									
Pumpkin leaves									
Puncture vine/Caltrops (*Tribulus terrestris*)									
Purslane (*Portulaca oleracea*)									
Roselle									
Sesanum spp. (Mlenda wima)									
Sweet potato leaves									
Swiss chard									
Taro leaves									
Water cress									
Water leaf (*Talinum* spp.)									
White Cabbage									
Fruit vegetables									
African eggplant									
Beans (pods) (*specify*)									
Bur gherkin (Maimbe)									
Cucumber, exotic									
Cucumber, wild									
Gourd									
Okra									
Peas (pods) (*specify*)									
Pumpkin									
Sweet pepper									
Tomato									
Root vegetables									
Carrot									
Onion									
Others (specify)									

**Linking Vegetable Diversity and Nutrition Security
Household Survey Tanzania June/July 2006**

Respondent Name ________________________________

No. of clinical test

8. Investigation of clinical data

Did you take any <u>Vitamin A</u> supplements recently (last 3 month)? MVIT

No [ ] 0   Yes [ ] 1

Did you take any <u>Iron</u> supplements recently (last 3 month)? MIRO

No [ ] 0   Yes [ ] 1

When you have the monthly bleeding, is it more severe or longer than 6 days? MBLE

No [ ] 0   Yes [ ] 1

Are you pregnant? - If yes, which trimester? MPRE

No [ ] 0   Yes [ ] 1   **Trimester:** 1st [ ] 1   2nd [ ] 2   3rd [ ] 3

Did you experience a loss of appetite recently? MAPP

No [ ] 0   Yes [ ] 1

   - If yes, do you have been diagnosed for Tuberculosis? MTUB

No [ ] 0   Yes [ ] 1

Did you take any medication to treat hookworms recently or at the moment? MHOO

No [ ] 0   Yes [ ] 1

Do you have any problem seeing in the daytime? MNB1

No [ ] 0   Yes [ ] 1

Do you have any problem seeing at nighttime? MNB2

No [ ] 0   Yes [ ] 1

   - If yes, is this problem different from others in your community? MNB3

No [ ] 0   Yes [ ] 1

Do you have night blindness (*use local term that describes the symptom*)? MNB4

No [ ] 0   Yes [ ] 1

Do you suffer from chronic diarrhoea for more than 1 month? MDIA

No ☐ 0 **Yes** ☐ 1   (major sign)

Do you suffer from prolonged fever for more than 1 month (intermittent or constant)? MFEV

No ☐ 0 **Yes** ☐ 1 | (major sign)

Did you loose weight recently (≥ 10%)? MWEI

No ☐ 0 **Yes** ☐ 1 | (major sign)

Do you suffer from a persistent cough (for more than 1 month)? MCOU

No ☐ 0 **Yes** ☐ 1 | (minor sign)

Do you suffer from an itching dermatitis? MDER

No ☐ 0 **Yes** ☐ 1 | (minor sign)

Do you suffer from recurrent painfull zoster / blistering dermatitis? MZOS

No ☐ 0 **Yes** ☐ 1 | (minor sign)

Do you suffer from fungal infection of the mouth and throat? MFUN

No ☐ 0 **Yes** ☐ 1 | (minor sign)

Do you suffer from swollen lymphatic glands? MGLA

No ☐ 0 **Yes** ☐ 1 | (minor sign)

9. Clinical data measured

Height in cm ☐ NHEI

BMI = ☐ NBMI

Weight in kg ☐ NWEI

Bitot's spots No ☐ 0 Yes ☐ 1   NBIT

Hb in g/dL ☐ NHAE

Linking Vegetable Diversity and Nutrition Security
Household Survey Tanzania Nov/Dec 2006

No. of participant ☐☐

1. Basic data

Date of survey (dd/mm/yyyy) ☐☐☐

Name of enumerator _______________________

Village _______________________ District _______________________

Respondent Name _______________________

2b. Vegetable consumption

What are the three most important criteria that you consider while choosing a vegetable?
Rank - do not tick but place "1", "2", "3" into boxes CCRI1, CCRI2, CCRI3

Taste ☐ 1   Fruit/leaf colour ☐ 5
Price ☐ 2   Fruit/leaf size ☐ 6
Availability ☐ 3   Other (specify) ☐ 7
Good for health ☐ 4   _______________________

What are the three most important factors influencing your decisions to consume many
<u>different</u> types of vegetables? *Rank - do not tick but place "1", "2", "3" into boxes* CDIV1, CDIV2, CDIV3

Different tastes of different vegetables ☐ 1
Prices fluctuate – always buy the cheapest vegetables ☐ 2
Availability – supply changes frequently (e.g. seasonality, quantities available) ☐ 3
A diverse diet is good for health ☐ 4
Vegetable diversity is high in own garden ☐ 5
Certain vegetables need to be cooked/mixed with others and certain foods
need to be consumed with certain vegetables ☐ 6
Availability of cash ☐ 7
Others (specify)_______________________ ☐ 8
I do not consume many different vegetables (less than 3 different per week) ☐ 9

3b. Vegetable production

Which vegetables (exotic and traditional) do you **produce** at the moment in your homegarden?
PPRO

No.	Local name	Kiswahili name	English name	Scientific name
1				
2				
3				
4				
5				
6				
7				
8				
9				
10				

Which naturally growing vegetables do you **gather** at the moment?   PGAT

No.	Local name	Kiswahili name	English name	Scientific name
1				
2				
3				
4				
5				
6				
7				
8				
9				
10				

This year in June/July you produced and gathered the following vegetables:

Produced	Gathered

Why do you **not** produce / gather these vegetables at the moment during this season but different ones? PNOP

Vegetables concerned

Temperature too high/low ☐ 1 _______________
Water availability ☐ 2 _______________
Pests and diseases stress ☐ 3 _______________
Input: availability of seeds, fertiliser, chemicals is lacking ☐ 4 _______________
Output: market prices and demand are low ☐ 5 _______________
Home consumption: not consumed at the moment ☐ 6 _______________
Other (please specify) _______________ ☐ 7 _______________

Do you **sell** any of your own produced vegetables at the moment (this time of the year)? PSEL

No ☐ 0   Yes ☐ 1

PSWE If yes, how many times a week? ☐ times
PSMO If yes, how many times a month? ☐ times
PSHO If yes, how do you normally sell?   Farm gate ☐ 1   Market ☐ 2
Walking from house to house ☐ 3   Others ☐ 4

Do you **buy** any vegetables at the moment (this time of the year)? PBUY1

No ☐ 0   Yes ☐ 1

If yes, which vegetables do you buy at the moment / at this time of the year? PBUY2

No.	Local name	Kiswahili name	English name	Scientific name
1				
2				
3				
4				
5				

If yes, how often do you buy vegetables at the moment / at this time of the year? PBUY3
☐ times per week
☐ times per month

If yes, where do you buy vegetables? PBUY4
Market ☐ 1
Farm gate / neighbours ☐ 2
From traders walking from house to house ☐ 3
Others ☐ 4

Why do you buy (additional) vegetables? _______________
PBUY5 _______________

Linking Vegetable Diversity and Nutrition Security
Household Survey Tanzania Nov/Dec 2006

No. of participant

1. Basic data

Date of survey (dd/mm/yyyy)

Name of enumerator _______________________________

Village _______________________________ District _______________________________

Respondent Name _______________________________

2a. Physical activity

Which typical positive and/or negative properties or qualities would you associate with a person being very big / corpulent? PHPO, PHNE

Positive	Negative

Visual analogue scale

Please make one cross on this scale between the two extremes of "sleeping" and "extremely strenuous physical activity" where you would set up your own daily activity (daily mean)? PHVI

sleeping

Extremely strenuous physical activity

3a. Knowledge about health and nutrition

Have you heard about <u>Vitamin A</u> as a nutrient? KNVI1   No ☐ 0   Yes ☐ 1

If yes, do you know why it is an important micronutrient for the body? KNVI2

No ☐ 0 Yes ☐ 1 Because _______________________________

Please name 3 - 5 foods that are good for sight and eye health. KNVI3
NOTE: food from all food groups, NOT only vegetables - and IF vegetables ask if indigenous or exotic!!

1. _______________________________
2. _______________________________
3. _______________________________
4. _______________________________
5. _______________________________

Have you heard about <u>Iron</u> as a nutrient? KNIR1   No ☐ 0   Yes ☐ 1

If yes, do you know why it is an important micronutrient for the body? KNIR2

No ☐ 0 Yes ☐ 1 Because _______________________________

Please name 3 - 5 foods that help the body to make red blood cells and allow the muscles and brain to work properly. KNIR3
NOTE: food from all food groups, NOT only vegetables - and IF vegetables ask if indigenous or exotic!!

1. _______________________________
2. _______________________________
3. _______________________________
4. _______________________________
5. _______________________________

4a. 24h-recall

Which <u>food and drinks</u> and how much of it did you (*alone <u>not</u> your household*) consume yesterday? Was the food item processed and where did you get it from?

<u>Main meal</u>: how many people shared this meal? How many older/younger 10 yrs?

Food item	Quantity - all foods - (piece, handful, cup, table spoon, etc.) (*show three containers - small, medium, large*)	Source - all foods - 1 = purchased 2 = produced 3 = collected 4 = gift 5 = other	Processing - vegetable only - (raw or dried, steamed, cooked, fried + for how many minutes) *e.g. steamed (10 min)*	No. pers. >10 yrs	No. pers. < 10 yrs
Breakfast					
Lunch					
Dinner					
Snacks					

5a. Seven-day recall of vegetables

(*Do not tick but place number for Source:* 1 = purchased; 2 = produced; 3 = collected; 4 = gift; 5 = other)

Vegetables	On which days of the last week did you consume a certain vegetable and where did you get it from? (*Use key above for source*)							No. of days	Source
	Mon	Tue	Wen	Thu	Fri	Sat	Sun		
Leafy vegetables									
African nightshade / Mnavu									
African spiderflower / Mgagani									
Amaranth leaves / Mchicha									
Baobab tree leaves / Majani ya mbuyu									
Bitter / hair lettuce / Mchunga									
Black jack / Kishonanguo									
Cassava leaves / Kisamvu mhogo									
Cassava leaves / Kisamvu mpira									
Chinese cabbage / Chainisi									
Cowpea leaves / Majani ya kunde									
Ethiopian kale / Sukuma wiki / Loshuu									
False sesame / Mlenda mbata									
Jute mallow / Mlenda wima									
Kibwando									
Malabar spinach / Nderema									
Mhilile									
Pumpkin leaves / Majani ya maboga									
Sweet potato leaves / Majani ya matembele									
Swiss chard / Spinachi									
Taro leaves / Majani ya magimbi									
Water cress / Saladi									
Wild simsim / Mlenda mwitu/wima									
White Cabbage / Kabichi									
Fruit vegetables									
African eggplant / Ngogwe									
Bean pods (*specify*)									
Bur gherkin (Maimbe)									
Cucumber, exotic									
Cucumber, wild									
Gourd									
Okra / Bamia									
Pea pods (*specify*)									
Pumpkin / Maboga									
Sweet pepper / Pilipili tamu									
Tomato / Njanja									
Root vegetables									
Carrot / Karotti									
Onion / Vitungu									
Others (specify)									

**Linking Vegetable Diversity and Nutrition Security
Household Survey Tanzania Nov/Dec 2006**

No. of participant

Respondent Name _______________________________

6. Investigation of clinical data

Did you take any <u>Vitamin A</u> supplements recently (last 3 month)? MVIT

No ☐ 0   Yes ☐ 1

Did you take any <u>Iron</u> supplements recently (last 3 month)? MIRO

No ☐ 0   Yes ☐ 1

When you have the monthly bleeding, is it more severe or longer than 6 days? MBLE

No ☐ 0   Yes ☐ 1

Are you pregnant? - If yes, which trimester? MPRE

No ☐ 0   Yes ☐ 1   **Trimester: 1st** ☐ 1   **2nd** ☐ 2   **3rd** ☐ 3

Did you experience a loss of appetite recently? MAPP

No ☐ 0   Yes ☐ 1

   - If yes, do you have been diagnosed for Tuberculosis? MTUB

No ☐ 0   Yes ☐ 1

Did you take any medication to treat hookworms recently or at the moment? MHOO

No ☐ 0   Yes ☐ 1

Do you have any problem seeing in the daytime? MNB1

No ☐ 0   Yes ☐ 1

Do you have any problem seeing at nighttime? MNB2

No ☐ 0   Yes ☐ 1

   - If yes, is this problem different from others in your community? MNB3

No ☐ 0   Yes ☐ 1

Do you have night blindness (kujigongagonga usiku / kutokuona vizuri usiku)? MNB4
(Je, unatatizo la kujigongagonga usiku wakati unatembea nyumbani / njiani?)

No ☐ 0   Yes ☐ 1

Do you suffer from chronic diarrhoea for more than 1 month? MDIA

No ☐ 0   Yes ☐ 1   (major sign)

Do you suffer from prolonged fever for more than 1 month (intermittent or constant)? MFEV

No ☐ 0   Yes ☐ 1   (major sign)

Did you loose weight recently (≥ 10%)? MWEI

No ☐ 0   Yes ☐ 1   (major sign)

Do you suffer from a persistent cough (for more than 1 month)? MCOU

No ☐ 0   Yes ☐ 1   (minor sign)

Do you suffer from an itching dermatitis? MDER

No ☐ 0   Yes ☐ 1   (minor sign)

Do you suffer from recurrent painfull zoster / blistering dermatitis? MZOS

No ☐ 0   Yes ☐ 1   (minor sign)

Do you suffer from fungal infection of the mouth and throat? MFUN

No ☐ 0   Yes ☐ 1   (minor sign)

Do you suffer from swollen lymphatic glands? MGLA

No ☐ 0   Yes ☐ 1   (minor sign)

7. Clinical data measured

Height in cm (June/July) ☐ NHEI

BMI = ☐ NBMI

Weight in kg ☐ NWEI

Bitot's spots   No ☐ 0   Yes ☐ 1   NBIT

Hb in g/dL ☐ NHAE

Linking Vegetable Diversity and Nutrition Security
Household Survey Tanzania March/April 2007

No. of participant

1. Basic data

Date of survey (dd/mm/yyyy)

Name of enumerator _______________________________

Village _______________________________District _______________________________

Respondent Name _______________________________

2b. Vegetable consumption

What are the three most important criteria that you consider while choosing a vegetable?
Rank - do not tick but place "1", "2", "3" into boxes CCRI1, CCRI2, CCRI3

Taste	1	Fruit/leaf size	6
Price	2	Easy to prepare	7
Availability	3	Family members preference	8
Good for health	4	Other (specify)	9
Fruit/leaf colour	5	_______________________	

What are the three most important factors influencing your decisions to consume many
__different__ types of vegetables? *Rank - do not tick but place "1", "2", "3" into boxes* CDIV1, CDIV2, CDIV3

Different tastes of different vegetables 1
Prices fluctuate – always buy the cheapest vegetables 2
Availability – supply changes frequently (e.g. seasonality, quantities available) 3
A diverse diet is good for health 4
Vegetable diversity is high in own garden 5
Certain vegetables need to be cooked/mixed with others and certain foods
need to be consumed with certain vegetables 6
Availability of cash 7
Preference of family members 8
Others (specify)_______________________________ 9
I do **not** consume many different vegetables (less than 3 different per week) 10

During the last survey we found that some women do __not__ consume vegetables every day. What
is the main reason for you __not__ to consume any vegetable during a whole day? CNOC

Seasonal availability 1
Prices too high 2
Preparation too time consuming/no time for preparation 3
Being not at home (but in the field, travelling, etc.) 4
Others (specify) _______________________________ 5

____

3b. Vegetable production

Which vegetables (exotic and traditional) do you **produce** at the moment in your homegarden?
PPRO

No.	Local name	Kiswahili name	English name	Scientific name
1				
2				
3				
4				
5				
6				
7				
8				
9				
10				

Which naturally growing vegetables do you **gather** at the moment? PGAT

No.	Local name	Kiswahili name	English name	Scientific name
1				
2				
3				
4				
5				
6				
7				
8				
9				
10				

Last year in November/December you produced and gathered the following vegetables:

Produced	Gathered

Why do you **not** produce / gather these vegetables <u>at the moment during this season</u> but different ones? PNOP

Vegetables concerned

Temperature too high/low	1	____________
Water availability	2	____________
Pests and diseases stress	3	____________
Input: availability of seeds, fertiliser, chemicals is lacking	4	____________
Output: market prices and demand are low	5	____________
Home consumption: not consumed at the moment	6	____________
It is not their season	7	____________
Flooding of garden/field	8	____________
Other (please specify) ____________	9	____________

Do you **sell** any of your own produced vegetables at the moment (this time of the year)? PSEL

No ☐ 0   Yes ☐ 1

PSWE If yes, how many times a week? ☐ times
PSMO If yes, how many times a month? ☐ times
PSHO If yes, how do you normally sell?   Farm gate ☐ 1   Market ☐ 2
Walking from house to house ☐ 3   Others ☐ 4

Do you **buy** any vegetables at the moment (this time of the year)? PBUY1

No ☐ 0   Yes ☐ 1

If yes, which vegetables do you buy at the moment / at this time of the year? PBUY2

No.	Local name	Kiswahili name	English name	Scientific name
1				
2				
3				
4				
5				

If yes, how often do you buy vegetables at the moment / at this time of the year? PBUY3

☐ times per week
☐ times per month

If yes, where do you buy vegetables? PBUY4
Market ☐ 1
Farm gate / neighbours ☐ 2
From traders walking from house to house ☐ 3
Others ☐ 4

Why do you buy (additional) vegetables? ____________________
PBUY5

Linking Vegetable Diversity and Nutrition Security
Household Survey Tanzania March/April 2007

1. Basic data

No. of participant ☐☐

Date of survey (dd/mm/yyyy) ☐☐☐

Name of enumerator ____________________

Village ____________________ District ____________________

Respondent Name ____________________

2a. Physical activity

<u>Visual analogue scale</u>
Please make one cross on this scale between the two extremes of "no physical activity" (sleeping) and "extremely strenuous physical activity". Where would you set up your own daily activity (<u>daily mean</u>)? PHVI
Tafadhali weka alama ya x katika mizani hii, kuonyesha ukubwa wa shughuli zako za kila siku.

hapana shughuli ●━━━━━━━━━━━━━● shughuli nyingi

3a. Attitude towards vegetable consumption

While vegetable consumption is important for women and children, it is also important for men. CATT1
Strongly agree ☐ 1   Agree ☐ 2   Not sure ☐ 3
Disagree ☐ 4   Strongly disagree ☐ 5

Vegetables are inferior foods that are good when one doesn't have much money or food at home. CATT2
Strongly agree ☐ 1   Agree ☐ 2   Not sure ☐ 3
Disagree ☐ 4   Strongly disagree ☐ 5

Fresh vegetables are likely to contain more nutrients than dried ones. CATT3
Strongly agree ☐ 1   Agree ☐ 2   Not sure ☐ 3
Disagree ☐ 4   Strongly disagree ☐ 5

4a. 24h-recall

Which <u>foods and drinks</u> and how much of it did you (*alone* <u>not</u> *your household*) consume yesterday? Was the food item processed and where did you get it from?

<u>Main meal</u>: how many people shared this meal? How many older/younger 10 yrs?

Food item	Quantity - all foods - (piece, handful, cup, table spoon, etc.) (*show three containers - small, medium, large*)	Source - all foods - 1 = purchased 2 = produced 3 = collected 4 = gift 5 = other	Processing - vegetable only - (raw or dried, steamed, cooked, fried + for how many minutes) *e.g. steamed (10 min)*	No. pers. >10 yrs	No. pers. < 10 yrs
Breakfast					
Lunch					
Dinner					
Snacks					

5a. Seven-day recall of vegetables

(*Do not tick but place number for Source:* 1 = purchased; 2 = produced; 3 = collected; 4 = gift; 5 = other)

Vegetables	On which days of the last week did you consume a certain vegetable and where did you get it from? (*Use key above for source*)							No. of days	Source
	Mon	Tue	Wen	Thu	Fri	Sat	Sun		
Leafy vegetables									
African nightshade / Mnavu									
African spiderflower / Mgagani									
Amaranth leaves / Mchicha									
Baobab tree leaves / Majani ya mbuyu									
Bitter/hair lettuce / Mchunga									
Black jack / Kishonanguo									
Cassava leaves / Kisamvu mhogo									
Cassava leaves / Kisamvu mpira									
Chinese cabbage / Chainisi									
Cowpea leaves / Majani ya kunde									
Cucumber leaves / Matango pori									
Ethiopian kale / Sukuma wiki / Loshuu									
False sesame / Mlenda mbata									
Jute mallow / Mlenda wima/pori									
Kibwando									
Malabar spinach / Nderema									
Mhilile									
Pumpkin leaves / Majani ya maboga									
Sweet potato leaves / Majani ya matembele									
Swiss chard / Spinachi									
Taro leaves / Majani ya magimbi									
Water cress / Saladi									
Wild simsim / Mlenda mwitu/mfuta/wima									
White Cabbage / Kabichi									
Fruit vegetables									
African eggplant / Ngogwe									
Cucumber fruit, exotic									
Cucumber fruit, wild / Matango pori									
Gourd									
Okra / Bamia									
Pea pods (*specify*)									
Pumpkin / Maboga									
Sweet pepper / Pilipili tamu									
Tomato / Njanja									
Root vegetables									
Carrot / Karoti									
Onion / Kitungu									
Others (specify)									
Vitamin A rich foods									
Papaya / papai									
Mango / embe									
Yellow/orange sweet potato / viazi vitamu									
Red palm oil (fresh, unbleached) / mafuta ya mawese (mbichi, haijakamuliwa)									
Liver / maini									
Egg (esp. egg yolk) / mayai									
Milk (butter, cheese) / maziwa (siagi, chizi)									
Blueband									

Linking Vegetable Diversity and Nutrition Security
Household Survey Tanzania March/April 2007

Respondent Name _______________________

No. of participant

6. Investigation on nutrition and health status

Did you take any <u>Vitamin A</u> supplements recently (last 3 month)? MVIT

No ☐ 0   **Yes** ☐ 1

Did you take any <u>Iron</u> supplements recently (last 3 month)? MIRO

No ☐ 0   **Yes** ☐ 1

When you have the monthly bleeding, is it more severe or longer than 6 days? MBLE

No ☐ 0   **Yes** ☐ 1

Are you pregnant? - If yes, which trimester? MPRE

No ☐ 0   **Yes** ☐ 1   **Trimester: 1st** ☐ 1   **2nd** ☐ 2   **3rd** ☐ 3

Did you experience a loss of appetite recently? MAPP

No ☐ 0   **Yes** ☐ 1

- If yes, do you have been diagnosed for Tuberculosis? MTUB

No ☐ 0   **Yes** ☐ 1

Did you take any medication to treat hookworms recently or at the moment? MHOO

No ☐ 0   **Yes** ☐ 1

Do you have any problem seeing in the daytime? MNB1

No ☐ 0   **Yes** ☐ 1

Do you have any problem seeing at nighttime? MNB2

No ☐ 0   **Yes** ☐ 1

- If yes, is this problem different from others in your community? MNB3

No ☐ 0   **Yes** ☐ 1

Do you have night blindness (kujigongagonga usiku / kutokuona vizuri usiku)? MNB4
(Je, unatatizo la kujigongagonga usiku wakati unatembea nyumbani / njiani?)

No ☐ 0   **Yes** ☐ 1

Do you suffer from chronic diarrhoea for more than 1 month? MDIA

No ☐ 0   **Yes** ☐ 1   (major sign)

Do you suffer from prolonged fever for more than 1 month (intermittent or constant)? MFEV

No ☐ 0   **Yes** ☐ 1   (major sign)

Did you loose weight recently (≥ 10%)? MWEI

No ☐ 0   **Yes** ☐ 1   (major sign)

Do you suffer from a persistent cough (for more than 1 month)? MCOU

No ☐ 0   **Yes** ☐ 1   (minor sign)

Do you suffer from an itching dermatitis? MDER

No ☐ 0   **Yes** ☐ 1   (minor sign)

Do you suffer from recurrent painfull zoster / blistering dermatitis? MZOS

No ☐ 0   **Yes** ☐ 1   (minor sign)

Do you suffer from fungal infection of the mouth and throat? MFUN

No ☐ 0   **Yes** ☐ 1   (minor sign)

Do you suffer from swollen lymphatic glands? MGLA

No ☐ 0   **Yes** ☐ 1   (minor sign)

Do you smoke? MSMO1

No ☐ 0   **Yes** ☐ 1   (minor sign)

- If yes, how many cigarettes per day? MSMO2

< 5 ☐   6-10 ☐   11-15 ☐   16-20 ☐   21-30 ☐   > 30 ☐

7. Clinical data measured

Weight in kg   ☐ NWEI   BMI = ☐ NBMI

Bitot's spots   No ☐ 0   Yes ☐ 1   NBIT

Hb in g/dL   ☐ NHAE

CIC	LEFT eye			RIGHT eye		
Cooperation in general	Good ☐	Interm. ☐	Bad ☐	Good ☐	Interm. ☐	Bad ☐
Blinking/Closing eye	No ☐	Yes ☐		No ☐	Yes ☐	
Tears	No ☐	Yes ☐		No ☐	Yes ☐	
If yes	Few ☐	Medium ☐	Many ☐	Few ☐	Medium ☐	Many ☐